The new Physics I
FIRST ALGEBRA OF MAGNITUDES

WE SAVED THE FALSE HYPOTHESIS OF THE INTERNATIONAL SYSTEM
OF UNITS WITH THE *DYADIC ALGEBRA OF MAGNITUDES* AND WE COME
TO THE «DISMETRIC» PHYSICS, WHICH RECOGNIZES NEW
PROPERTIES IN THE EMPTY SPACE

ENGLISH VERSION

PRIMERA ÁLGEBRA DE MAGNITUDES

SALVAMOS LA HIPÓTESIS FALSA DEL SISTEMA
INTERNACIONAL DE UNIDADES CON EL *ÁLGEBRA
DIÁDICA DE MAGNITUDES* Y LLEGAMOS A LA FÍSICA
«DISMÉTRICA», QUE RECONOCE NUEVAS PROPIEDADES
EN EL ESPACIO VACÍO

VERSIÓN ORIGINAL EN ESPAÑOL

J. M. Arnaiz
cfejma@gmail.com

The good thing if brief twice good;
and even the bad, if little, not so bad

Lo bueno, si breve, dos veces bueno;
y aun lo malo, si poco, no tan malo

Baltasar Gracián

Todos los derechos reservados
Autor: J. M. Arnaiz
© Ediciones Go Beyond, 2017
ISBN: 9781974491292
Cubiertas diseñadas por Freepik

To curious spirits who yearn to understand the principles of science, to facilitate their work; and also in honor of the absent friends who inspired everything with altruistic spirit.

And especially to Maria, my wife, without whose inspiration «dysmetry» would not exist.

A los espíritus curiosos que anhelan comprender los principios de la ciencia, para facilitarles su quehacer; y también en honor de los amigos ausentes que lo inspiraron todo con ánimo altruista.

Y especialmente a María, mi esposa, sin cuya inspiración la «dismetría» no existiría.

ILATIVE SCHEME

EXORDIUM
Physics «arithmetization» paradox
False hypothesis of the International System of Units 13

Section I
Definitions of magnitude and quantity 33

Section II
Definition of measurement of
the quantity of a magnitude 40

Section III
Definition of concrete entity or physical dyad
Dyadic sets 44

Section IV
Definitions of homogeneity,
uniformity and equality 60

Section V
Definition of dyadic addition 64

Section VI
Commutative properties and
associative of the dyadic addition 74

Section VII
Existence of neutral and symmetrical
elements for dyadic addition 78

Section VIII
Definition of dyadic subtration 82

Section IX
Definition of multiplication
of a real number for a dyad 89

Section X
Distributive properties of the
dyadic multiplication by a scalar 95

Section XI
Definition of division between
homogenous dyads 104

Section XII
Definition of multiplication
dyadic of scalar dyads 112

Section XIII
Commutative and associative properties
of the multiplication of scalar dyads 125

Section XIV
Inexistence of unit or reverse elements
for the multiplication of scalar dyads 131

Section XV
Distributive ownership of the multiplication
on the addition of the dyadic algebra 139

Section XVI
Definition of dyadic division
between scalar measurements 143

Section XVII
Definition of power and
radication of scalar dyads 165

Section XVIII
Definition of dyadic logarithmation
scalar and the legendary calculation rule 169

Section XIX
Definition of scalar products
and vector of vector dyads 187

Section XX
Product definition between
scalar and vector dyads 203

Section XXI
Definition of dyadic entities imaginaries
and their laws of composition 207

Section XXII
Effects of the principle
of symbolic economy 215

Section XXIII
Classes of magnitudes 227

Section XXIV
Equality, identity, equation and physical law 239

Section XXV
Definition of dimensions
of the physical magnitudes 275

Section XXVI
The physical constants 291

Section XXVII
Philosophical consequences 311

Section XXVIII
Dyadic or physical algebra compendium
Need and basic development 335

Section XXIX
The black lagoon of math
Origin of the «arithmetization» of physics 449

Bibliography 499

ESQUEMA ILATIVO

EXORDIO
Paradoja de «aritmetización» de la Física
Hipótesis falsa del Sistema Internacional de Unidades 23

Apartado I
Definiciones de magnitud y cantidad 36

Apartado II
Definición de medición de la
cantidad de una magnitud 42

Apartado III
Definición de ente concreto o díada física
Conjuntos diádicos 52

Apartado IV
Definiciones de homogeneidad,
uniformidad e igualdad 62

Apartado V
Definición de adición diádica 69

Apartado VI
Propiedades conmutativa y
asociativa de la adición diádica 76

Apartado VII
Existencia de elementos neutro y
simétrico para la adición diádica 80

Apartado VIII
Definición de sustracción diádica 86

Apartado IX
Definición de multiplicación
de un escalar por una díada 92

Apartado X
Propiedades distributivas de la
multiplicación diádica por un escalar 100

Apartado XI
Definición de división entre
díadas homogéneas 108

Apartado XII
Definición de multiplicación
diádica de concretos escalares 118

Apartado XIII
Propiedades conmutativa y asociativa de
la multiplicación de díadas escalares 128

Apartado XIV
Inexistencia de elementos unidad ni inverso
para la multiplicación de díadas escalares 135

Apartado XV
Propiedad distributiva de la multiplicación
sobre la adición del álgebra diádica 141

Apartado XVI
Definición de división diádica
entre mediciones escalares 154

Apartado XVII
Definición de potenciación y
radicación de díadas escalares 167

Apartado XVIII
Definición de logaritmación diádica
escalar y la legendaria regla de cálculo 178

Apartado XIX
Definición de los productos escalar
y vectorial de díadas vectoriales 195

Apartado XX
Definición de producto diádico entre
un concreto escalar y otro vectorial 205

Apartado XXI
Definición de entes diádicos o concretos
imaginarios y de sus leyes de composición 211

Apartado XXII
Efectos del principio
de economía simbólica 221

Apartado XXIII
Clases de magnitudes 233

Apartado XXIV
Igualdad, identidad, ecuación y ley física 257

Apartado XXV
Definición de dimensiones
de las magnitudes físicas 283

Apartado XXVI
Las constantes físicas 301

Apartado XXVII
Consecuentes filosóficos 323

Apartado XXVIII
Compendio de álgebra diádica o física
Necesidad y desarrollo básico 340

Apartado XXIX
La laguna negra de la matemática
Origen de la «aritmetización» de la física 452

Bibliografía 499

EXORDIUM

Physics «arithmetization» paradox
False hypothesis of the International System of Units

Does anyone know the answers to these seemingly elementary questions: How do you multiply a kilogram by a meter? What is the multiplier, the kilogram or the meter? How does one second square? divide the product kg×m by one second squared? What is the full meaning of compound units such as the newton, unit of force, or the joule, unit of energy?

Dear reader, do not feel ignorant or self-conscious if you cannot pin down the answers, because, incredible as it may sound, no one currently knows how to answer them at all. Neither the Nobel Laureates in Physics nor the most eminent physicists or mathematicians, no one can fully and rationally justify such questions, and they do not even realize it. Well, that is the object of the First Algebra of Magnitudes, which solves and explains all these questions with considerable didactic effort and, furthermore, as the first great achievement of this **new non-arithmetic algebra**, reveals the «dysmetry» of space, exhibited in volume II of this work.

Since we started in the elementary study of Physics we used to make use of operations with entities that indicate specific quantities of magnitudes and, due to the subliminal influence of arithmetic operations with abstract numbers such as real ones, we naturally believe that concrete operations they must follow the same calculation rules. Thus, for example, if a mobile travels a distance of 100 meters, using a time of 25 seconds, we say that the speed of the movement is $100/25=4$ meters per second, and we will briefly write this speed with the letter $4\ m/s$. Likewise, if in 3 seconds the velocity changes from an initial value of $10\ m/s$

to 70 m/s, we firmly affirm that the mobile would have experienced an acceleration of (70-10)/3 = 20 meters per second per second, writing it 20 m/s^2. With this we will have unconsciously assumed that with the symbols of the units, such as the meter or the second of the examples, one must operate with the same composition laws that we have established for abstract real numbers, that is, as if the units were mere variables formal algebraics; a happy assumption that already openly bothered Fourier or Maxwell, and so many other eminent physicists, given the absence of any justification for it, which implies upsetting scientific knowledge, depriving it of one of its fundamental pillars, because without fully clarifying this point It should not go to the second page of any Physics book; but it turns out that the texts themselves induce this mess, because they all forget to plumb a fundamental pillar of science: the algebra of magnitudes.

So what unscrupulously we get used to and they teach us to do unconsciously from a young age, seeming to us so natural, in reality, it is not only not obvious, but it is totally wrong, since it dispenses with something capital: the epistemic definitions of the laws of composition between entities that represent measurements or quantities of magnitudes and their units.

So it is not surprising that the effects of this omission are worrying and that they continue to disturb the wise men of Physics of all times; what is more, what is striking is that it must be the prominent ones who philosophize about it and that no one else discusses the tradition, because the root of the problem is very elementary. Perhaps the first to recognize it was Fourier in his Analytical Theory of Heat (1822), where he introduced the important concept of dimension inherent in all physical magnitudes. Another distinguished thinker who warned about this issue was Clerk Maxwell, author of the unification of electromagnetism, who alluded to the typical case of teachers of all educational levels, who explain the unit of work without more than referring to the formula one joule is equal to a kilogram multiplied by a square meter and divided by a second squared, according to the well-known expression:

$$1\,joule = \frac{1\,kg \times 1\,m^2}{1\,s^2}$$

And he continued saying, not without a certain grace, that teachers do not realize that they would find themselves in a bind if some inquisitive student asked them what it is to multiply a kilogram by a square meter and how to divide the result by a second squared, or even what is the point of a second squared.

And it is that, indeed, we happily admit that 1 s×1 s=1 s², because it seems elementary algebra; however, we must take a closer look at expressions like this. It is clear that any multiplication of abstract numbers such as 3×4 means, by definition of multiplication, adding the multiplicand 3 as many times as indicated by the multiplier 4, that is, we have the abbreviated sum 3×4=3+3+3+3=12. So, if we applied this same algorithm to the product 3 s×4 s, establishing as multiplying the first factor and the second as multiplier, or vice versa, we would have 3 s×4 s=3 s+3 s+3 s+3 s=12 s. Why, then, do we so easily admit that 3 s×4 s=12 s²? Just because it has the written form of an algebraic expression, do we ignore the arithmetic definition of multiplication? Because obviously 12 s cannot be the same as 12 s², since we can easily understand what a second is, since we have clocks that measure that quantity of time; Now, what kind of entity will a second squared be? It does not seem that such a thing can be observed in nature, which would seem to delegitimize compound units like this one.

There is therefore a gap pending to be resolved in this of operations with physical magnitudes, which causes the proliferation of diverse and contradictory opinions regarding their nature and formulation, discussions that would be ended simply by defining the appropriate composition laws. A group of authors such as RC Tolman attribute to the symbols of dimensional expressions a certain impenetrable or mystical character and consider that «the true essence of magnitudes, from the physical point of view, is represented by their dimensional formula» (Physics Review, p 25, 1917). This hypothesis does not seem to be true, because it would suppose that such disparate magnitudes as the momentum of a force and its work, which can both be expressed in «newton×meter», were essentially manifestations of the same magnitude, energy, which seems clearly a madness, as we justify in section XXVI of the First Algebra of Magnitudes. Great authors such as Planck indicate that «it is as meaningless to speak of the "real" dimension of a magnitude as it is of the "real" name of an object», which would mean that physical magnitudes should be hidden from the understanding. Planck seems to indicate to us that we must not forget that physical quantities are mental entities and that, like any other name that indicates an extramental object, they are the result of the arbitrariness of thought. The positivist faction of the Vienna Circle, led by Bridgman, states that «dimensions do not have absolute value at all, but must be defined precisely from the process used to measure the respective magnitude», which makes us to suggest that, indeed, in the field of magnitudes there must be a good deal of arbitrariness; which bothered Planck so much that he criticized positivism thus: «The views of the positivists cannot be fought from a purely logical point of view. And yet a careful examination of them reveals that they are inadequate and sterile, because they dispense with a circumstance that is of decisive importance for scientific progress. As much as positivism boasts of being free from prejudice, it has to start from a fundamental premise if it is not to degenerate into an unintelligible solipsism. This premise is that every physical measurement can be reproduced in such a way that the result is independent of the observer's personality, the place and time in which the measurement is made, and any other circumstance. All this simply reveals that the decisive factor for the measurement result lies outside the observer and that consequently the measurements pose problems involving causal connections in an objective reality independent of the observer».

And in the same way as in these notable examples, different beliefs about the nature of physical quantities proliferate, forming a kind of intellectual pandemonium in this matter, so that everyone conforms to the usual way of operating with quantities of magnitudes, although each one has their own subjective notion of them and without being aware of the problem that the lack of epistemic definition of their composition laws raises, so that for the majority the vice does not exist and it is common not even to ask why it must operate with such physical entities as it is done and not otherwise.

Moreover, all authors versed in dimensional analysis take it for granted that unit abbreviations operate with the same algebra of abstract numbers, and on this tacit assumption and without justifying it in any way they elaborate their respective theories, which completely omit all specific algebra. for the magnitudes. And the same happens in the educational field, where the philosophical problems related to magnitudes and their laws of composition are ignored, as if they did not exist, teaching concrete operations in an intuitive, subjective and arbitrary way, leaving students with, Even without knowing it, a residue of uncertainty that vitiates all the knowledge acquired with this gap pending to be clarified, degrading the teaching quality, because the key to perfect understanding is not to advance at all without first having precisely defined all of the foregoing, and more, if possible, in the case of something so fundamental to understand and develop natural laws, such as magnitudes, their measurements and their operations.

In our view, the reason for this inertia common to science and education could be due to the intuition fed by the algebra of geometric segments, which more or less we all know to some extent since adolescence. We know that the multiplication of segments does not follow the model of the arithmetic product, but is conceived in geometry as a new magnitud, called surface, with two factors, or called volume, if there are three, so that the multiplication of lengths gives rise to two new derived magnitudes, the surface and the volume, in accordance with certain suitably defined laws of composition, and whose commutative, associative and distributive properties retain the form of the arithmetic. We will analyze all this in more detail throughout this work.

Thus, in the same way that there are algebras for numbers and abstract vectors, universally accepted, an algebra of concrete entities, representatives of quantities of magnitudes, should be established, because only in this way would the prevailing confusion be ended and they would be better clarified the meanings of the various composite magnitudes. And this is precisely what is humbly carried out in the monographic memory that we have entitled First Algebra of Magnitudes: starting as the foundation of the algebra of geometric segments, one immediately has that of lengths and, by mere generalization, it is easy to arrive at the convenient definition of the laws of composition for any magnitudes. This reveals the hidden frameworks of the derived units and the meanings that can be attributed to them can be judged more accurately. Thus we will see that such mythical magnitudes as, for example, the so-called energy, surprisingly, are rather a kind of ether fruit of the arbitrary nature of scientific thought, rather than real entities or qualities. This does not mean that the magnitudes do not correspond to some aspect of the physical realm, but rather that we must be prudent when we derive conclusions about the world from dimensional formulations, pertaining to the mental realm, in order not to end up entangled in childish disquisitions wrong. And for this it is very convenient to understand where the compositions of the magnitudes that intervene in any phenomenon have arisen, since, otherwise, the entire analysis carried out on the case under examination will be mutilated or corrupted without possible remission and subject to capricious speculation.

So the notion of dimension of all magnitude has to be considered after and not before having conceived an algebra of magnitudes, whose mathematical expression is concrete entities. Hence, the method followed in the work of the First Algebra of Magnitudes, faithful to the step-by-step sequence that characterizes us, has to be presented according to the following sequence: first, the basic concepts of physical magnitudes are established in general; then, they are assigned a mathematical entity and the concrete entities or physical dyads are created; An algebra is defined below for such special entities, which are adopted as precise representatives of the quantities of natural magnitudes; then the meaning of the definition equations, of the universal laws and of other physical entities is investigated; and, finally, the basis for dimensional analysis is finished. Without trying at all to describe this matter with perfect exhaustiveness, but with enough detail so that its unfailing character can be appreciated, leaving for successive editions the development of more abstract structures tailored to the various scientific theories.

The capital elements of Physics are the measurements or quantities of magnitudes, associated through certain invariant relations that operate with them; so, considering that simple algebraic entities are not composed, because they always carry a unit, it is not admissible that the laws of composition that define exactly how these physical features of nature should be composed are lacking. It would be something like if Mathematics established the algebraic formulas in the abstract without having specified the tables for the addition or multiplication of natural numbers. We would certainly be shocked by it. And yet, with Physics, this is what has been tacitly done for centuries with all naturalness and indifference on the part of the majority, except for a few wise men who have been upset by the tradition of simply admitting that magnitudes operate. as elements of ordinary algebra, without their concern having resulted in a definitive solution, adding to the confusion that many refer to as the «mysterious problem of dimensions» or the «puzzling character of compound units».

With the First Algebra of Magnitudes, the foundations are laid so that the forgotten fundamental pillar of science is properly installed and, when we are before a physical law such as Newton's second law $\overline{F} = m \times \overline{a}$, we do not believe that we are facing an expression of vector algebra Rather, we must understand that $\overline{F}$, m y $\overline{a}$ indicate quantities of physical magnitudes or measurements, represented by concrete mathematical entities, and that to operate with them it is necessary to have defined first of all how to do it in a convenient way with epistemic composition laws without the least ambiguity. With this, any subjective controversy about the interpretation or natural meaning of the composite magnitudes should decline, because it will be the definitions themselves that allow it to be elucidated objectively. And this will undoubtedly result in a substantial improvement in the quality of teaching and the intellectual performance of students.

A good revelation of the First Algebra of Magnitudes is the **doubling theorem**, which determines how every physical law is decomposed into its two basic elements, the pure algebraic equation and the dimensional relationship between the units of its two members, a process in which they appear the constants of homogeneity, which show the difference between the equations of the mathematical metric, in which said constants are always the unit, hence the doubling does not produce any change in their formulations, and the physical measurements, in which in general you do not have to comply with this restriction. And with this result, in our opinion very remarkable and transcendent, culminates what with some license of the language and without much property we could consider as a certain dyadic algebra sui generis applied to physical magnitudes, since ultimately the measurements are nothing but dyads formed by pairs of closely related elements: an algebraic primary and a dimensional secondary.

But the most spectacular and striking achievement of the algebra of magnitudes is the tremendous discovery of the

«dysmetry» of space, which springs naturally from the rigorous logic of the First Algebra of magnitudes. The «dysmetric» spaces are called to revolutionize the future Physics. The «dysmetric» mathematical structures are in the innovative research phase and their first fundamental development will be the subject of a separate publication. The «dysmetric» observation is unobjectionable and inalienable, and involves giving a dazzling Copernican twist to modern Physics, as opposed to the elemental and invisible isometry that has prevailed since the origins of science. The First Algebra of Magnitudes in principle selflessly follows the dogmatic isometric tradition, practically limiting itself to discovering, describing and solving the **paradox of «arithmetization» of Physics** and the **false hypothesis of the International System of Units**, but soon it inevitably meets the revolutionary **«dysmetry» of space**, which marks a new course for science, with the consequent appearance of an inexhaustible hotbed of physical innovations, which many enterprising researchers will undoubtedly appreciate. It is an inevitable discovery for logic and unavoidable for Physics.

The paradox of «arithmetization» of Physics is covered in a pernicious way with the false hypothesis of the International System of Units, admitted by everyone by tacit and comfortable conventionalism, most of the time unconscious, which gives the appearance of rigor where only there is arbitrariness: it is universally admitted by hypothesis that physical quantities form an abelian multiplicative group, which cannot be sustained with the rigorous algebra here exposed.

It is not possible to operate seriously with magnitudes only through arithmetic, because operations with abstract numbers are internal composition laws and, as shown throughout this text, especially from section XII, operations with magnitudes require new **generating external composition laws**, which produce the composite magnitudes born from the fundamentals, the only way to give physical meaning to the laws and equations of Physics. And these generating operations are not much like the arithmetic ones, displaying very notable differences such as, for example, the non-existence of inverse elements for the quantities of magnitudes, for which reason unitary dyads or inverses and inverses cannot exist for the generating laws, therefore, there cannot be the inverse of any unit, as explained in section XIV. This supposes that, just as the inverse of the number 2 is the rational $1/2$, the inverses of a meter, a kilogram or a second cannot be found.

Here we do not limit ourselves to simply denouncing this false hypothesis, but the problem is identified and solved with the first coherent algebra in the history of Physics. Nobody well informed and impartial will maintain after studying the matter, that this absurd hypothesis must be maintained because Physics has given great results, because they should be reproached for how much more fruit it would bear without that falsehood in its principles.

In all the informative efforts to convince of the relevance of what is explained, we have tried to use a mathematical language that is not too abstruse or abstract, because our intention is that this algebra and its spatial «dysmetry» be accessible to most of the students intellects, including those at the college level. To do this, we have endeavored to make clear the algebraic logical method, which reasons step by step, based solely on the previously established definitions and properties, thus saving any blinding prejudice of understanding. We hope we have managed to explain ourselves with the simplicity necessary for this and at the same time with sufficient significance to convey the philosophical and practical significance for the future of Physics of the algebra of magnitudes, necessary for the discovery of «dysmetry», which is an impressive physical-mathematical truth.

J. M. Arnaiz

EXORDIO

Paradoja de «aritmetización» de la Física
Hipótesis falsa del Sistema Internacional de Unidades

¿Alguien conoce las respuestas a estas preguntas aparentemente tan elementales?: ¿Cómo se multiplica un kilogramo por un metro?, ¿cuál es el multiplicador, el kilogramo o el metro?, ¿cómo se eleva un segundo al cuadrado?, ¿cómo se divide el producto $kg \times m$ entre un segundo al cuadrado?, ¿cuál es el significado completo de unidades compuestas tales como el newton, unidad de fuerza, o el julio, unidad de energía?

Querido lector, no se sienta ignorante ni acomplejado si no puede precisar las respuestas, porque, por increíble que parezca, nadie sabe actualmente responderlas en absoluto. Ni los premios nobeles de física ni los más eminentes físicos ni matemáticos, nadie puede justificar cabal y racionalmente tales interrogantes, y ni siquiera se dan cuenta de ello. Pues bien, ese es el objeto de la *Primera álgebra de magnitudes*, que resuelve y explica con notable esfuerzo didáctico todas esas cuestiones y, además, como primera gran conquista de esa **nueva álgebra no aritmética**, revela la «dismetría» del espacio, expuesta en el volumen II de esta obra.

Desde que nos iniciamos en el estudio elemental de la Física acostumbramos a servirnos de las operaciones con entes que indican cantidades concretas de magnitudes y, por la influencia subliminal de las operaciones aritméticas con números abstractos como los reales, creemos con toda naturalidad que las operaciones concretas deban seguir las mismas reglas de cálculo. Así, por ejemplo, si un móvil recorre una distancia de 100 metros, empleando en ello un tiempo de 25 segundos, decimos que la velocidad del movimiento sea de $100/25 = 4$ metros por segundo, y abreviadamente escribiremos tal velocidad con la grafía $4\ m/s$.

Asimismo, si en 3 segundos la velocidad cambia de un valor inicial de $10\,m/s$ a $70\,m/s$, afirmamos rotundamente que el móvil habría experimentado una aceleración de $(70-10)/3=20$ metros por segundo y por segundo, escribiéndolo $20\,m/s^2$. Con ello habremos supuesto inconscientemente que con los símbolos de las unidades, como el metro o el segundo de los ejemplos, deba operarse con las mismas leyes de composición que tenemos establecidas para los números reales abstractos, es decir, como si las unidades fuesen meras variables algebraicas formales; alegre suposición que ya incomodaba abiertamente a Fourier o Maxwell, y a tantos otros eminentes físicos, dada la ausencia de toda motivación que la justifique, lo que implica descabalar el conocimiento científico, privándolo de uno de sus pilares fundamentales, porque sin aclarar del todo este punto no debería pasarse a la segunda página de ningún libro de Física; pero resulta que los propios textos inducen este desaguisado, porque todos olvidan aplomar un pilar fundamental de la ciencia: el álgebra de magnitudes.

Así que lo que sin escrúpulos nos acostumbramos y nos enseñan a hacer desde pequeños inconscientemente, pareciéndonos tan natural, en realidad, no solo no es nada evidente, sino que es totalmente incorrecto, puesto que se prescinde de algo capital: de las definiciones epistémicas de las leyes de composición entre entes que representen mediciones o cantidades de magnitudes y de sus unidades. Así que no es extraño que los efectos de esta omisión preocupasen y que sigan inquietando a los sabios de la Física de todos los tiempos; es más, lo llamativo es que hayan de ser los prominentes quienes filosofen al respecto y que nadie más discuta la tradición, porque la raíz del problema es muy elemental.

Quizá el primero en reconocerlo fuese Fourier en su *Teoría analítica del calor* (1822), donde introdujo el importante concepto de dimensión inherente a toda magnitud física. Otro insigne pensador que advirtió sobre esta cuestión fue Clerk Maxwell, autor de la unificación del electromagnetismo, que aludía al caso típico de los profesores de todos los niveles educativos, que explican la unidad de trabajo sin más que aludir a la fórmula un julio es igual a un kilogramo multiplicado por un metro cuadrado y dividido por un segundo al cuadrado, de acuerdo con la conocida expresión:

$$1\,julio = \frac{1\,kg \times 1\,m^2}{1\,s^2}$$

Y continuaba diciendo, no sin cierta gracia, que no caen en la cuenta los maestros de que se verían en un aprieto si algún alumno inquisitivo les preguntase qué es eso de multiplicar un kilogramo por un metro cuadrado y cómo se divide el resultado por un segundo al cuadrado, o incluso qué sentido tiene un segundo al cuadrado.

Y es que, en efecto, admitimos alegremente que $1\,s \times 1\,s = 1\,s^2$, porque parece álgebra elemental; sin embargo, debemos observar con más atención expresiones como esta. Está claro que toda multiplicación de números abstractos como 3×4 significa, por definición de multiplicación, sumar el multiplicando 3 tantas veces como indique el multiplicador 4, es decir, tenemos la suma abreviada $3 \times 4 = 3 + 3 + 3 + 3 = 12$. Así que, si aplicásemos este mismo algoritmo al producto $3\,s \times 4\,s$, estableciendo como multiplicando el primer factor y el segundo como multiplicador, o viceversa, tendríamos $3\,s \times 4\,s = 3\,s + 3\,s + 3\,s + 3\,s = 12\,s$. ¿Por qué, entonces, admitimos con tanta facilidad que $3\,s \times 4\,s = 12\,s^2$?, ¿solo porque tiene la forma escrita de una expresión algebraica desoímos la definición aritmética de multiplicación?, porque obviamente no puede ser lo mismo $12\,s$ que $12\,s^2$, ya que podemos entender sin problemas lo que sea un segundo, pues disponemos de relojes que miden esa cantidad de tiempo; ahora bien, ¿qué clase de ente habrá de ser un segundo elevado al cuadrado?, no parece que tal cosa pueda observarse en la naturaleza, lo cual parecería deslegitimar unidades compuestas como esta.

Existe entonces una laguna pendiente de resolver en esto de las operaciones con magnitudes físicas, que provoca la proliferación de opiniones diversas y contradictorias respecto a su naturaleza y formulación, discusiones a las que se pondría fin simplemente definiendo las leyes de composición convenientes. Un grupo de autores como R. C. Tolman atribuyen a los símbolos de las expresiones dimensionales cierto carácter impenetrable o místico y consideran que «la verdadera esencia de las magnitudes, desde el punto de vista físico, está representada por su fórmula dimensional» (*Physics Review*, p. 25, 1917). Esta hipótesis no parece que pueda ser cierta, porque supondría que magnitudes tan dispares como el momento de una fuerza y su trabajo, que pueden expresarse ambas en «newton×metro», fuesen esencialmente manifestaciones de la misma magnitud, la energía, lo cual parece a todas luces un desvarío, como justificamos en el apartado XXVI de la *Primera álgebra de magnitudes*. Grandes autores como Planck indican que «tan falto de sentido es hablar de la dimensión "real" de una magnitud como del nombre "real" de un objeto», lo que supondría que las magnitudes físicas habrían de ocultarse al entendimiento. Planck parece indicarnos que no hemos de olvidar que las magnitudes físicas son entes mentales y que, como cualquier otro nombre que señale a un objeto extramental, son fruto de la arbitrariedad del pensamiento. La facción positivista del Círculo de Viena, encabezada por Bridgman, dispone que «las dimensiones no tienen en modo alguno valor absoluto, sino que han de definirse, precisamente, a partir del proceso que se utilice para medir la magnitud respectiva», que nos vuelve a sugerir que, en efecto, en el ámbito de las magnitudes debe de haber una buena dosis de arbitrariedad; lo que incomodaba tanto a Planck, que criticó el positivismo así: «Las opiniones de los positivistas no pueden ser combatidas desde un punto de vista puramente lógico. Y, sin embargo, un examen detenido de las mismas revela que son inadecuadas y estériles, porque prescinden de una circunstancia que tiene importancia decisiva para el progreso científico. Por mucho que alardee el positivismo de estar exento de prejuicios, tiene que partir de una premisa fundamental si no quiere degenerar en un solipsismo ininteligible. Tal premisa consiste en que toda medida física puede ser reproducida de tal modo que el resultado es independiente de la personalidad del observador, del lugar y tiempo en que se efectúa la medición, y de cualquier otra circunstancia. Todo esto revela simplemente que el factor decisivo para el resultado de la medición está fuera del observador y que, en consecuencia, las medidas plantean problemas que implican conexiones causales en una realidad objetiva independiente del observador».

Y de la misma manera que en estos ejemplos notables, proliferan las diferentes creencias sobre la naturaleza de las magnitudes físicas, formando una especie de pandemónium intelectual en esta materia, de modo que todo el mundo se conforma con la manera usual de operar con las cantidades de magnitudes, aunque cada cual tenga su propia noción subjetiva de ellas y sin tomar conciencia del problema que suscita la falta de definición epistémica de sus leyes de composición, por lo que para la mayoría el vicio ni existe y es común ni siquiera preguntarse por qué debe operarse con tales entes físicos como se hace y no de otro modo.

Es más, todos los autores versados en análisis dimensional dan por sentado que las abreviaturas de unidades operen con la misma álgebra de los números abstractos, y sobre este presupuesto tácito y sin justificarlo en modo alguno elaboran sus respectivas teorías, que omiten absolutamente toda álgebra específica para las magnitudes. Y lo mismo sucede en el ámbito educativo, donde se pasan por alto, como si no existiesen, los problemas filosóficos atinentes a las magnitudes y sus leyes de composición, enseñando las operaciones concretas de manera intuitiva, subjetiva y arbitraria, dejando en los alumnos, aun sin saberlo, un poso de incertidumbre que vicia todo el conocimiento adquirido con esta laguna pendiente de ser clarificada, envileciendo la calidad docente, porque la clave del perfecto entendimiento es no avanzar en absoluto sin antes haber definido con precisión todo lo precedente, y más, si cabe, tratándose de algo tan fundamental para comprender y desarrollar las leyes naturales, como lo son las magnitudes, sus mediciones y sus operaciones.

A nuestro entender el motivo de esta inercia común a la ciencia y a la educación podría deberse a la intuición alimentada por el álgebra de los segmentos geométricos, que más o menos todos conocemos hasta cierto punto desde la adolescencia. Sabemos que la multiplicación de segmentos no sigue el modelo del producto aritmético, sino que se concibe en geometría como una nueva magnitud, denominada superficie, con dos factores, o llamada volumen, si fuesen tres, de modo que la multiplicación de longitudes da lugar a dos nuevas magnitudes derivadas, la superficie y el volumen, de conformidad con ciertas leyes de composición convenientemente definidas, y cuyas propiedades conmutativa, asociativa y distributiva conservan la forma de las aritméticas. Todo esto lo analizaremos con más detalle a lo largo de este trabajo.

Así, pues, de la misma manera que existen álgebras para los números y vectores abstractos, aceptadas universalmente, debería asentarse un álgebra de los entes concretos, representantes de las cantidades de magnitudes, porque solo así se acabaría con la confusión imperante y quedarían mejor aclarados los significados de las distintas magnitudes compuestas. Y esto es precisamente lo que humildemente se lleva a cabo en la memoria monográfica que hemos titulado *Primera álgebra de magnitudes*: partiendo como fundamento del álgebra de los segmentos geométricos, se tiene inmediatamente la de longitudes y, por mera generalización, se llega con facilidad a la definición conveniente de las leyes de composición para cualesquiera magnitudes. Con ello quedan al descubierto los entramados ocultos de las unidades derivadas y pueden juzgarse con más acierto los significados que se les pueda atribuir. Así veremos que magnitudes tan míticas como, por ejemplo, la denominada energía, sorprendentemente, son más bien una especie de éter fruto de la arbitrariedad del pensamiento científico, antes que entes o cualidades reales. Sin que ello quiera decir que las magnitudes no se correspondan con algún aspecto del ámbito físico, sino que debemos ser prudentes cuando derivemos conclusiones acerca del mundo a partir de las formulaciones dimensionales, pertenecientes al ámbito mental, a fin de no acabar enredados en pueriles disquisiciones erróneas. Y para ello es muy conveniente entender de dónde han surgido las composiciones de las magnitudes que intervengan en cualquier fenómeno, pues, de otro modo, quedará mutilado o corrompido sin posible remisión y sujeto a caprichosas especulaciones todo el análisis realizado sobre el caso sujeto a examen.

De modo que la noción de dimensión de toda magnitud ha de considerarse después y no antes de haber concebido un álgebra de magnitudes, cuya expresión matemática son los entes concretos. De ahí que el método seguido en del trabajo de la *Primera álgebra de magnitudes*, fieles a la ilación paso a paso que nos caracteriza, ha de presentarse según la siguiente secuencia: en primer lugar, se asientan los conceptos básicos propios de las magnitudes físicas en general; luego, se les asigna entidad matemática y se crean los entes concretos o díadas físicas; a continuación se define un álgebra para tales entes especiales, que se adoptan como representantes precisos de las cantidades de magnitudes naturales; se investiga después el significado de las ecuaciones de definición, de las leyes universales y de otros entes físicos; y, finalmente, se termina con la base del análisis dimensional. Sin pretender en absoluto describir esta materia con perfecta exhaustividad, sino con suficiente detalle para que se pueda apreciar su carácter indefectible, dejando para sucesivas ediciones el desarrollo de estructuras más abstractas a la medida de las diversas teorías científicas.

Los elementos capitales de la Física son las mediciones o cantidades de magnitudes, asociadas mediante ciertas relaciones invariantes que operan con ellas; así que, considerando que no se componen simples entes algebraicos, porque llevan aparejada siempre una unidad, no es admisible que falten las leyes de composición que definan con exactitud cómo deban componerse esos rasgos físicos de la naturaleza. Sería algo así como si la Matemática estableciese las fórmulas algebraicas en abstracto sin haber concretado las tablas de la adición ni de la multiplicación de números naturales. Sin duda nos escandalizaríamos por ello. Y, sin embargo, con la Física es eso mismo lo que se está haciendo tácitamente durante siglos con toda naturalidad e indiferencia por parte de la mayoría, salvo unos pocos sabios que se han visto trastornados por la tradición de admitir sin más que las magnitudes operen como elementos del álgebra ordinaria, sin que su preocupación haya cuajado en una solución definitiva, abonando la confusión que muchos refieren como el «misterioso problema de las dimensiones» o el «carácter desconcertante de las unidades compuestas».

Con la *Primera álgebra de magnitudes* se sientan las bases para que el olvidado pilar fundamental de la ciencia sea instalado debidamente y, cuando estemos ante una ley física como la *segunda ley de Newton* $\overline{F} = m \times \overline{a}$, no creamos hallarnos frente a una expresión del álgebra vectorial, sino que comprendamos que $\overline{F}$, m y $\overline{a}$ indican cantidades de magnitudes físicas o mediciones, representadas por entes matemáticos concretos, y que para operar con ellos es preciso haber definido antes que nada cómo hacerlo de manera conveniente con leyes de composición epistémicas sin la menor ambigüedad. Con ello deberá decaer toda controversia subjetiva sobre la interpretación o significado natural de las magnitudes compuestas, porque serán las propias definiciones las que permitan dilucidarlo con objetividad. Y ello redundará sin duda en la mejora sustancial de la calidad docente y del rendimiento intelectual de los estudiosos.

Una buena revelación de la *Primera álgebra de magnitudes* es el **teorema del desdoblamiento**, que determina cómo se descompone toda ley física en sus dos elementos básicos, la ecuación algebraica pura y la relación dimensional entre las unidades de sus dos miembros, proceso en el que aparecen las constantes de homogeneidad, que evidencian la diferencia entre las ecuaciones de la métrica matemática, en las que dichas constantes son siempre la unidad, de ahí que el desdoblamiento no produzca ningún cambio en sus formulaciones, y las mediciones físicas, en las que en general no se tiene por qué cumplir dicha restricción. Y con este resultado, a nuestro juicio muy notable y trascendente, culmina lo que con alguna licencia del lenguaje y sin mucha propiedad podríamos considerar como una cierta álgebra diádica sui géneris aplicada a las magnitudes físicas, toda vez que en definitiva las mediciones no son sino díadas formadas por parejas de elementos estrechamente vinculados entre sí: un primario algebraico y un secundario dimensional.

Pero el logro más espectacular y llamativo del álgebra de magnitudes es el apoteósico descubrimiento de la **«dismetría» del espacio**, que brota con naturalidad desde la lógica rigurosa de la *Primera álgebra de magnitudes*. Los espacios **«dismétricos»** están llamados a transformar la Física. Las estructuras matemáticas «dismétricas» están en fase de investigación innovadora y su primer desarrollo fundamental será objeto de una publicación aparte. La observación «dismétrica» es inobjetable e irrenunciable, y supone dar un deslumbrante giro copernicano a la Física moderna, frente a la elemental e invisible isometría imperante desde los orígenes de la ciencia. La *Primera álgebra de magnitudes* sigue en principio abnegadamente la dogmática tradición isométrica, limitándose prácticamente a descubrir, describir y resolver la **paradoja de «aritmetización» de la Física** y la **hipótesis falsa del Sistema Internacional de Unidades**, pero pronto se encuentra inevitablemente con la revolucionaria **«dismetría» del espacio**, que marca un nuevo rumbo para la ciencia, con la consiguiente aparición de un inagotable semillero de innovaciones físicas, que sin duda sabrán apreciar muchos investigadores emprendedores. Se trata de un descubrimiento inevitable para la lógica e irrenunciable para Física.

La paradoja de «aritmetización» de la Física se encubre de modo pernicioso con la hipótesis falsa del Sistema Internacional de Unidades, admitida por todo el mundo por convencionalismo tácito y cómodo, la mayoría de las veces inconsciente, que da la apariencia de rigor donde solo hay arbitrariedad: se admite universalmente por hipótesis que las magnitudes físicas forman un grupo multiplicativo abeliano, lo que no se puede sostener con el álgebra rigurosa aquí expuesta.

No es posible operar seriamente con magnitudes solo mediante la aritmética, porque las operaciones con números abstractos son leyes de composición internas y, como se muestra a lo largo de este texto, especialmente a partir del apartado XII, las operaciones con magnitudes requieren nuevas **leyes de composición externas generatrices**, que producen las magnitudes compuestas nacidas de las fundamentales, única forma de dotar de sentido físico a la leyes y ecuaciones de la Física. Y estas operaciones generatrices no son ni mucho menos como las aritméticas, ostentando diferencias muy notables como, por ejemplo, la inexistencia de elementos inversos para las cantidades de magnitudes, por lo que no pueden existir para las leyes generatrices las díadas unitarias ni las inversas y, por tanto, no puede existir el inverso de ninguna unidad, como se explica en el apartado XIV. Ello supone que, así como el inverso del número 2 es el racional 1/2, no se pueden encontrar los inversos de un metro, de un kilogramo ni de un segundo.

Aquí no nos limitamos a denunciar sin más dicha hipótesis falsa, sino que se identifica y se resuelve el problema con la primera álgebra coherente de la historia de la Física. Nadie bien informado e imparcial sostendrá después de estudiar la materia, que debe mantenerse esa hipótesis absurda porque la Física haya dado grandes frutos, porque a estos habría que reprocharles que cuántos más frutos daría sin esa falsedad en sus principios.

En todos los esfuerzos divulgativos para convencer de la relevancia de lo que se explica, se ha procurado usar un lenguaje matemático no demasiado abstruso ni abstracto, porque nuestra intención es que esta álgebra y su «dismetría» espacial sean accesibles a la mayor parte de los intelectos, incluidos los de nivel preuniversitario. Para ello, nos hemos esforzado en dejar patente el método lógico algebraico, que razona paso a paso, fundamentándose únicamente en las definiciones y propiedades previamente asentadas, salvando así cualquier prejuicio cegador del entendimiento. Esperamos haber conseguido explicarnos con la llaneza necesaria para ello y a la vez con suficiente significación para transmitir la transcendencia filosófica y práctica para el futuro de la Física del álgebra de magnitudes, necesaria para el descubrimiento de la «dismetría», que es una impresionante verdad físico-matemática.

J. M. Arnaiz

Section I

DEFINITIONS OF MAGNITUDE AND QUANTITY[1]

The first obstacle to overcome in the analysis of physical phenomena is to specify what is meant by magnitude. The contemplation of nature inspires the mental concepts of length, area, volume, time, speed, acceleration, mass, force, energy, and many others. Each of these entities does not always present itself with the same intensity, resulting in relationships of «equality» or «greater than» or «less than» between their different portions. Thus, by factual comparison, we can establish whether a certain portion of length is equal to, greater than, or less than another, just as two portions of areas, volumes, or time intervals, etc., can be determined. This fact allows us to conceive the definition of **physical magnitude** as any property that allows relationships of equality and inequality to be established between its different portions. On the other hand, the **quantity implicit** in any portion of a magnitude is the extent, degree, or intensity with which it manifests itself[2]. In the case of the magnitude of length, any straight line represents it; a portion would be any segment of the straight line; the amount of length implicit in a segment would be non-numerical and refers to the true extension of the segment. Since every quantity of magnitude is non-numérical, that is, it cannot be reduced to an abstract number, to express it analytically and be able to operate with various quantities, it must be symbolized with a mathematical object that indirectly represents it. This mathematical

---

[1] In English, it is common to consider the terms quantity and magnitude as synonyms. This poses a problem for English speakers in understanding the substantial difference between the meanings of these two concepts in Physics. Magnitude refers to a measurable natural property. In contrast, the quantity implicit in a portion of magnitude is precisely what symbolizes that measurement. Therefore, readers of that language are advised to rigorously understand and differentiate these two fundamental concepts for dyadic algebra and «dysmetry».

[2] A more elementary version of the algebra of magnitudes, although equally significant, can be found in another title by the same author, «Lesson 3» of Matematize 3.

element arises from the measurement process and is formed by a pair consisting of an abstract number or other multiplying entity and a portion of any magnitude that is taken as the unit. The physical unit is non-numerical, so it is replaced by an abstract symbol. It is vital to differentiate the physical or real quantity implicit in a portion of magnitude from the mathematical pair that represents it.

Let's take an example. When we say that the distance between two points is four meters, we write it as 4 m. In this nomenclature we have the following meanings: the magnitude we are not referring to is the length; the abbreviation m means that **we have taken as a reference the quantity of non-numerical length implicit in the portion we call the standard unit or meter** to measure the distance, therefore m means an amount of length that is not explicit because it cannot be expressed numerically, which is why it is symbolized with a letter. In turn, the abstract number 4 acts as a multiplier of m and means that the established distance is such that it corresponds to four times the amount of length implicit in the standard unit m. It is common to confuse 4 with a quantity of length, which is a fundamental conceptual error, because 4 is only an abstract number that by itself does not indicate magnitude; only by associating it with a unit like m does the pair thus formed acquire in this case the meaning of the quantity of the magnitude length. Thus, the difference between magnitude and quantity is clear: 4 m means a quantity that refers to the magnitude of length. Since the unit m is not countable, the quantity 4 m is not either, but this non-numerical quantity is represented by the mathematical pair 4 m. If the same number is associated with another quantity of magnitude, for example, the kilogram, the pair formed by the abstract number 4 and the quantity of non-numerical mass implicit in the standard unit, symbolized kg, acquires the meaning of a quantity four times the quantity implicit in the kilogram, expressed as 4 kg, of the magnitude we call mass. Here, the quantity is the meaning of the pair 4 kg, and the magnitude referred to is mass. The quantity of mass is not the pair 4 kg; this is merely the mathematical symbol for the real quantity implicit in a certain portion of mass.

Why do we say that every quantity of any magnitude is non-numerical? Let's take length. The distance between two points in space is a quantity of real or true length. How can we express it? No matter how much we search, we cannot associate it with any number. Therefore, what we do is invent measurement. We take a portion of magnitude or segment that we assume includes a certain amount of length and call it a standard unit. We take it for granted that the segment implicitly contains a certain quantity of length. Since we cannot express it numerically, we assign it a symbol to represent it, for example, m. Thus, to measure the distance between two points, we simply need to determine how many times the standard segment fits juxtaposed within the distance to be established. We tacitly admit that the quantity of length implicit in the standard segment does not vary along the distance to be measured, and we can now say that the quantity of length existing between the two points in question is that many times the quantity of length present in the standard segment, which seems constant to us. We have thus established a measurement consisting of a pair of elements: a number and an non-numerical standard unit. We say that this pair symbolically represents the real value of the measured quantity of length, $4\ m$ in the example. In this way, we overcome the fact that we cannot directly number the magnitude of length and assume that measurement replaces it. This impossibility arises for any other magnitude, which is why there is always the need for measurement and for establishing standard units associated with the magnitude whose quantities we wish to establish. Thus, we symbolically express the quantities of magnitude that seem true to us through pairs of measurements or dyads, defined in section III, which consist of a multiplier number and an non-numerical portion or physical unit that appears to produce constant measurements. It is important to recognize that $4\ m$ is the symbol or dyad associated with the quantity of physical, real, or true length that we wish to represent. **Understanding the difference between symbol and meaning, that is, between dyad and quantity of real magnitude, is crucial to appreciating the "dysmetric" phenomenon, where the same standard can indicate different quantities.**

When studying operations with physical quantities, it will be understood that the coherent definition of magnitude refers to **any physical property affine to length,** in order to provide a logical connection to the algebra of magnitudes. Throughout this work, this affinity will gradually become more specific, allowing us to define the various laws of composition that provide an algebraic structure to the quantities of physical quantities.

Apartado I

DEFINICIONES DE MAGNITUD Y CANTIDAD [3]

El primer obstáculo a superar en el análisis de los fenómenos físicos es concretar qué ha de entenderse por magnitud. La contemplación de la naturaleza inspira los conceptos mentales de longitud, superficie, volumen, tiempo, velocidad, aceleración, masa, fuerza, energía y otros muchos. Cada uno de estos entes no se presentan siempre con la misma intensidad, resultando que se pueden establecer entre sus distintas porciones relaciones de «igualdad» o de «mayor» o «menor que». De modo que por comparación fáctica podemos establecer si una cierta porción de longitud sea igual, mayor o menor que otra, al igual que dos porciones de superficies o de volúmenes o de intervalos de tiempo, etc. Este hecho nos permite concebir la definición de **magnitud física** como toda propiedad que permita establecer entre sus diferentes porciones relaciones de igualdad y desigualdad. Por otra parte, la **cantidad implícita** en toda porción de una magnitud es la extensión, grado o intensidad con que se manifiesta[4]. En el caso de la magnitud longitud, cualquier recta la representa, una porción sería todo segmento de la recta, la cantidad de longitud implícita

---

[3] En inglés es común considerar los términos cantidad y magnitud como sinónimos. Y esto es un problema para que los angloparlantes entiendan la sustancial diferencia para la Física entre los significados de esos dos conceptos. La magnitud en Física se refiere a una propiedad natural medible. En cambio, la cantidad implícita en una porción de magnitud es la que simboliza precisamente esa medición. Por tanto, se recomienda a los lectores en esa lengua que procuren entender y diferenciar con rigor esos dos conceptos fundamentales para el álgebra diádica y la «dismetría».

[4] Una versión más elemental del álgebra de magnitudes, aunque igualmente significativa, puede encontrarse en otro título del mismo autor, «Lección 3» de *Matematizar 3*.

en un segmento sería innúmera y se refiere a la extensión verdadera del segmento. Como toda cantidad de magnitud es innúmera, es decir, no se puede reducir a un número abstracto, para expresarla analíticamente y poder operar con diversas cantidades, hay que simbolizarla con un objeto matemático que indirectamente la represente. Ese elemento matemático surge del proceso de medición y está formado por un par integrado por un número abstracto u otro ente multiplicador y una porción de magnitud cualquiera que se toma como unidad. La unidad física es innúmera, por lo que se sustituye por un símbolo abstracto. Es vital diferenciar la cantidad física o real implícita en una porción de magnitud frente al par matemático que la representa.

Pongamos un ejemplo. Cuando decimos que la distancia entre dos puntos es de cuatro metros, lo escribimos abreviadamente 4 *m*. En esta nomenclatura tenemos los siguientes significados: la magnitud a que no referimos es la longitud, la abreviatura *m* significa que **hemos tomado como referencia la cantidad de longitud innúmera implícita en la porción que llamamos unidad patrón o metro** para medir la distancia, por tanto *m* significa una cantidad de longitud que no es explícita porque no se puede expresar numéricamente, por lo que se simboliza con una letra. A su vez, el número abstracto 4 hace las veces de multiplicador de *m* y significa que la distancia establecida es tal que se corresponde con cuatro veces la cantidad de longitud implícita en la unidad patrón *m*. Es frecuente confundir el 4 con una cantidad de longitud, lo cual es un error de concepto fundamental, porque el 4 solo es un número abstracto que por sí solo no indica magnitud, solo al asociarlo con una unidad como *m* el par así formado adquiere en este caso el significado de cantidad de la magnitud longitud. Por tanto, la diferencia entre magnitud y cantidad es clara, 4 *m* significa una cantidad que se refiere a la magnitud longitud. Como la unidad *m* no es numerable, la cantidad 4 *m* tampoco lo es, pero esa cantidad innúmera queda representada por el par matemático 4 *m*. Si el mismo número se asocia con otra cantidad de magnitud, por ejemplo, el kilogramo, el par que forman el número abstracto 4 y la cantidad de masa innúmera implícita en la unidad patrón, simbolizada *kg*, adquiere el significado de cantidad de cuatro veces la cantidad implícita en el kilogramo, expresada 4 *kg*, de la magnitud que llamamos masa. Aquí la cantidad es el significado del par 4 *kg* y la magnitud aludida es la masa. La cantidad de masa no es el par 4 *kg*, este no es más que el símbolo matemático de la cantidad real implícita en una cierta porción de masa.

¿Por qué decimos que toda cantidad de cualquier magnitud es innúmera? Tomemos la longitud. La distancia entre dos puntos del espacio es una cantidad de longitud real o verdadera. ¿Cómo expresarla? Por mucho que indaguemos no la podemos asociar con ningún número. Por tanto, lo que hacemos es inventar la medición. Tomamos una porción de magnitud o segmento que suponemos incluye una cierta cantidad de longitud y lo llamamos unidad patrón. Damos por obvio que el segmento contiene implícita una cierta cantidad de longitud. Como no la podemos expresar numéricamente, le asignamos un símbolo para representarla, por ejemplo, *m*. Así, para medir la distancia entre dos puntos, nos basta con averiguar cuántas veces cabe yuxtapuesto el segmento patrón en la distancia a establecer. Admitimos tácitamente que la cantidad de longitud implícita en el segmento patrón no varía a lo largo de la distancia a medir, y ya podemos decir que la cantidad de longitud existente entre los dos puntos en cuestión es esas veces la cantidad de longitud presente en el segmento patrón, que nos parece constante. Hemos establecido así una medición formada por un par de elementos: un número y una unidad patrón innúmera. Este par decimos que representa simbólicamente el valor real de la cantidad de longitud medida, 4 *m* en el ejemplo. De este modo salvamos el hecho de no poder numerar directamente la magnitud longitud y damos por sentado que la medición la sustituye. Esta imposibilidad se presenta para cualquier otra magnitud, por lo que siempre existe la necesidad de la medición y de establecer unidades patrón asociadas a la magnitud cuyas cantidades queremos establecer. Así expresamos simbólicamente las cantidades de magnitud que nos parecen verdaderas mediante pares de medición o díadas, definidas en el apartado III, que constan de un número multiplicador y una porción o unidad física innúmera que parece producir mediciones constantes. Es importante reconocer que 4 *m* es el símbolo o díada que se asocia a la cantidad de longitud física, real o verdadera que se quiere representar. **Entender la diferencia entre símbolo y significado, es decir, entre díada y cantidad de magnitud real, es crucial para apreciar el fenómeno «dismétrico», donde un mismo patrón puede indicar cantidades diferentes.**

Cuando se estudien las operaciones con cantidades físicas se comprenderá que la definición coherente de magnitud se refiere a **toda propiedad física afín a la longitud**, con el objeto de dar conexión lógica al álgebra de magnitudes. A lo largo de este trabajo se irá concretando paulatinamente en qué consiste esa afinidad, que nos permitirá definir las diversas leyes de composición para dotar de estructura algebraica a las cantidades de magnitudes físicas.

Quien siga atentamente los razonamientos expuestos en el texto descubrirá poco a poco la **riqueza que esconden las magnitudes**. Por ejemplo, provisionalmente en el *Álgebra diádica de magnitudes* se asume, como siempre se ha hecho de manera automática y sin tomar conciencia de ello, que toda porción de magnitud idéntica a una misma unidad patrón parece que siempre se asociará con la misma cantidad de magnitud real de forma absolutamente constante y en toda situación o circunstancia. Sin embargo, nada impide, y además es necesario para la lógica y la ciencia, formular la previsión genérica, esto es, que porciones iguales a una misma unidad patrón puedan identificarse con diferentes cantidades de magnitud verdaderas en función de diversas causas, por ejemplo, la posición, el tiempo o el entorno material. Esta opción más amplia, que llamaremos **variante «dismétrica»**, supera los límites de la actual **hipótesis isométrica**, admitida tácitamente, y es objeto de estudio específico en la segunda parte de esta obra, cuyo atento examen se recomienda encarecidamente, porque introduce al lector en los originales **espacios «dismétricos»**, novedad nacida del orden algebraico diádico y fecunda herramienta matemática que producirá inagotables innovaciones científicas.

Section II

DEFINITION OF MEASUREMENT OF
THE QUANTITY OF A MAGNITUDE

The reasoning in section I allows us to define the measurement of a quantity of magnitude in this way: **we will call measurement the application of any procedure that allows the quantity implicit in a portion of magnitude to be represented by the pair formed by two heterogeneous elements, one mathematical and the other physical, the first serving as a multiplier of the second, which is any symbol that represents a quantity of non-numerical magnitude.** The order in which these two entities are written is irrelevant, but we agree to write the mathematical element first, followed by the physical element.

By definition of magnitude, their quantities can be treated as quantities of length, given the affinity we have postulated. Therefore, the various quantities will be comparable in terms of «equality», «greater than» or «less than» as if they were geometric segments. What quantity of length is implicit in a given segment? Obviously, this quantity is not visible to direct observation, so we must quantify it indirectly; we must determine it in some relative way. Any segment can be fragmented into equal, that is, geometrically congruent, segments. If we assume that the quantity of length implicit in each of these smaller segments is the same, we have the traditional **isometric hypothesis,** as opposed to the more general option we called «**dysmetry**» in the second part of the text, which allows for different implicit lengths in congruent segments. It is always possible to arbitrarily choose any segment as a unit of length to construct multiples and submultiples of it that allow us to compare that unit with other segments and observe how many units or fractions of the unit a given segment contains. We have called this action measurement, and given the affinity of magnitudes with length, we accept that it is valid for any magnitude. The result of the measurement will be a real number or other mathematical entity, indicating the number of units or fractions of a unit contained in the quantity of magnitude being measured. A heterogeneous pair is thus obtained: the physical element ϕ, which represents the reference quantity of magnitude, whose quantity is innumerable, so it is replaced by this abstract symbol; and the mathematical multiplying element μ of the multiplicand ϕ. We will say that the heterogeneous pair thus formed represents the real quantity of the magnitude considered, and we will express it using any of the forms $\mu\,\phi$, $(\mu\,\phi)$, (μ,ϕ), or $\mu\times\phi$, which we will call the significant **measurement** of that real quantity. We define **measure** as the mathematical multiplying element resulting from a measurement, coinciding with Newton. And this is not capricious, because, if we have $\mu\times\phi=\varphi$, where φ is the quantity of magnitude to be measured with the unit ϕ, the division $\varphi/\phi=\mu$ is clear. So, if the measure is defined as the whole or fractional times that the quantity φ contains its unit ϕ, it turns out that the measure is the ratio φ/ϕ, which is the multiplier μ. This division operation is detailed in section XI. Measurement symbolizes the quantity of real magnitude implicit in the measured phenomenon, showing its relative value to a certain quantity of magnitude taken as a reference standard, whose true value is non-numerical, so it is replaced by an abstract symbol. The comparison is made by affine geometric congruence. Assuming that congruent segments implicitly include the same quantities of length is the common isometric hypothesis. Recognizing the more general observation that congruent segments can implicitly contain different quantities of length is essential to appreciating the «dysmetric» phenomenon, the subject of the second part of this work, starting with section XXX. This means that **mathematical congruence is not synonymous with physical equality**, with important consequences, which are outlined in said second part.

Apartado II

DEFINICIÓN DE MEDICIÓN DE LA
CANTIDAD DE UNA MAGNITUD

Los razonamientos del apartado I nos permiten definir la medición de una cantidad de magnitud de esta forma: **llamaremos medir a la aplicación de cualquier procedimiento que permita representar la cantidad implícita en una porción de magnitud mediante el par formado por dos elementos heterogéneos, uno matemático y otro físico, el primero con la función de multiplicador del segundo, que es cualquier símbolo que represente una cantidad de magnitud innúmera.** El orden de escritura de estos dos entes es indiferente, pero convenimos en escribir primero el elemento matemático seguido del elemento físico.

Por definición de magnitud, sus cantidades podrán ser tratadas como cantidades de longitud, dada la afinidad que hemos postulado. Por tanto, las diversas cantidades serán comparables en términos de «igualdad», «mayor que», o «menor que» como si fuesen segmentos geométricos. ¿Qué cantidad de longitud está implícita en un segmento determinado? Es obvio que esa cantidad no se muestra a la observación directa, por lo que hemos de cuantificarla indirectamente, debemos fijarla de algún modo relativo. Cualquier segmento se puede fragmentar en segmentos iguales, es decir, geométricamente congruentes. Si se supone que la cantidad de longitud implícita en cada uno de estos segmentos menores es la misma, tenemos la **hipótesis isométrica** tradicional, frente a la opción más general que llamamos **«dismetría»** en la segunda parte del texto, que permite diferentes longitudes implícitas en segmentos congruentes. Siempre es posible elegir de manera arbitraria un segmento cualquiera como unidad de longitud para construir múltiplos y submúltiplos suyos que permitan comparar esa unidad con otros segmentos y observar cuántas unidades o fracciones de la unidad contiene un segmento dado. Esta acción la hemos llamado medición y, dada la afinidad de las magnitudes con la longitud, admitimos que es válida para cualquier magnitud. El resultado de la medición será un número real u otro ente matemático, que indique el número de unidades o fracciones de la unidad que contiene la cantidad de magnitud que se mide. Se obtiene así un par heterogéneo: el elemento físico ϕ que representa la cantidad de magnitud de referencia, cuya cuantía es innúmera, por lo que se sustituye por dicho símbolo abstracto; y el elemento matemático multiplicador μ del multiplicando ϕ. El par heterogéneo así formado diremos que representa la cantidad real de la magnitud considerada y la expresaremos mediante cualquiera de las formas $\mu\,\phi$, $(\mu\,\phi)$, (μ,ϕ) o $\mu\times\phi$, que llamaremos **medición** significante de esa cantidad real. Definimos la **medida** como el elemento matemático multiplicador resultante de una medición, coincidiendo con Newton. Y esto no es caprichoso, porque, si tenemos $\mu\times\phi=\varphi$, siendo φ la cantidad de magnitud a medir con la unidad ϕ, es clara la división $\varphi/\phi=\mu$. De modo que, si la medida se define como las veces enteras o fraccionarias que la cantidad φ contiene a su unidad ϕ, resulta que la medida es la razón φ/ϕ, que es el multiplicador μ. En el apartado XI se detalla esta operación de división.

La medición simboliza la cantidad de magnitud real implícita en el fenómeno medido, mostrando su valor relativo a cierta cantidad de magnitud tomada como patrón de referencia, cuyo valor verdadero no es numerable, por lo que es sustituido por un símbolo abstracto. La comparación se hace por congruencia geométrica afín. Suponer que segmentos congruentes incluyen implícitas las mismas cantidades de longitud es la hipótesis isométrica común. Reconocer la observación más general sobre que segmentos congruentes puedan contener implícitas cantidades de longitud diferentes resulta esencial para apreciar el fenómeno «dismétrico», objeto de la segunda parte de este trabajo a partir del apartado XXX. Esto significa que **congruencia matemática no es sinónimo de igualdad física**, con importantes consecuencias, que se esbozan en dicha segunda parte.

Section III

DEFINITION OF CONCRETE ENTITY OR PHYSICAL DYAD
DYADIC SETS

Let's consider the most common measurements expressed with real numbers R or vectors of R^3. We have called measurement the quantity of a magnitude expressed with the form $q\,U$, as a symbol of the times q that a unitary quantity U is present in a phenomenon, calling q multiplier or measure with the unit U of the observed magnitude. And analogously if the measurement were a vector. The term «measure» or multiplier should not be confused with the «measured quantity» corresponding to a measurement. Here the word «measure» is the participle of the verb «to measure». Measurement is the product of the measure by the unit, which can be expressed with the common multiplication symbol $q{\times}U$, and the quantity it represents refers to the value that the measurement yields based on the supposedly true value of the quantity of magnitude implicit in the unit U, **an non-numerical quantity but which is taken into account at all times with the abstraction of symbolizing it with a sign that replaces it.** The expression $q{\times}U$ as an abstract formula for any quantity of magnitude associated with the unit U reveals the presence of the pair (q,U) formed by the elements q and U, and for convenience nothing prevents it from being written more briefly as $q\,U$, representing a mathematical entity formed by a multiplier, real number, vector or other mathematical object, and a multiplicand or dimensional quantity associated with a certain magnitude. This newborn entity that alludes to physical **measurement** can also be given a mathematical name, for example, **concrete entity** or **physical dyad**, and its elements will be called primary, measure, mathematical element or multiplier q or , and unit U, secondary, physical element, dimensional part or multiplicand. The primary is the mathematical part of the dyad. The secondary is the physical or dimensional part. Perhaps the name unit number or physical number, or some other suggestive title, would be appropriate for the secondary; But, since the name does not make things, we will not waste time on this trifle, but rather we will attend to what is important, which is the nature of the concrete entity, truly born from the action of counting and joining a certain number of whole or fractional times the pattern U, an operation that, we repeat, can be indicated as the product of a number by a quantity of determined magnitude noted $q \times U$ or $q \times (1\,U)$; or, if the measurement is vector, $\overline{q} \times U$ or $\overline{q} \times (1\,U)$, where $(1\,U)$, although not specified, corresponds to a supposed true quantity of the magnitude in question, established by its empirical definition and taken as an elementary unitary pattern of said magnitude. And there is no problem in admitting, by definition, that the indicated quantity does not depend on the order of writing, so the same quantity will be $q \times (1\,U)$ as $(1\,U) \times q$, which is equivalent to axiomatizing the commutative property of this symbology. In short, it is necessary to establish a principle that allows for the construction of further reasoning, and we will call it the **fundamental postulate to be kept in mind in operations with dyads**, whose statement is that the measurement symbol $q\,U$ means that the real quantity is q whole or fractional times the quantity implicit in the unit U, which is indicated by the three forms of the following definition:

$$q\,U = q \times (1\,U) = (1\,U) \times q \qquad [3.1]$$

As we will define in the following section, here we advance that the equal sign means that all members symbolize the same quantity of magnitude, so they are substitutable for each other.

In turn, in the case of a vector primary $\overline{q}$, the concrete $\overline{q}\,U$ must symbolize the quantity of a vector magnitude that has the same direction and sense as $\overline{q}$ and whose module is the number of whole or fractional times equal to the module of $|\overline{q}|$ the quantity of the magnitude contained in the unit U. As in the scalar

45

case, the following three notations will be accepted as indicative of this meaning:

$$\overline{q}\, U = \overline{q} \times (1\, U) = (1\, U) \times \overline{q} \qquad [3.2]$$

Therefore, in view of [3.1] and [3.2] it is not possible to distinguish between scalar and vector units, because, for both, every unit or quantity of magnitude U must be admitted by algebraic axiom to be identified with the scalar dyad $(1\, U)$ and that the numerical element that acts as multiplier of the dyad is q or $|\overline{q}|$, depending on whether it is associated with a magnitude of scalar nature in R or vector in R^3, or with another mathematical entity that serves as multiplier.

Magnitudes whose multipliers are such that $q \in R$ and that can take any value are called continuous, whereas those in which the multipliers can only be integers, with $q \in Z$, are called discrete. It is observed that operations with discrete magnitudes are included in the continuous ones, since their primaries will be represented by integers, which is a subset of the real numbers, so that the continuous magnitudes present greater generality than the discrete magnitudes; and the continuous magnitudes will be explained in the abstract in any case by means of the affinity with length, which fictitiously represents them all, because any of them can be assimilated to the real line, resulting in any case in the same reasoning scheme.

The choice of units for any magnitude is arbitrary. Therefore, **the broad definition of a physical dyad is any pair formed by a primary mathematical multiplier, number, vector, or other, and a secondary consisting of any symbol or symbols that designate a certain quantity of unspecified and non-numerical magnitude.**

In section I, we defined magnitude as **any physical property affine to length.** We said that this means that quantities of magnitudes can be treated as if they were geometric segments. And this helps us **formulate the concept of dyad in the abstract.**

Thus, any quantity f of a magnitude can be associated with a segment of arbitrary length, which can be added to itself by juxtaposition as many times as a multiplying element indicates, or decomposed into equal segments of lesser length as many times as a divisor element indicates, in accordance with the elementary geometric operations, which we assume the reader is familiar with. In this way, we can form, with the quantity of magnitude ϕ, other quantities defined by a multiplying element μ, integer or fractional, which we symbolize with the multiplicative form $\mu \times \phi$. This operation is developed in section IX, assigning it the symbol of the law of composition «○». The factor μ is generally an element of the set of real numbers R or a vector of R^3. Once we have obtained the homogeneous quantity $\mu \times \phi$ of ϕ, it is obvious that $\mu \times \phi$ can be observed as a pair of heterogeneous elements μ and ϕ. Let us remember that μ in isolation is not a quantity of length, it is merely an abstract mathematical entity. Now, nothing prevents us from naming this pair with the term abstract dyad, choosing symbols such as $\mu \phi$, $(\mu \phi)$, (μ,ϕ) or $\mu \times \phi$ to denote these pairs. As we have already seen, we will say that μ is the primary, mathematical element or multiplier of the dyad, and we will call the quantity of magnitude ϕ secondary, physical unit, dimensional part, or multiplicand. The dyad is, therefore, the mathematical reflection of the measurement process, by which a unit, divided into other smaller and equal units, allows us to formulate a measurement in the form of pairs of the type (μ,ϕ). Every measurement is a dyad, but an abstract dyad does not have to be a measurement. For operations on dyads, which represent quantities of magnitudes, to verify the associative-commutative and distributive properties, we will see in the following sections that the set of multipliers $\{\mu\}$ must have a field structure, like R. If we designate μ_1 as the multiplicative unitary element of said set $\{\mu\}$, we postulate that $\mu_1 \times \phi = \phi$ for any ϕ, and we can extend the ways of expressing the same quantity of a magnitude indicated by a dyad (μ,ϕ) with $\mu \times (\mu_1,\phi)$ and $(\mu_1,\phi) \times \mu$. The set of all dyads formed with the set of multipliers $\{\mu\}$ and associated with the quantity of magnitude f could be represented abstractly with the graph $\{\{\mu\}, \phi\}$, which we will call a dyadic set. Thus, by definition, the dyadic set f over the field of multipliers $\{\mu\}$ is, by definition, $\{\{\mu\}, \phi\} = \{(\mu, \phi) \mid \mu \in \{\mu\}\}$.

Once the definition of physical dyads and dyadic sets has been established, they would be of no use if the composition laws that allow us to operate with them were not formulated, providing them with an algebraic structure. And this is the crux of the key task for resolving the gap described at the beginning of this work, as is properly justifying operations with quantities of different physical magnitudes, which is the objective of the first part of this work, in which the terms **concrete** and **physical dyad** will be used interchangeably to name the basic elements that symbolically represent every quantity of magnitude.

The name concrete is retained due to the historical weight of this concept, which has long served to differentiate abstract numbers, those that indicate a quantity without specifying any unit and formed by a single mathematical element, from classical concrete numbers, which indicate a quantity associated with the unit to which they refer. However, the name **dyad** is our preference for heterogeneous pairs associated with any physical quantity, which is why the name dyadic algebra is reserved for the various structures that arise from the laws of composition that will be defined for dyads, symbolized indistinctly with the forms $q\ U$, $(q\ U)\ \underline{o}\ (q, U)$. And analogously for vector numbers, replacing q with $\bar{q}$.

To formally develop dyadic algebra, it is necessary to epistemically define the addition of dyads, the multiplication of a dyad by a scalar, for example, a real number, the subtraction of dyads, the multiplication and division of scalar dyads, and the scalar and vector products of vector dyads. Once these operations are established, other derived operations such as exponentiation, radicalization, or logarithm of scalar dyads can be deduced by logical means, for rather theoretical purposes, in the same way that subtraction and division are derived from addition and multiplication.

Particularizing for the field R, let (μ,ϕ) be a dyad, which represents a quantity of a certain magnitude, where μ and ϕ can be any; if the magnitude is a scalar, $\mu=q$ will be a real number with $q \in R$. Any quantity of the given magnitude can be taken as the unit $\phi=U$, and we will denote the universal set of all of them with $\{U\}$; therefore, we will have that every unit U will be in the total set of quantities $\{U\}$ and we will write $U \in \{U\}$. It is concluded with this notation that every dyad (q,U) will be an element of the set of all possible dyads, which we will denote $\{R,U\}$ and we will call it the dyadic set of the magnitude considered relative to the quantity U; this set in turn is evidently constructed with the Cartesian product of R and $\{U\}$, that is, $\{R,U\}=R \times \{U\}$. We reiterate that U represents any quantity of the magnitude in question, which in any case could be taken as a pattern. We observe that $\{R,U\}=\{U\}$. Obviously, for any quantity $U_o \in \{U\}$ the dyadic set constructed with it $\{R,U_o\}$ is complete, it includes all quantities of magnitude, so it coincides with $\{R,U\}$, so $\{R,U_o\}=\{R,U\}$. Dyads can be composed among themselves by means of internal composition laws, defined by establishing applications of the Cartesian product $\{R,U\} \times \{R,U\}$ in the same $\{R,U\}$; and they can also be composed with the elements of other sets, such as R, by means of external composition laws such as applications of $R \times \{R,U\}$ in $\{R,U\}$, so the task of establishing an adequate algebra for them must be addressed. We will verify that the field conditions of R are necessary for the dyadic structure to maintain the associative, commutative and distributive properties in its operations. Instead, we will see that dyadic entities lack unitary and inverse elements in the sense that refers to the internal composition laws inherent in the group structure. In summary, for scalar quantities, we have the following fundamental analytic definitions:

$\{U\}=\{$set of all quantities U of a magnitude$\}$
Dyadic set of U over R: $\{R,U\}=\{(q,U)\,|\,q \in R\}=R \times \{U\}$
Every dyad is $(q,U) \in \{R,U\}$ with $q \in R$ y $U \in \{U\}$

In turn, the vector dyads form a set that can be symbolized $\{R^3,U\}$ or if preferred $\{\mathbf{V}^3,U\}$ or $\{E^3,U\}$, which are equivalent notations used interchangeably. They are susceptible to being composed among themselves by means of internal composition laws, with applications of the Cartesian product $\{R^3,U\}\times\{R^3,U\}$ in $\{R^3,U\}$; and they can also be composed with the elements of other sets, such as R, by means of external composition laws, with applications of $R\times\{R^3,U\}$ in $\{R^3,U\}$, so the task of establishing an adequate algebra for them will also be addressed, which we will have to ensure is as isomorphic as possible with the structure of the vector space R^3 over R.

We will see that the above composition laws do not present too much difficulty for their analytical formulation, because they are all **additive operations** built on the same magnitude, as defined in sections V to XI, despite which they reveal important properties, resolving, for example, the historical mysteries of the dimensionless nature of angular magnitudes, and in turn promoting the development of new and important concepts such as «dysmetric» density.

We will find more resistance in the foundation of **multiplicative operations** based on two or more equal or different magnitudes with any units U_1 and U_2 for the case of two factors, whose definitions and properties are established in sections XII to XVII. However, it will be possible to establish coherent applications between sets such as $\{R,U_1\}\times\{R,U_2\}$ and $\{R,U_c\}$, where U_c indicates a new composite unit produced when operating on U_1 and U_2 for scalar magnitudes and analogously for vector magnitudes. And this notable capacity to generate new magnitudes is precisely the characteristic note of such multiplicative operations. The sets $\{R,U_1\}$ and $\{R,U_2\}$ may be equal, when they refer to the same magnitude, but **the dyadic set $\{R,U_c\}$ will always be different from the previous two, because its elements are quantities of a different magnitude, born by multiplying magnitudes in accordance with what is explained in section XII.** Thus it turns out that, even when $\{R,U_1\}$ and $\{R,U_2\}$ are the same set, for example, if both referred to the magnitude length, it will turn out that $\{R,U_c\}$ represents all the quantities that an area can take, because **the product of two lengths gives rise to a new magnitude composed with them that we call surface or area**. In this way, what we could call **generating external composition laws** appear, which have the special quality of producing new magnitudes from the multiplication of any others. Once this is done, it will be understood that this greater difficulty of such multiplicative operations is what has caused everyone, including the International System of Units, to ignore them and replace them with an easy, arbitrary and illusory undesirable hypothesis of «arithmetization» of magnitudes, an error that we have assumed by allowing ourselves to be deceived by arithmetic symbology and creating erroneous concepts. For example, physical units and any quantity of magnitude cannot have multiplicative inverses as if they were internal operations, because the multiplication of magnitudes is externally generative. Thus, notations such as U^{-1} must be reformulated, as explained in section XIV and in the appendix. Symbologies such as m^{-1}, s^{-1} or kg^{-1} must be expressly defined for physical quantities. Note also that not every external law is a generating law, although every generating law must be external. For example, applications of $R \times \{R,U\}$ to $\{R,U\}$, being external, do not generate a new quantity, because the set $\{R,U\}$, the image of that external law, is one of the initial sets. Generating laws are essential laws for the algebra of quantities, since every physical formulation is built with them, giving rise to new composite quantities. For example, Newton's second law involves composing mass and acceleration, relating them to another different magnitude: force. In turn, by composing length and time, a new magnitude arises, which is speed, or by operating with mass and volume, we generate another magnitude, which is density. In any case, it is clear that **the algebra of magnitudes must obey operational criteria with dyads**, so we must establish specific composition laws that allow us to build sui generis structures and take into account the dyadic nature of physical phenomena, avoiding illusory simplifications.

Apartado III

DEFINICIÓN DE ENTE CONCRETO O DÍADA FÍSICA
CONJUNTOS DIÁDICOS

Consideremos las mediciones más comunes expresadas con números reales de R o vectores de R^3. Hemos llamado medición a la cantidad de una magnitud expresada con la forma $q\,U$, como símbolo de las veces q que una cantidad unitaria U esté presente en un fenómeno, denominando a q multiplicador o medida con la unidad U de la magnitud observada. Y análogamente si la medida fuese un vector $\overline{q}$. No hay que confundir el término «medida» o multiplicador, con la «cantidad medida» correspondiente a una medición. Aquí la palabra «medida» es el participio del verbo «medir». La medición es el producto de la medida por la unidad, que se puede expresar con el símbolo de multiplicación común $q \times U$, y la cantidad que representa se refiere al valor que arroje la medición en función del valor supuestamente verdadero de la cantidad de magnitud implícita en la unidad U, **cuantía innúmera pero que se tiene en cuenta en todo momento con la abstracción de simbolizarla con un signo que la sustituya.** La expresión $q \times U$ como fórmula abstracta de cualquier cantidad de magnitud asociada a la unidad U revela la presencia del par (q, U) formado por los elementos q y U, y por comodidad nada impide escribirla más brevemente con la forma $q\,U$, en representación de un ente matemático formado por un multiplicador, número real, vector u otro objeto matemático, y un multiplicando o cantidad dimensional asociada a cierta magnitud. Este ente recién nacido que alude a la **medición** física, también puede recibir un nombre matemático, por ejemplo, **ente concreto** o **díada física**, y a sus elementos los llamaremos primario, medida, elemento matemático o multiplicador q o $\overline{q}$, y unidad U, secundario, elemento físico, parte dimensional o multiplicando. El primario es la parte matemática de la díada. El secundario es la parte física o

52

dimensional. Quizá fuese adecuado para el secundario el nombre de número unitario o número físico, o cualquier otro título sugestivo; pero, como el nombre no hace a las cosas, no perderemos el tiempo en esta pequeñez, sino que atenderemos a lo importante, que es la naturaleza del ente concreto, nacido realmente de la acción de contar y juntar cierto número de veces enteras o fraccionarias el patrón U, operación que, repetimos, puede indicarse como producto de un número por una cantidad de magnitud determinada notada $q \times U$ o $q \times (1\ U)$; o, si la medida es vectorial, $\overline{q} \times U$ o $\overline{q} \times (1\ U)$, donde $(1\ U)$, aunque no se especifique, corresponde a una supuesta cantidad verdadera de la magnitud interesada, establecida por su definición empírica y tomada como patrón unitario elemental de dicha magnitud. Y la cantidad indicada no hay problema en admitir, por definición, que no dependa del orden de escritura, por lo que la misma cantidad será $q \times (1\ U)$ que $(1\ U) \times q$, lo que equivale a axiomatizar la propiedad conmutativa de esta simbología. En resumen, es necesario fijar un principio que permita construir razonamientos ulteriores, y lo llamaremos **postulado fundamental a tener presente en las operaciones con díadas**, cuyo enunciado es que el símbolo de la medición $q\ U$ significa que la cantidad real es q veces enteras o fraccionarias la cantidad implícita en la unidad U, lo cual se indica mediante las tres formas de la definición siguiente:

$$q\ U = q \times (1\ U) = (1\ U) \times q \qquad [3.1]$$

Como definiremos en el apartado siguiente, aquí adelantamos que el signo igual significa que todos los miembros simbolizan la misma cantidad de magnitud, por lo que son sustituibles entre sí.

A su vez, en el caso de un primario vectorial $\overline{q}$, el concreto $\overline{q}\ U$ debe simbolizar la cantidad de una **magnitud vectorial** que tenga la misma dirección y sentido que $\overline{q}$ y cuyo módulo sea el número de veces enteras o fraccionarias igual al módulo de $|\overline{q}|$ la cantidad de la magnitud contenida en la unidad U. Como en el caso escalar, se admitirán como indicativas de este significado las tres notaciones siguientes:

$$\overline{q}\ U = \overline{q} \times (1\ U) = (1\ U) \times \overline{q} \qquad [3.2]$$

Por tanto, a la vista de [3.1] y [3.2] no cabe distinguir entre unidades escalares y vectoriales, porque, tanto para unas como para las otras, toda unidad o cantidad de magnitud U debe admitirse por axioma algebraico que se identifique con la díada escalar $(1\ U)$ y que el elemento numérico que actúe como multiplicador de la díada sea q o $|\overline{q}|$, según se asocie con una magnitud de índole escalar en R o vectorial en R^3, o con otro ente matemático que sirva de multiplicador.

Las magnitudes cuyos multiplicadores sean tales que $q \in \mathrm{R}$ y que puedan tomar cualquier valor se denominan continuas, en cambio, aquellas en que los multiplicadores solo puedan ser números enteros, con $q \in \mathrm{Z}$, se llaman discretas. Se observa que las operaciones con magnitudes discretas quedan comprendidas en las continuas, puesto que sus primarios vendrán representadas por números enteros, que es un subconjunto de los números reales, por lo que las continuas presentan mayor generalidad que las magnitudes discretas; y las continuas quedarán explicadas en abstracto en todo caso por medio de la afinidad con la longitud, que las representa ficticiamente a todas, porque cualquiera de ellas se puede asimilar a la recta real, resultando en todo caso el mismo esquema de razonamiento.

La elección de unidades para cualquier magnitud es arbitraria. Por tanto, **la definición amplia de díada física es la que corresponde a todo par formado por un primario matemático multiplicador, número o vector u otro, y un secundario integrado por cualquier símbolo o símbolos que designen una cierta cantidad de magnitud no especificada e innúmera.**

En el apartado I hemos definido la magnitud como **toda propiedad física afín a la longitud.** Hemos dicho que esto significa que las cantidades de magnitudes pueden tratarse como si fueran segmentos geométricos. Y esto nos sirve para **formular en abstracto el concepto de díada.** Así, cualquier cantidad ϕ de una magnitud se puede asociar con un segmento de longitud arbitraria, que puede sumarse consigo mismo por yuxtaposición las veces que marque un elemento multiplicador o descomponerse en segmentos iguales de menor extensión en tantos como indique un elemento divisor, de acuerdo con las operaciones geométricas elementales, que suponemos conoce el lector. De este modo podemos formar con la cantidad de magnitud ϕ otras cantidades definidas por un elemento multiplicador μ, entero o fraccionario, que simbolizamos con la forma multiplicativa $\mu \times \phi$. Esta operación es la que se desarrolla en el apartado IX, asignándola el símbolo de ley de composición «∘». El factor μ generalmente es un elemento del conjunto de los números reales R o un vector de R^3. Una vez obtenida la cantidad homogénea $\mu \times \phi$ de ϕ, es obvio que $\mu \times \phi$ podemos observarla como un par de elementos heterogéneos μ y ϕ. Recordemos que μ aislado no es una cantidad de longitud, solo es un ente matemático abstracto. Ahora nada nos impide dar nombre a ese par con la denominación **díada abstracta**, eligiendo para notar estos pares simbologías como $\mu\,\phi$, $(\mu\,\phi)$, (μ,ϕ) o $\mu \times \phi$. Como ya hemos visto, diremos que μ es el primario, elemento matemático o multiplicador de la díada, y llamaremos a la cantidad de magnitud ϕ secundario, unidad física, parte dimensional o multiplicando. La díada es, por tanto, el reflejo matemático del proceso de medir, mediante el cual, una unidad fraccionada en otras menores e iguales entre sí, permite formular una medición con la forma de pares del tipo (μ,ϕ). Toda medición es una díada, pero una díada abstracta no tiene por qué ser una medición. Para que las operaciones con díadas, que representan cantidades de magnitudes, verifiquen las propiedades asociativa conmutativa y distributiva, veremos en los apartados siguientes que el conjunto de los multiplicadores $\{\mu\}$ ha de tener estructura de cuerpo, como R. Si designamos μ_1 al elemento unitario multiplicativo de dicho conjunto $\{\mu\}$, postulamos que $\mu_1 \times \phi = \phi$ para cualquier ϕ, y podemos ampliar las formas de expresar la misma cantidad de una magnitud señalada por una díada (μ,ϕ) con $\mu \times (\mu_1,\phi)$ y $(\mu_1,\phi) \times \mu$. El conjunto de todas las díadas formadas con el conjunto de multiplicadores $\{\mu\}$ y asociados a la cantidad de magnitud ϕ se podría representar en abstracto con la grafía $\{\{\mu\},\phi\}$, que llamaremos **conjunto diádico**. Así, por definición, el conjunto diádico de ϕ sobre el cuerpo de multiplicadores $\{\mu\}$ es, por definición, $\{\{\mu\},\phi\} = \{(\mu,\phi) \mid \mu \in \{\mu\}\}$.

Establecida así la definición de las díadas físicas y los conjuntos diádicos, de nada servirían si no se diera forma a las leyes de composición que permitan operar con ellos, dotándolos de estructura algebraica. Y este es el meollo de la tarea capital para resolver la laguna descrita al principio de este trabajo, como lo es justificar debidamente las operaciones con cantidades de las diferentes magnitudes físicas, que es el objeto de la primera parte de esta obra, en la que se utilizarán indistintamente los términos **concreto** o **díada física** para nombrar los elementos básicos que representan simbólicamente toda cantidad de magnitud.

Se mantiene el nombre de concreto por el peso histórico de este concepto, que durante mucho tiempo ha servido para diferenciar los números abstractos, aquellos que indican una cantidad sin especificar ninguna unidad y formados por un único elemento matemático, de los números concretos clásicos, que indican una cantidad asociados a la unidad a que se refieren. No obstante, el nombre **díada** es nuestro preferido para los pares heterogéneos asociados a toda cantidad física, razón por la que se reserva el nombre de álgebra diádica a las diversas estructuras que surgen de las leyes de composición que se van a definir para las díadas, simbolizadas indistintamente con las formas $q\ U$, $(q\ U)$ o (q,U). Y análogamente para las vectoriales, sustituyendo q por $\overline{q}$.

Para desarrollar formalmente el álgebra diádica es preciso definir epistémicamente la adición de díadas, la multiplicación de una díada por un escalar, por ejemplo, un número real, la sustracción de díadas, la multiplicación y división de díadas escalares, y los productos escalar y vectorial de díadas vectoriales. Establecidas estas operaciones se podrán deducir por medios lógicos, con fines más bien teóricos, otras operaciones derivadas como la potenciación, la radicación o la logaritmación de díadas escalares, de la misma manera que la sustracción y la división las derivamos de la adición y de la multiplicación.

Particularizando para el cuerpo R, sea una díada (μ,ϕ), que representa una cantidad de cierta magnitud, donde μ y ϕ pueden ser cualesquiera; si la magnitud es escalar, $\mu=q$ será un número real con $q \in$ R. Cualquier cantidad de la magnitud dada se puede tomar como unidad $\phi = U$, al conjunto universal de todas ellas lo señalaremos con $\{U\}$; por tanto, tendremos que toda unidad U estará en el conjunto total de cantidades $\{U\}$ y escribiremos $U \in \{U\}$. Se concluye con esta notación que toda díada (q, U) será un elemento del conjunto de todas las díadas posibles, que indicaremos $\{R, U\}$ y lo llamaremos **conjunto diádico** de la magnitud considerada relativo a la cantidad U; este conjunto a su vez es evidente que se construye con el producto cartesiano de R y $\{U\}$, es decir, $\{R, U\} = R \times \{U\}$. Reiteramos que U representa cualquier cantidad de la magnitud en cuestión, que en todo caso podría tomarse como patrón. Observamos que $\{R, U\} = \{U\}$. Obviamente, para toda cantidad $U_0 \in \{U\}$ el conjunto diádico construido con ella $\{R, U_0\}$ es completo, comprende todas las cantidades de magnitud, por lo que coincide con $\{R, U\}$, conque $\{R, U_0\} = \{R, U\}$. Las díadas pueden componerse entre sí mediante leyes de composición internas, definidas estableciendo aplicaciones del producto cartesiano $\{R, U\} \times \{R, U\}$ en el mismo $\{R, U\}$; y también pueden componerse con los elementos de otros conjuntos, como por ejemplo R, mediante leyes de composición externas tales como aplicaciones de $R \times \{R, U\}$ en $\{R, U\}$, por lo que hay que abordar la tarea de establecer para ellos un álgebra adecuada. Comprobaremos que las condiciones de cuerpo de R son necesarias para que la estructura diádica mantenga las propiedades asociativa, conmutativa y distributiva en sus operaciones. En cambio, veremos que los entes diádicos carecen de elementos unitario e inverso en el sentido que se refiere a las leyes de composición internas propias de la estructura de grupo. En resumen, para las magnitudes escalares tenemos las siguientes definiciones analíticas fundamentales:

$\{U\} = \{$conjunto de todas las cantidades U de una magnitud$\}$
Conjunto diádico de U sobre R: $\{R, U\} = \{(q, U) \,|\, q \in R\} = R \times \{U\}$
Toda díada es $(q, U) \in \{R, U\}$ con $q \in R$ y $U \in \{U\}$

A su vez, las díadas vectoriales forman un conjunto que se puede simbolizar $\{R^3, U\}$ o si se prefiere $\{V^3, U\}$ o $\{E^3, U\}$, que son notaciones equivalentes usadas indistintamente, son susceptibles de componerse entre sí mediante leyes de composición internas, con aplicaciones del producto cartesiano $\{R^3, U\} \times \{R^3, U\}$ en $\{R^3, U\}$; y también pueden componerse con los elementos de otros conjuntos, como por ejemplo R, mediante leyes de composición externas, con aplicaciones de $R \times \{R^3, U\}$ en $\{R^3, U\}$, por lo que también se abordará la tarea de establecer para ellos un álgebra adecuada, que habremos de procurar sea lo más isomorfa posible con la estructura del espacio vectorial R^3 sobre R.

Las leyes de composición anteriores veremos que no ofrecen demasiada dificultad para su formulación analítica, porque son todas **operaciones aditivas** construidas sobre una misma magnitud, tal como se definen en los apartados V a XI, a pesar de lo cual revelan importantes propiedades, resolviendo, por ejemplo, los misterios históricos de la naturaleza adimensional de las magnitudes angulares, y propiciando a su vez el desarrollo de nuevos e importantes conceptos como la densidad «dismétrica».

Encontraremos más resistencia en la fundamentación de las **operaciones multiplicativas** basadas en dos o más magnitudes iguales o diferentes con unidades cualesquiera U_1 y U_2 para el caso de dos factores, cuyas definiciones y propiedades se establecen en los apartados XII a XVII. Se logrará, no obstante, establecer aplicaciones coherentes entre conjuntos como $\{R, U_1\} \times \{R, U_2\}$ y $\{R, U_C\}$, donde U_C indica una nueva unidad compuesta producida cuando se opera sobre U_1 y U_2 para las magnitudes escalares y análogamente para las vectoriales. Y esta notable capacidad generadora de nuevas magnitudes es precisamente la nota característica de tales operaciones multiplicativas. Los conjuntos $\{R, U_1\}$ y $\{R, U_2\}$ pueden ser iguales, cuando se refieren a la misma magnitud, pero **el conjunto diádico $\{R, U_C\}$ siempre será distinto de los dos anteriores, porque sus elementos son cantidades de una magnitud diferente, nacida al multiplicar magnitudes de acuerdo con lo expuesto en el apartado XII.** Así resulta que, incluso cuando $\{R, U_1\}$ y $\{R, U_2\}$ sean el mismo conjunto, por ejemplo, si ambos se refirieran a la magnitud longitud, resultará que $\{R, U_C\}$ representa todas las cantidades que puede tomar un área, porque

el producto de dos longitudes da lugar a una nueva magnitud compuesta con ellas que llamamos superficie o área. De este modo aparecen las que podríamos llamar **leyes de composición externas generatrices**, que tienen la cualidad especial de producir nuevas magnitudes a partir de la multiplicación de otras cualesquiera. Hecho esto, se comprenderá que esa mayor dificultad de tales operaciones multiplicativas es la que ha provocado que todos, incluido el Sistema Internacional de Unidades, las hayamos ignorado y sustituido por una fácil, arbitraria e ilusoria hipótesis indeseable de «aritmetización» de las magnitudes, error que hemos asumido dejándonos engañar por la simbología aritmética y creando conceptos erróneos. Por ejemplo, las unidades físicas y cualquier cantidad de magnitud no pueden tener inversos multiplicativos como si se tratase de operaciones internas, porque la multiplicación de magnitudes es externa generatriz. Así que notaciones como U^{-1} deben reformularse, como se expone en el apartado XIV y en el anexo. Simbologías como m^{-1}, s^{-1} o kg^{-1} hay que definirlas ex profeso para las magnitudes físicas. Nótese también que no toda ley externa es generatriz, aunque toda ley generatriz ha de ser externa. Por ejemplo, aplicaciones de $R \times \{R, U\}$ en $\{R, U\}$, siendo externas, no generan una nueva magnitud, porque el conjunto $\{R, U\}$, imagen de esa ley externa, es uno de los conjuntos iniciales. Las generatrices son leyes imprescindibles para el álgebra de magnitudes, pues toda formulación física se construye con ellas, dando lugar a nuevas magnitudes compuestas. Por ejemplo, la *segunda ley de Newton* implica componer masa y aceleración, relacionándolas con otra magnitud diferente que es la fuerza. A su vez, componiendo la longitud y el tiempo surge una nueva magnitud que es la velocidad, u operando con la masa y el volumen generamos otra magnitud que es la densidad.

En todo caso, es claro que **el álgebra de magnitudes ha de obedecer a criterios operacionales con díadas**, por lo que hemos de establecer leyes de composición específicas que permitan construir estructuras sui géneris y tengan en cuenta la naturaleza diádica de los fenómenos físicos, evitando simplificaciones ilusorias.

Section IV

DEFINITIONS OF HOMOGENEITY, UNIFORMITY AND EQUALITY

Two scalar dyads $a_1 U_1$ and $a_2 U_2$, each formed by a real number and a unit, we will say that they are **homogeneous** if and only if they refer to quantities of the same magnitude, that is, if their units U_1 and U_2 symbolize quantities uncountable empirical values of the same magnitude.

In turn, the concretes or dyads whose unit is the same we will say that they are **uniform**. So all elements of a set like $\{R,U\}$ are concrete uniform scalars.

On the other hand, two concrete or homogeneous scalar dyads $a_1 U_1$ and $a_2 U_2$ we will say that they are equal if and only if, by definition, they describe the same quantity of the associated magnitude, and we will symbolize the **equality** with the usual equal sign with an expression such as the next:

$$a_1 U_1 = a_2 U_2 \qquad\qquad [4.1]$$

Together $\{R,U_1\}$ determines all quantities of the associated magnitude with reference to the unit U_1. In turn, the set $\{R,U_2\}$ also determines all the quantities of the same magnitude linked to the unit U_2. The equal elements of both sets are related by equation [4.1]. Under these conditions, the criterion of equality of two concretes or uniform scalar dyads cannot be established in a more convenient way than this: we will say that two uniform concretes $a_1 U$ and $a_2 U$ are equal if and only if they have the same numerical or measured part, that is, if the real numbers $a_1 = a_2$ are equal.

What has been said for the dyad scalars with respect to homogeneity and uniformity must be analogous for the vector dyads $\overline{a}_1 U_1$ and $\overline{a}_2 U_2$, and regarding equality it must mean that both dyads refer to the same quantity of the vector magnitude in question. The analytic expression for equality must logically respond to the dyadic equation:

$$\overline{a}_1 U_1 = \overline{a}_2 U_2 \tag{4.2}$$

If the dyads were uniform, it will result $U_1 = U_2 = U$, and it will be said that two uniform vector dyads $\overline{a}_1 U$ and $\overline{a}_2 U$ are equal if and only if the vector equality $\overline{a}_1 = \overline{a}_2$ is verified.

Given two homogeneous units U_1 and U_2, that is, associated to the same magnitude, it is necessary to axiomatize, because physical observations so advise, that there is a real number k such that:

$$U_2 = (1, U_2) = k\, U_1 \tag{4.3}$$

We will call this statement the **axiom of continuity** and it will promote the transformation of quantities of homogeneous magnitudes, linking them to the same unit, which will allow us to add homogeneous concrete entities, as we will see later. In particular, if the units are uniform, we will have $k = 1 \in R$. We must note that the algebraic meaning of [4.3] will be completed with the definition [9.1] of multiplication of a scalar by a dyadic entity.

Note an essential condition of the definition of equality, which is that only homogeneous quantities can be compared, that is, of the same magnitude, so that **every physical equation of equality establishes a law that relates the quantities of magnitudes specified by predefined dyadic algebra operations. and both members must be homogeneous.**

At the end of sections IX and XI we complete the concept of dyadic equality once the operations that allow it to be established with algebraic rigor have been defined: the multiplication of a scalar by a dyad and the homogeneous dyadic division.

Apartado IV

DEFINICIONES DE HOMOGENEIDAD, UNIFORMIDAD E IGUALDAD

Dos díadas escalares $a_1 U_1$ y $a_2 U_2$, formadas cada una de ellas por un número real y una unidad, diremos que son **homogéneas** si y solo si se refieren a cantidades de la misma magnitud, es decir, si sus unidades U_1 y U_2 simbolizan cantidades empíricas no numerables de la misma magnitud.

A su vez, los concretos o díadas cuya unidad sea la misma diremos que son **uniformes**. Así que todos los elementos de un conjunto como $\{R, U\}$ son concretos escalares uniformes.

Por otra parte, dos concretos o díadas escalares homogéneos $a_1 U_1$ y $a_2 U_2$ diremos que son iguales si y solo si, por definición, describen la misma cantidad de la magnitud asociada, y la **igualdad** la simbolizaremos con el signo igual usual con una expresión como la siguiente:

$$a_1 U_1 = a_2 U_2 \qquad [4.1]$$

El conjunto $\{R, U_1\}$ determina todas las cantidades de la magnitud asociada con referencia a la unidad U_1. A su vez, el conjunto $\{R, U_2\}$ determina también todas las cantidades de la misma magnitud vinculadas a la unidad U_2. Los elementos iguales de ambos conjuntos quedan relacionados por la ecuación [4.1]. En estas condiciones, el criterio de igualdad de dos concretos o díadas escalares uniformes no puede establecerse de otro modo más conveniente que este: diremos que dos concretos uniformes $a_1 U$ y $a_2 U$ son iguales si y solo si tienen la misma parte numérica o medida, es decir, si son iguales los números reales $a_1 = a_2$.

Lo dicho para los concretos escalares respecto a la homogeneidad y uniformidad ha de ser análogo para las díadas vectoriales $\overline{a}_1\ U_1$ y $\overline{a}_2\ U_2$, y en cuanto a la igualdad ha de significar que ambas díadas se refieran a la misma cantidad de la magnitud vectorial en cuestión. La expresión analítica de igualdad debe responder lógicamente a la ecuación diádica:

$$\overline{a}_1\ U_1 = \overline{a}_2\ U_2 \qquad [4.2]$$

Si los concretos fuesen uniformes, resultará $U_1 = U_2 = U$, y se dirá que dos concretos vectoriales uniformes $\overline{a}_1\ U$ y $\overline{a}_2\ U$ son iguales si y solo si se verifica la igualdad vectorial $\overline{a}_1 = \overline{a}_2$.

Dadas dos unidades homogéneas U_1 y U_2, es decir, asociadas a una misma magnitud, es preciso axiomatizar, porque las observaciones físicas así lo aconsejan, que existe un número real k tal que:

$$U_2 = (1, U_2) = k\ U_1 \qquad [4.3]$$

Este enunciado lo denominaremos **axioma de continuidad** y propiciará la transformación de cantidades de magnitudes homogéneas, vinculándolas a una misma unidad, lo que permitirá sumar entes concretos homogéneos, como luego veremos. En particular, si las unidades son uniformes, se tendrá $k = 1 \in \mathrm{R}$. Debemos advertir que el significado algebraico de [4.3] quedará completo con la definición [9.1] de multiplicación de un escalar por un ente diádico.

Obsérvese una condición esencial de la definición de igualdad, cual es que solo pueden compararse cantidades homogéneas, es decir de una misma magnitud, por lo que **toda ecuación física de igualdad establece una ley que relaciona las cantidades de magnitudes especificadas mediante operaciones del álgebra diádica predefinida y ambos miembros han de ser homogéneos.**

Al final de los apartados IX y XI completamos el concepto de igualdad diádica una vez definidas las operaciones que permiten establecerla con rigor algebraico: la multiplicación de un escalar por una díada y la división diádica homogénea.

Section V

DEFINITION OF DYADIC ADDITION

A first observation to bear in mind when defining operations with dyadic entities is that certain rules of an axiomatic nature must be respected, based on rational observations of the facts. Thus, to add concretes it is required that the addends be homogeneous, that is, that they refer to the same magnitude, although it would be admissible that the units expressed in the addends were not the same. It would not make sense to add meters with kilograms, because the result could not be indicated in any of the units of the addends; but it does have coherence to add seconds and hours, because the magnitude associated with both units is time, so they can be added and express the sum in seconds or hours, although for this one of the addends should be converted to the unit of the other, otherwise the sum of different units would also be meaningless and should be rejected, because the dyadic addition consists in fact in counting the elements of the addends and those that can be said to be equal.

Therefore, it is necessary to admit as a prior and necessary axiom for the dyadic addition to be valid that in every sum of concrete entities the addends must refer to the same magnitude, that is, the addends must be homogeneous, and before adding them they must be represented in the same unit, a statement that we could call the **axiom of uniformity** of addition. Such a transformation will always be possible by virtue of the axiom of continuity.

Let's start with adding scalar dyads. The sum cannot be conceived in any other way than by establishing a law of internal composition called addition between the uniform scalar concretes, through an application of the Cartesian product

$\{R,U\} \times \{R,U\}$ in $\{R,U\}$. In this way, when the addends are already expressed in the same unit U of a certain magnitude, that is, when they are uniform, the sum of two scalar dyads $a\,U$ and $b\,U$ can be written $a\,U + b\,U$, with the meaning of counting the number of units U that accept the two addends at the same time; and here there is no other option but to admit as a result of the sum the statement that it is equal to $(a+b)$ units U, which would be written with the dyadic form $(a+b)\,U$, because what is added are quantities of elements all equal to the quantity symbolized by the letter U, which is elementary arithmetic, with which it is fully based on an application of the Cartesian product set $\{R,U\} \times \{R,U\}$ in $\{R,U\}$, characterized by the abstract formula that describes the **addition definition equation for scalar diads**:

$$a\,U + b\,U = (a+b)\,U \qquad [5.1]$$

Observing the previous definition, it should be emphasized, even at the risk of seeming repetitive, that it represents the addition of elements equal to the quantity of the unit considered, so it should be read with a meaning like the following: the addition of a quantities equal to the quantity of the unit U added to b quantities equal to the quantity of the unit U equals the sum of real numbers $(a+b)$ quantities equal to the quantity of the unit U; which reduces the addition of dyads to the addition of real numbers with perfect precision.

In order not to reiterate heavily the expression «quantities equal to the quantity of the unit U», we simply substitute the letter U, and thus we will speak simply of «a units U», of «b units U» or of «$(a+b)$ units U».

We must note that the **principle of symbolic economy** leads us to identify different laws of composition with the same symbol: in effect, the definition equation of the concrete addition [5.1] includes in the first member the plus sign in $a\,U + b\,U$, with the meaning of addition of scalar dyads, while the same sign of the second member in $(a+b)\,U$ refers to the addition of the field of real numbers.

In a more abstract and symbolic way but equivalent in result, it could be observed that the addition of scalar dyads behaves analytically as if U were a number, since it could be considered that it reflects the distributive form, which allows to consider that to operate with the addition of scalar concretes it is enough to do it symbolically as if the symbol of the unit of the addends were one more algebraic element and then read the result with the meaning that the addition of concretes is a dyadic entity with a primary equal to the addition in R of the real parts of the addends, associated to the same unit as these.

Let's see an example of addition: if you wanted to add the amount of time of 2 minutes, abbreviated 2 *min*, and the amount of time of 15 seconds, expressed 15 s, sum that symbolically is 2 *min* +15 s, both addends are homogeneous, because they refer to the same magnitude, time, then they can be added; but first they must be expressed in the same unit of time, given the axiom of uniformity of addition; be this the second, by definition of minute, we will have 2 *min* =120 s, so the sum to be calculated is 120 s +15 s, and now, since the addends specify quantities of the same unit, so they are uniform, it is enough add the numerical parts according to the definition of addition to state that the sum is 135 s. It could have been reasoned by operating symbolically with the distributive property for the letter s in the following way, first applying the axiom of uniformity of addition to put the minutes in seconds:

$$2 \text{ } min +15 \text{ } s =120 \text{ } s +15 \text{ } s =(120 +15) \text{ } s =135 \text{ } s$$

And the result would be read with the meaning that the sum of 2 minutes and 15 seconds equals 135 seconds. Therefore, the addition of uniform scalar dyads allows operating in abstract analytical terms with the unit symbols as if they were algebraic elements, although without losing sight of the proper meaning of the indicated notation. If you think about it, this circumstance is not strange, because when 15 s is indicated the meaning is 15×1 s, that is, that 15 s actually represents 15 times the quantity of the time magnitude contained in one second, **a quantity that**

cannot be expressed numerically it is symbolized by the letter s, that is, the product of the real number 15 by s; so the distributive behavior of the symbol s in the above reasoning scheme is not illogical, but undeniable. Simply, the number 15 acts with respect to the quantiity of time s as a **multiplier**.

In the case of the addition of vector dyads, the reasoning scheme to define this operation is totally similar to that of scalars, with the exception that the supporting structure is that of the vector space R^3 or $\mathbf{V}^3$ and, therefore, The addition to which it is reduced is not that of R but the vector, according to the following equation for the **definition of addition of vector dyads**:

$$\overline{a}\,U+\overline{b}\,U=(\overline{a}+\overline{b})\,U \qquad [5.2]$$

Note that the addition of the second member of [5.2] is not that of R, as in the case of [5.1], but the vector addition of R^3 or $\mathbf{V}^3$; while the addition sign of the first member indicates the sum of vector dyads defined here. The same addition sign for two different composition laws.

If you want to be more precise in the differentiation of the composition laws involved, even if only for didactic purposes, to better explain the precise meanings of the definitions [5.1] and [5.2] they must be identified with those of the exact equations, which they differentiate the operations and they could be written with «⊕» for scalar or vector dyadic additions, as well as «+» for sums of scalars or vectors. This is how the explicit analytic expressions of scalar and vector dyadic additions result:

$$a\,U\oplus b\,U=(a+b)\,U \qquad [5.3]$$

$$\overline{a}\,U\oplus\overline{b}\,U=(\overline{a}+\overline{b})\,U \qquad [5.4]$$

And even so, as with the «+» sign between scalars or vectors, which refers to two different operations, we would not be distinguishing in the spellings all those involved, although they are easy to interpret due to the nature of the elements that connect, thus, if the operation sign «⊕» were placed between scalar dyads, the addition would be the scalar dyad; and, if you linked vector dyads, the sum would be the vector dyadic. And so analogously with all the operations involved.

In what precedes, in order to materialize the dyadic addition we have required that the units of the addends be the same, which we have called the axiom of uniformity; however, there is an exception that does allow the addition of non-uniform amounts without reducing them to a common unit: the case where the primaries are equal. In this case, nothing prevents the addition from being analytically expressed when the homogeneous units of the addends are different. Thus, given two units of the same magnitude U_1 and U_2, by the affinity postulate and the isomorphism with the geometric addition of segments described in section XXVIII, article 13, it turns out that adding quantities of magnitudes corresponds biunivocally by affinity with the addition of segments, with which the dyad in explicit notation $(q,U_1 \oplus U_2)$, is equivalent to the sum $(q,U_1) \oplus (q,U_2)$, so it can be indicated:

$$(q,U_1 \oplus U_2) = (q,U_1) \oplus (q,U_2)$$

Or with the classical notation that we have been using, we can also express the above with the form:

$$q\, U_1 \oplus U_2 = q\, U_1 \oplus q\, U_2$$

And the same exception can be established for vector magnitudes, resulting indistinct for both notations:

$$(\overline{q},U_1 \oplus U_2) = (\overline{q},U_1) \oplus (\overline{q},U_2)$$

$$\overline{q}\, U_1 \oplus U_2 = \overline{q}\, U_1 \oplus \overline{q}\, U_2$$

Apartado V

DEFINICIÓN DE ADICIÓN DIÁDICA

Una primera observación a tener presente al definir las operaciones con entes diádicos es que deben respetarse ciertas reglas de índole axiomática, a tenor de las observaciones racionales de los hechos. Así, para sumar concretos se requiere que los sumandos sean homogéneos, es decir, que se refieran a la misma magnitud, aunque sería admisible que las unidades expresadas en los sumandos no fueran la misma. No tendría sentido sumar metros con kilogramos, porque el resultado no se podría indicar en ninguna de las unidades de los sumandos; pero sí que tiene coherencia sumar segundos y horas, porque la magnitud asociada a ambas unidades es el tiempo, por lo que pueden sumarse y expresar la suma en segundos u horas, aunque para ello uno de los sumandos debería convertirse a la unidad del otro, pues si no la suma de unidades distintas carecería también de sentido y debería rechazarse, porque la adición diádica consiste de hecho en contar los elementos de los sumandos y de los que se pueda afirmar que son iguales.

Por tanto, hay que admitir como axioma previo y necesario para que la adición diádica sea válida que en toda suma de entes concretos los sumandos deben referirse a la misma magnitud, es decir, los sumandos deben ser homogéneos, y antes de sumarlos deben representarse en la misma unidad, enunciado que podríamos denominar el **axioma de uniformidad** de la adición. Dicha transformación siempre será posible en virtud del axioma de continuidad.

Comencemos con la adición de díadas escalares. La suma no puede concebirse de otro modo que estableciendo una ley de composición interna llamada adición entre los concretos escalares uniformes, mediante una aplicación del producto cartesiano

$\{R, U\} \times \{R, U\}$ en $\{R, U\}$. De este modo, cuando los sumandos estén ya expresados en la misma unidad U de cierta magnitud, esto es, cuando sean uniformes, la suma de dos díadas escalares $a\,U$ y $b\,U$ se puede escribir $a\,U + b\,U$, con el significado de contar el número de unidades U que acogen los dos sumandos a la vez; y aquí ya no cabe más remedio que admitir como resultado de la suma la afirmación de que sea igual a $(a+b)$ unidades U, que se escribiría con la forma diádica $(a+b)\,U$, porque lo que se suman son cantidades de elementos iguales todos a la cantidad simbolizada con la letra U, que es aritmética elemental, con lo que se llega con pleno fundamento a una aplicación del conjunto producto cartesiano $\{R, U\} \times \{R, U\}$ en $\{R, U\}$, caracterizada por la fórmula abstracta que describe la ecuación de **definición de adición de concretos escalares**:

$$a\,U + b\,U = (a+b)\,U \qquad [5.1]$$

Observando la definición anterior, se debe hacer énfasis, aun a riesgo de parecer reiterativos, en que representa la adición de elementos iguales a la cantidad de la unidad considerada, por lo que debería leerse con un significado como el siguiente: la adición de a cantidades iguales a la cantidad de la unidad U sumadas a b cantidades iguales a la cantidad de la unidad U es igual a la suma de números reales $(a+b)$ cantidades iguales a la cantidad de la unidad U; con lo cual se reduce la adición de concretos a la adición de números reales con perfecta precisión.

Para no reiterar con pesadez la expresión «cantidades iguales a la cantidad de la unidad U», la sustituimos simplemente por la letra U, y así hablaremos simplemente de «a unidades U», de «b unidades U» o de «$(a+b)$ unidades U».

Debemos advertir que el **principio de economía simbólica** nos lleva a identificar con el mismo símbolo leyes de composición diferentes: en efecto, la ecuación de definición de la adición concreta [5.1] incluye en el primer miembro el signo más en $a\,U + b\,U$, con el significado de adición de concretos escalares, mientras que el mismo signo del segundo miembro en $(a+b)\,U$ se refiere a la adición del cuerpo de los números reales.

De una manera más abstracta y simbólica pero equivalente en resultado se podría observar que la adición de concretos escalares se comporta analíticamente como si U fuese un número, ya que podría considerarse que refleja la forma distributiva, lo que permite considerar que para operar con la adición de concretos escalares baste hacerlo simbólicamente como si el símbolo de la unidad de los sumandos fuese un elemento algebraico más y luego leer el resultado con el significado de que la adición de concretos sea un ente diádico con un primario igual a la adición en R de las partes reales de los sumandos, asociada a la misma unidad que estos.

Veamos un ejemplo de adición: si quisieran sumarse la cantidad de tiempo de 2 minutos, abreviadamente 2 min, y la cantidad de tiempo de 15 segundos, expresado 15 s, suma que simbólicamente es 2 $min + 15$ s, ambos sumandos son homogéneos, porque se refieren a la misma magnitud, el tiempo, luego se pueden sumar; pero antes deben expresarse en la misma unidad de tiempo, dado el axioma de uniformidad de la adición; sea esta el segundo, por definición de minuto, tendremos 2 $min = 120$ s, conque la suma a calcular es 120 $s + 15$ s, y ahora, puesto que los sumandos especifican cantidades de la misma unidad, por lo que son uniformes, basta sumar las partes numéricas de acuerdo con la definición de adición para afirmar que la suma es 135 s. Se podría haber razonado operando simbólicamente con la propiedad distributiva para la letra s de la siguiente manera, aplicando en primer lugar el axioma de uniformidad de la adición para poner los minutos en segundos:

$$2\ min + 15\ s = 120\ s + 15\ s = (120 + 15)\ s = 135\ s$$

Y el resultado se leería con el significado de que la suma de 2 minutos y 15 segundos resulta igual a 135 segundos. Por tanto, la adición de díadas escalares uniformes admite operar en términos analíticos abstractos con los símbolos de las unidades como si de elementos algebraicos se tratase, aunque sin perder de vista el significado propio de la notación indicada. Si se piensa bien, no es extraña esta circunstancia, porque cuando se indica 15 s el significado es 15×1 *s*, es decir, que 15 *s* representa realmente 15 veces la cantidad de la magnitud tiempo contenida en un segundo, **cantidad esta no expresable numéricamente que se simboliza con la letra *s***, es decir, el producto del número real 15 por *s*; así que el comportamiento distributivo del símbolo *s* en el esquema de razonamiento anterior no es ilógico, sino innegable. Simplemente, el número 15 actúa respecto de la cantidad de tiempo *s* como **multiplicador**.

En el caso de la adición de concretos vectoriales el esquema de razonamiento para definir esta operación es totalmente similar al de los escalares, con la salvedad de que la estructura que le sirve de soporte es la del espacio vectorial $\mathbf{R}^3$ o $\mathbf{V}^3$ y, por tanto, la adición a la que se reduce no es la de R sino la vectorial, de acuerdo con la ecuación de **definición de adición de concretos vectoriales** siguiente:

$$\overline{a}\ U + \overline{b}\ U = (\overline{a} + \overline{b})\ U \qquad [5.2]$$

Nótese que la adición del segundo miembro de [5.2] no es la de R, como en el caso de [5.1], sino la adición vectorial de $\mathbf{R}^3$ o $\mathbf{V}^3$; mientras que el signo de adición del primer miembro señala la suma de concretos vectoriales aquí definida. El mismo signo de adición para dos leyes de composición diferentes.

Si se quiere ser más precisos en la diferenciación de las leyes de composición que intervienen, aunque solo sea a efectos didácticos, para explicitar mejor los significados precisos de las definiciones [5.1] y [5.2] deben identificarse con los de las ecuaciones exactas, que diferencian las operaciones y que se podrían escribir con «⊕» para las adiciones diádicas escalar o vectorial, así como « + » para las sumas de escalares o de vectores. Así resultan las expresiones analíticas explícitas de las adiciones diádicas escalar y vectorial:

$$a\ U \oplus b\ U = (a + b)\ U \qquad [5.3]$$

$$\overline{a}\ U \oplus \overline{b}\ U = (\overline{a} + \overline{b})\ U \qquad [5.4]$$

Y aún así, como ocurre con el signo « + » entre escalares o vectores, que alude a dos operaciones distintas, no estaríamos distinguiendo en las grafías todas las que intervienen, aunque son fáciles de interpretar por la naturaleza de los elementos que conectan, de modo que, si el signo de operación «$\oplus$» se situase entre díadas escalares, la adición sería la diádica escalar; y, si enlazara díadas vectoriales, la suma sería la diádica vectorial.

En lo que precede, para poder materializar la adición diádica hemos exigido que las unidades de los sumandos sean la misma, lo que hemos denominado axioma de uniformidad; sin embargo, existe una excepción que sí permite la adición de cantidades no uniformes sin reducirlas a una unidad común: el caso en que los primarios sean iguales. En este supuesto, nada impide expresar analíticamente la adición cuando las unidades homogéneas de los sumandos sean distintas. Así, dadas dos unidades de la misma magnitud U_1 y U_2, por el postulado de afinidad y el isomorfismo con la adición geométrica de segmentos descritos en el apartado **XXVIII**, artículo 13, resulta que sumar cantidades de magnitudes se corresponde biunívocamente por afinidad con la adición de segmentos, con lo cual la díada en notación explícita $(q, U_1 \oplus U_2)$, equivale a la suma $(q, U_1) \oplus (q, U_2)$, por lo que se puede indicar:

$$(q, U_1 \oplus U_2) = (q, U_1) \oplus (q, U_2)$$

O con la notación clásica que venimos utilizando también podemos expresar lo anterior con la forma:

$$q \ U_1 \oplus U_2 = q \ U_1 \oplus q \ U_2$$

Y la misma excepción puede establecerse para las magnitudes vectoriales, resultando para ambas notaciones indistintas:

$$(\overline{q}, U_1 \oplus U_2) = (\overline{q}, U_1) \oplus (\overline{q}, U_2)$$

$$\overline{q} \ U_1 \oplus U_2 = \overline{q} \ U_1 \oplus \overline{q} \ U_2$$

Section VI

COMMUTATIVE PROPERTIES AND
ASSOCIATIVE OF THE DYADIC ADDITION

First, let's look at adding scalar dyads. Once this internal law has been defined on the set of the concrete scalars $\{R,U\}$, it is worth asking whether $a\ U$ and $b\ U$ will turn out to be commutative for any two of its elements. The definition of concrete addition [5.3] allows us to write the equality:

$$a\ U \oplus b\ U = (a+b)\ U$$

The commutative property of the addition of the real numbers in R determines that $a+b=b+a$ then, in effect, the dyad $(a+b)\ U$ is the same as $(b+a)\ U$, which in turn is $b\ U \oplus a\ U$, and with it the scalar dyadic addition verifies the commutative property:

$$a\ U \oplus b\ U = b\ U \oplus a\ U$$

Furthermore, this dyadic addition is associative because, starting from the triple addition $(a\ U \oplus b\ U) \oplus c\ U$, the definition [5.3] of dyadic addition allows us to write the equality without problems:

$$(a\ U \oplus b\ U) \oplus c\ U = [(a+b)\ U \oplus c\ U] = [(a+b)+c]\ U \qquad [6.1]$$

Since the addition of real numbers is associative, we will have the equality in the additive group of R:

$$(a+b)+c = a+(b+c)$$

Therefore, the relationship between the dyads scalars indicated below is justified:

$$[(a+b)+c]\ U = [a+(b+c)]\ U$$

The definition of addition [5.3] allows the second member to be decomposed into the dyadic sum of concretes:

$$[a+(b+c)]\,U=a\,U\oplus(b+c)\,U=a\,U\oplus(b\,U\oplus c\,U) \qquad [6.2]$$

So the initial dyad $(a\,U+b\,U)+c\,U$ of [6.1] is the same as the concrete of the second member in [6.2], which is $a\,U\oplus(b\,U\oplus c\,U)$, a result we can call associative property of dyadic addition, written analytically with equality:

$$(a\,U\oplus b\,U)\oplus c\,U=a\,U\oplus(b\,U\oplus c\,U)$$

Regarding the addition of vector dyads, with an identical reasoning to that of scalars, but with the only exception that, instead of the commutative and associative properties of the real numbers of R, based on the commutative and associative properties of the addition of vectors in R^3 or $\mathbf{V}^3$, the commutative and associative properties of the addition [5.4] of vector dyads are concluded, whose analytical forms are:

$$\overline{a}\,U\oplus\overline{b}\,U=\overline{b}\,U\oplus\overline{a}\,U$$

$$(\overline{a}\,U\oplus\overline{b}\,U)\oplus\overline{c}\,U=\overline{a}\,U\oplus(\overline{b}\,U\oplus\overline{c}\,U)$$

Apartado VI

PROPIEDADES CONMUTATIVA Y
ASOCIATIVA DE LA ADICIÓN DIÁDICA

En primer lugar, analicemos la adición de díadas escalares. Una vez definida esta ley interna sobre el conjunto de los concretos escalares $\{R, U\}$, cabe preguntarse si resultará ser conmutativa para dos de sus elementos cualesquiera $a\ U$ y $b\ U$. La definición de adición concreta [5.3] permite escribir la igualdad:

$$a\ U \oplus b\ U = (a+b)\ U$$

La propiedad conmutativa de la adición de los números reales en R determina que $a+b=b+a$ luego, en efecto, la díada $(a+b)\ U$ es la misma que $(b+a)\ U$, que a su vez es $b\ U \oplus a\ U$, y con ello la adición diádica escalar verifica la propiedad conmutativa:

$$a\ U \oplus b\ U = b\ U \oplus a\ U$$

Además, esta adición concreta es asociativa, porque, partiendo de la adición triple $(a\ U \oplus b\ U) \oplus c\ U$, la definición [5.3] de adición diádica permite escribir sin problemas la igualdad:

$$(a\ U \oplus b\ U) \oplus c\ U = [(a+b)\ U \oplus c\ U] = [(a+b)+c]\ U \qquad [6.1]$$

Como la adición de números reales es asociativa, tendremos en el grupo aditivo de R la igualdad:

$$(a+b)+c = a+(b+c)$$

Por lo que está justificada la relación entre los concretos escalares que se indican a continuación:

$$[(a+b)+c]\ U = [a+(b+c)]\ U$$

La definición de adición [5.3] permite descomponer el segundo miembro en la suma diádica de concretos:

$$[a+(b+c)] \ U = a \ U \oplus (b+c) \ U = a \ U \oplus (b \ U \oplus c \ U) \qquad [6.2]$$

Así que la díada inicial $(a \ U + b \ U) + c \ U$ de [6.1] es la misma que el concreto del segundo miembro en [6.2], que es $a \ U \oplus (b \ U \oplus c \ U)$, resultado que podemos llamar propiedad asociativa de la adición diádica, escrita analíticamente con la igualdad:

$$(a \ U \oplus b \ U) \oplus c \ U = a \ U \oplus (b \ U \oplus c \ U)$$

En cuanto a la adición de díadas vectoriales, con un razonamiento idéntico al de los escalares, pero con la única salvedad de que, en vez de las propiedades conmutativa y asociativa de los números reales de **R**, basándose en las propiedades conmutativa y asociativa de la adición de vectores en $\mathbf{R}^3$ o $\mathbf{V}^3$, se concluyen las propiedades conmutativa y asociativa de la adición [5.4] de díadas vectoriales, cuyas formas analíticas son:

$$\overline{a} \ U \oplus \overline{b} \ U = \overline{b} \ U \oplus \overline{a} \ U$$

$$(\overline{a} \ U \oplus \overline{b} \ U) \oplus \overline{c} \ U = \overline{a} \ U \oplus (\overline{b} \ U \oplus \overline{c} \ U)$$

Section VII

EXISTENCE OF NEUTRAL AND SYMMETRICAL
ELEMENTS FOR DYADIC ADDITION

Let's see that the addition of scalar dyads defined on $\{R,U\}$ is a law of internal composition such that there is a neutral element. Indeed, we can easily observe that the dyad $0\,U$, where 0 is the real zero, is such that added to any dyad $a\,U$ it verifies the following reasoning:

$$a\,U \oplus 0\,U = 0\,U \oplus a\,U \qquad [7.1]$$

The above equality is a consequence of the commutative property. In turn, the definition [5.3] of addition of scalar dyads allows us to write:

$$a\,U \oplus 0\,U = (a+0)\,U \text{ y } 0\,U \oplus a\,U = (0+a)\,U$$

The zero or neutral element of the real numbers is such that $a+0$ is the same as $0+a$ and equal in both cases to a, with what we have:

$$a+0 = 0+a = a$$

Therefore, the two members of the first relation [7.1] are equal to $a\,U$:

$$a\,U \oplus 0\,U = 0\,U \oplus a\,U = a\,U$$

And this means, by definition of a neutral element, that the dyad $0\,U$ is one of the law of internal composition called dyadic addition and defined in $\{R,U\}$ by [5.3].

Furthermore, for every concrete scalar entity or dyad $a\,U$ can always form the dyad $(-a)\,U$, because in R there exists the opposite $-a$ of every $a \in R$, by virtue of the additive group structure of the set R of real numbers, and so that it is $a+(-a)=0$; so that, adding the two indicated dyads, applying the definition of addition and operating with the real numbers, we have the chain of equalities:

$$a\,U \oplus (-a)\,U = [a+(-a)]\,U = 0\,U$$

And so it turns out that $(-a)\,U$ is the scalar dyad opposite to the right of $a\,U$. The commutative property makes it unnecessary to check the condition of neutral element from the left $(-a)\,U \oplus a\,U$, which is also satisfied; and, in sum, for every concrete scalar to $a\,U$ there exists another with the form $(-a)\,U$ such that, added both from the right or from the left, give the neutral element $0\,U$, which means that the set of dyads scalars $\{R,U\}$, formed with the reals R and the unit U, and endowed with the internal composition law of addition defined by [5.3], has the abelian group structure, because the addition verifies the commutative and associative properties, there is a neutral element and for every dyad there is its opposite[5].

In turn, for the vector concretes $\{R^3,U\}$ and, given the group structure for the vector addition of R^3 or $\mathbf{V}^3$, which are the same space symbolized in two different ways, there are the neutral or null $\overline{0}$ and symmetric or opposite vectors $-\overline{a}$ of every vector $\overline{a}$, so that, by means of a reasoning scheme identical to the previous one, with $\overline{a}+\overline{0}=\overline{0}+\overline{a}=\overline{a}$ and with $\overline{a}+(-\overline{a})=(-\overline{a})+\overline{a}=\overline{0}$, it can be concluded that there are null and opposite vector dyads, and that are those indicated by the symbols $\overline{0}\,U$, for the null vector concrete, and $(-\overline{a})\,U$, for the opposite vector dyad of any other $\overline{a}\,U$, because with [5.4] the following reasonings are spun:

$$\overline{a}\,U \oplus \overline{0}\,U = \overline{0}\,U \oplus \overline{a}\,U = (\overline{a}+\overline{0})\,U = (\overline{0}+\overline{a})\,U = \overline{a}\,U$$

$$\overline{a}\,U \oplus (-\overline{a})\,U = [\overline{a}+(-\overline{a})]\,U = [(-\overline{a})+\overline{a}]\,U = (-\overline{a})\,U \oplus \overline{a}\,U = \overline{0}\,U$$

An introduction to algebraic structures can be found in the same author's syllabus, «Lesson 37» of Mathematize I.

Apartado VII

EXISTENCIA DE ELEMENTOS NEUTRO Y
SIMÉTRICO PARA LA ADICIÓN DIÁDICA

Veamos que la adición de díadas escalares definida sobre $\{R, U\}$ es una ley de composición interna tal que existe elemento neutro. En efecto, observamos con facilidad que el concreto $0\ U$, siendo 0 el cero real, es tal que sumado a cualquier concreto $a\ U$ verifica el siguiente razonamiento:

$$a\ U \oplus 0\ U = 0\ U \oplus a\ U \qquad [7.1]$$

La igualdad anterior es consecuencia de la propiedad conmutativa. A su vez, la definición [5.3] de adición de díadas escalares permite escribir:

$$a\ U \oplus 0\ U = (a+0)\ U \text{ y } 0\ U \oplus a\ U = (0+a)\ U$$

El cero o elemento neutro de los números reales es tal que $a+0$ es lo mismo que $0+a$ e igual en ambos casos a a, con lo que se tiene:

$$a+0 = 0+a = a$$

Por tanto, los dos miembros de la primera relación [7.1] son iguales a $a\ U$:

$$a\ U \oplus 0\ U = 0\ U \oplus a\ U = a\ U$$

Y ello significa, por definición de elemento neutro, que la díada $0\ U$ lo es de la ley de composición interna llamada adición diádica y definida en $\{R, U\}$ mediante [5.3].

Además, para todo ente concreto escalar $a\ U$ se puede formar siempre el $(-a)\ U$, porque en R existe el opuesto $-a$ de todo $a \in R$, en virtud de la estructura de grupo aditivo del conjunto R de los números reales, y de tal suerte que es $a+(-a)=0$; de modo que,

80

sumando las dos díadas indicadas, aplicando la definición de adición y operando con los números reales, se tiene la cadena de igualdades:

$$a \ U \oplus (-a) \ U = [a + (-a)] \ U = 0 \ U$$

Y así resulta que $(-a) \ U$ es el concreto escalar opuesto por la derecha del $a \ U$. La propiedad conmutativa hace innecesario comprobar la condición de elemento neutro por la izquierda $(-a) \ U \oplus a \ U$, que también se satisface; y, en suma, para todo concreto escalar $a \ U$ existe otro con la forma $(-a) \ U$ tal que, sumados ambos por la derecha o por la izquierda, dan el elemento neutro $0 \ U$, lo que significa que el conjunto de las díadas escalares $\{R, U\}$, formado con los reales R y la unidad U, y dotado de la ley de composición interna de la adición definida por [5.3], tiene la estructura de grupo abeliano, porque la adición verifica las propiedades conmutativa y asociativa, existe elemento neutro y para todo concreto existe su opuesto[6].

A su vez, para los concretos vectoriales $\{R^3, U\}$ y, dada la estructura de grupo para la adición vectorial de R^3 o $\mathbf{V}^3$, que son el mismo espacio simbolizado de dos formas distintas, existen los vectores neutro o nulo $\overline{0}$ y simétrico u opuesto $-\overline{a}$ de todo vector $\overline{a}$, por lo que, mediante un esquema de razonamiento idéntico al anterior, con $\overline{a} + \overline{0} = \overline{0} + \overline{a} = \overline{a}$ y con $\overline{a} + (-\overline{a}) = (-\overline{a}) + \overline{a} = \overline{0}$, se puede concluir que existen las díadas vectoriales nula y opuesta, y que son las indicadas por los símbolos $\overline{0} \ U$, para el concreto vectorial nulo, y $(-\overline{a}) \ U$, para la díada vectorial opuesta de cualquier otra $\overline{a} \ U$, porque con [5.4] se hilan los siguientes razonamientos:

$$\overline{a} \ U \oplus \overline{0} \ U = \overline{0} \ U \oplus \overline{a} \ U = (\overline{a} + \overline{0}) \ U = (\overline{0} + \overline{a}) \ U = \overline{a} \ U$$

$$\overline{a} \ U \oplus (-\overline{a}) \ U = [\overline{a} + (-\overline{a})] \ U = [(-\overline{a}) + \overline{a}] \ U = (-\overline{a}) \ U \oplus \overline{a} \ U = \overline{0} \ U$$

Una introducción a las estructuras algebraicas se puede encontrar en el temario del mismo autor, «Lección 37» de *Matematizar 1*.

Section VIII

DEFINITION OF DYADIC SUBTRATION

The generic operation called subtraction derives from addition, therefore, to be consistent with this algebraic criterion, it must be admitted that the subtraction of dyadic entities must satisfy the same axiom of uniformity of operating with equal units, and this because the definition of concrete subtraction cannot have any other foundation than the dyadic addition, in which all the terms are associated to the same unit, which leads us to the following formulation: **the subtraction of a minuend and a subtrahend, both being dyads scalars of the set $\{R,U\}$ or dyads vectors of the set $\{R^3,U\}$, is said equal to the difference if and only if the subtrahend added to the difference equals the minuend.**

So that addition and subtraction require by the condition of law of internal composition that the addends, in one case, or the minuend and the subtrahend, in the other, refer to the same unit, that is, that they be uniform. Given the abelian group structure of $\{R,U\}$ and $\{R^3,U\}$, we have no problem in defining the subtraction of concrete entities as the sum of the minuend and the opposite of the subtrahend, which always exists, as we have established in the previous section, and thus we will have a result that added to the subtrahend will be equal to the minuend, according to the generic, convenient and usual definition that has been accepted for the subtraction of numerical entities. So we agree to indicate the subtraction of any two scalar dyads from $a\,U$ and $b\,b\,U$ as the application of $\{R,U\} \times \{R,U\}$ to $\{R,U\}$ defined with the following formula, which assumes, as the addition, that uniform quantities referring to the same unit are subtracted, because it operates as a law of internal composition on the set of the concrete scalars

{R,U}, and written applying the symbolic economy of operations is as follows:

$$a\,U - b\,U = [a + (-b)]\,U = (a - b)\,U \qquad [8.1]$$

With this definition the concrete or dyadic scalar subtraction is reduced to that of R and, as we pointed out for the addition, the same subtraction sign with the hyphen is used with different meanings, because in [8.1] the minus of $a\,U - b\,U$ refers to the subtraction of scalar dyads and the minus of $(a - b)\,U$ indicates the subtraction of real numbers.

In turn, the subtraction or subtraction of any two vector dyads $\overline{a}\,U$ and $\overline{b}\,U$ is defined as the scalar with the application of {R³,U}×{R³,U} in {R³,U} such that, like addition, it assumes that uniform quantities referred to the same unit are subtracted, because it operates as a law of internal composition on the set of vector dyads {R³,U}, and this according to the definition equation:

$$\overline{a}\,U - \overline{b}\,U = [\overline{a} + (-\overline{b})]\,U = (\overline{a} - \overline{b})\,U \qquad [8.2]$$

Here it should also be noted, as we can see in the addition, that the definitions are such that the abbreviation or symbol of the unit behaves for the purposes of formal writing with the distributive property, as if the unit symbol were another algebraic element.

Therefore, the different meanings that correspond to the same sign with which the different operations are indicated, depending on the elements between which it is located, should be noted, so the precise meanings of definitions [8.1] and [8.2] should be understood as the of the exact equations, which differentiate the composition laws and which could be written with «⊖» for scalar or vector dyadic differences, as well as «−» for subtractions of scalars or vectors, in this way:

$$a\,U \ominus b\,U = [a + (-b)]\,U = (a - b)\,U \qquad [8.3]$$

$$\overline{a}\,U \ominus \overline{b}\,U = [\overline{a} + (-\overline{b})]\,U = (\overline{a} - \overline{b})\,U \qquad [8.4]$$

As we have already indicated, the symbol $-b$ denotes the opposite real number of b, and with $-\overline{b}$ the opposite vector of $\overline{b}$.

The dyadic subtraction can be deduced from the addition and based on the generic subtraction criterion. To do this, consider the following scalar sum:

$$d\,U \oplus s\,U = m\,U \qquad\qquad [8.5]$$

The symbology of addition has simply been adapted to indicate with the letters m a minuend, s a subtrahend and d for the difference that corresponds to them. The usual subtraction criterion, as an operation that, given an addition, allows one of the addends to be obtained as a function of the sum and the other by adding, authorizes the establishment of the dyadic difference, distinguished with the sign «$\ominus$», by means of the equation:

$$m\,U \ominus s\,U = d\,U \qquad\qquad [8.6]$$

The definition [5.3] of dyadic addition applied to [8.5], allows us to write:

$$(d+s)\,U = m\,U \qquad\qquad [8.7]$$

The equality criterion of section IV applied to expression [8.7] gives us the relation $(d+s)=m$, and the subtraction in R leads us to write $d=m-s$. So, substituting d in [8.6], we have:

$$m\,U \ominus s\,U = (m-s)\,U \qquad\qquad [8.8]$$

Conclusion [8.8] is identical to definition [8.1], and it means that the dyadic difference between two scalar concretes, called minuend and subtrahend, is a dyad called difference whose primary is the subtraction in R of the primaries and with the same secondary that they.

The reasoning for the subtraction of vector dyads is completely analogous to the previous one with scalars, given the additive and abelian group structure of R³, which presents the same formal properties for the sum of vectors that are given with real numbers.

Apartado VIII

DEFINICIÓN DE SUSTRACCIÓN DIÁDICA

La operación genérica llamada sustracción deriva de la adición, por lo que, para ser coherentes con este criterio algebraico, habrá que admitir que la sustracción de entes diádicos deba satisfacer el mismo axioma de uniformidad de operar con unidades iguales, y ello porque la definición de sustracción concreta no puede tener otro fundamento que la adición diádica, en la que todos los términos están asociados a la misma unidad, lo que nos lleva a la siguiente formulación: **la sustracción de un minuendo y un sustraendo, siendo ambos concretos escalares del conjunto $\{R, U\}$ o vectoriales del conjunto $\{R^3, U\}$, se dice igual a la diferencia si y solo si el sustraendo sumado a la diferencia iguala al minuendo.**

De modo que la adición y la sustracción exigen por la condición de ley de composición interna que los sumandos, en un caso, o el minuendo y el sustraendo, en el otro, se refieran a la misma unidad, es decir, que sean uniformes. Dada la estructura de grupo abeliano de $\{R, U\}$ y de $\{R^3, U\}$, no tenemos problema en definir la resta de entes concretos como la suma del minuendo y el opuesto del sustraendo, que siempre existe, como hemos asentado en el apartado anterior, y así se tendrá un resultado que sumado al sustraendo será igual al minuendo, de acuerdo con la definición genérica, conveniente y usual que viene admitiéndose para la resta de entes numéricos. De modo que estaremos de acuerdo en indicar la **resta de dos díadas escalares** cualesquiera $a\,U$ y $b\,U$ como la aplicación de $\{R, U\} \times \{R, U\}$ en $\{R, U\}$ definida con la fórmula siguiente, que asume, como la adición, que se restan cantidades uniformes referidas a la misma unidad, porque opera como ley de composición interna sobre el conjunto de los concretos escalares $\{R, U\}$, y escrita aplicando la economía simbólica de operaciones queda así:

$$a \ U - b \ U = [a + (-b)] \ U = (a - b) \ U \qquad [8.1]$$

Con esta definición se reduce la sustracción concreta o diádica escalar a la de R y, como apuntamos para la adición, el mismo signo de resta con el guión se utiliza con significados diferentes, porque en [8.1] el menos de $a \ U - b \ U$ se refiere a la sustracción de díadas escalares y el menos de $(a - b) \ U$ señala la sustracción de números reales.

A su vez, la **resta o sustracción de dos díadas vectoriales** cualesquiera $\overline{a} \ U$ y $\overline{b} \ U$ queda definida como la escalar con la aplicación de $\{R^3, U\} \times \{R^3, U\}$ en $\{R^3, U\}$ tal que, como la adición, asume que se restan cantidades uniformes referidas a la misma unidad, porque opera como ley de composición interna sobre el conjunto de los concretos vectoriales $\{R^3, U\}$, y ello de acuerdo con la ecuación de definición:

$$\overline{a} \ U - \overline{b} \ U = [\overline{a} + (-\overline{b})] \ U = (\overline{a} - \overline{b}) \ U \qquad [8.2]$$

Aquí también hay que notar, como apreciamos en la adición, que las definiciones son tales que la abreviatura o símbolo de la unidad se comporta a efectos de escritura formal con la propiedad distributiva, como si el símbolo unitario fuese un elemento algebraico más.

Por tanto, deben advertirse los diferentes significados que corresponden al mismo signo con que se indican las distintas operaciones, según los elementos entre los que se sitúe, por lo que los significados precisos de las definiciones [8.1] y [8.2] deben entenderse como los de las ecuaciones exactas, que diferencian las leyes de composición y que se podrían escribir con «⊖» para las diferencias diádicas escalar o vectorial, así como «−» para las restas de escalares o de vectores, de esta manera:

$$a \ U \ominus b \ U = [a + (-b)] \ U = (a - b) \ U \qquad [8.3]$$

$$\overline{a} \ U \ominus \overline{b} \ U = [\overline{a} + (-\overline{b})] \ U = (\overline{a} - \overline{b}) \ U \qquad [8.4]$$

Como ya hemos indicado, con el símbolo $-b$ se denota el número real opuesto de b, y con $-\overline{b}$ el vector opuesto de $\overline{b}$.

La resta diádica se puede deducir a partir de la adición y en función del criterio genérico de sustracción. Para ello, consideremos la suma escalar siguiente:

$$d\ U \oplus s\ U = m\ U \qquad [8.5]$$

Simplemente se ha adaptado la simbología de la adición para indicar con las letras m a un minuendo, s a un sustraendo y d para la diferencia que les corresponda. El criterio usual de sustracción, como operación que, dada una adición, permite obtener uno de los sumandos en función de la suma y del otro sumando, autoriza a establecer la diferencia diádica, distinguida con el signo «$\ominus$», mediante la ecuación:

$$m\ U \ominus s\ U = d\ U \qquad [8.6]$$

La definición [5.3] de adición diádica aplicada a [8.5], nos permite escribir:

$$(d+s)\ U = m\ U \qquad [8.7]$$

El criterio de igualdad del apartado IV aplicado a la expresión [8.7] nos proporciona la relación $(d+s)=m$, y la sustracción en R nos lleva a escribir $d=m-s$. De modo que, sustituyendo d en [8.6], se tiene:

$$m\ U \ominus s\ U = (m-s)\ U \qquad [8.8]$$

La conclusión [8.8] es idéntica a la definición [8.1], y significa que la diferencia diádica entre dos concretos escalares, llamados minuendo y sustraendo, es un concreto llamado diferencia cuyo primario es la sustracción en R de los primarios y con el mismo secundario que ellos.

El razonamiento para la sustracción de díadas vectoriales es completamente análogo al anterior con escalares, dada la estructura de grupo aditivo y abeliano de R^3, que presenta las mismas propiedades formales para la suma de vectores que se dan con los números reales.

Section IX

DEFINITION OF MULTIPLICATION
OF A REAL NUMBER FOR A DYAD

If a quantity $a\,U$ or (a,U) of a certain scalar magnitude, from the set of uniform concretes $\{R,U\}$, is multiplied by a real number p, by definition, we will establish that the result is a dyad such that Its primary or measure is the real product $a \times p$. For now we will use the same multiplication sign for the new operation, but knowing that it is not the product of real numbers, but the multiplication of a scalar by a quantity of magnitude. Then we will detail our own symbology to highlight the difference between these two operations. In analytical terms the definition of this product is:

$$(a\,U) \times p = (a \times p)\,U \qquad [9.1]$$

If in the product $(a\,U) \times p$ its factors are commuted to form the multiplication $p \times (a\,U)$, we must establish by convenient definition that both quantities coincide, so we can axiomatically admit the **commutative property** of the product of a real number times a concrete scalar, expressing it analytically using the expression:

$$(a\,U) \times p = p \times (a\,U) \qquad [9.2]$$

In the interest of mathematical precision, it is worth clarifying that the definition of the product of a scalar by a dyad, defined here analytically with the definition equations [9.1] and [9.2], does not represent more than a law of external composition or application of the Cartesian product $R \times \{R,U\}$ on $\{R,U\}$ on the left and the symmetric on the right of $\{R,U\} \times R$ on $\{R,U\}$.

It should be idle and remember that the same multiplication sign, generally the cross «×» or a blank space, are used to symbolize different laws of composition, depending on the pairs of elements between which they appear. However, let's clarify it: indicating «×» the product of R, designating «•» the product of a scalar by a vector, and indicating with the sign «∘» the product of a real number by a scalar or vector dyadic element, the definitions [9.1] and [9.2] must be interpreted according to the explicit expressions:

$$(a\ U)\circ p = p\circ(a\ U) = (a{\times}p)\ U = (p{\times}a)\ U$$

$$(\overline{a}\ U)\circ p = p\circ(\overline{a}\ U) = (\overline{a}{\bullet}p)\ U = (p{\bullet}\overline{a})\ U$$

Likewise, it is observed that the previous definitions allow the unit U to be manipulated symbolically as if it were one more algebraic element, such that formally in writing and together with the other elements they appear to be commutative and associative, although the operations on each member are different.

Multiplying the null element of R or zero by any concrete scalar to U of {R,U}, we will have:

Multiplicando el elemento nulo de R o cero por cualquier concreto escalar $a\ U$ de {R,U}, se tendrá:

$$0\circ(a\ U) = (a\ U)\circ 0 = (0{\times}a)\ U = (a{\times}0)\ U = 0\ U$$

That is, any concrete of {R,U} multiplied by the zero scalar, $0{\in}R$, the null element of the addition in R, from the right and from the left, is equal to the null element $0\ U{\in}\{R,U\}$.

In turn, the unit element for the multiplication of R, which is usually symbolized by the number 1, is such that, when compounded with this new external law, it leaves any scalar dyad unchanged, which we can verify without more than taking the generic $a\ U$ and compose it with the unit of R, according to the reasoning that is based on [9.1], [9.2] and on the condition of 1 as a unit element of the product in R, which is such that $a{\times}1 = 1{\times}a = a$, all of which motivates the following reasoning of analytic dyadic algebra:

$$1 \circ (a\ U) = (a\ U) \circ 1 = (1 \times a)\ U = (a \times 1)\ U = a\ U$$

So, in effect, the unit $1 \in R$ of the multiplicative group R is such that it operates as a unit scalar for the external law «∘» of $R \times \{R,U\}$ in $\{R,U\}$ or $\{R,U\} \times R$ in $\{R,U\}$.

In turn, for the uniform vector dyads of $\{R^3,U\}$, the multiplication by a scalar «∘» must refer to the external law «•» of the vector space R^3 or $\mathbf{V}^3$ over R, with the definition having the same form of [9.1] and [9.2], although with the proper meaning of this algebraic structure:

$$(\overline{a}\ U) \circ p = (\overline{a} \bullet p)\ U \qquad\qquad [9.3]$$

$$(\overline{a}\ U) \circ p = p \circ (\overline{a}\ U) \qquad\qquad [9.4]$$

These definition equations represent an external composition law «∘» or application of the Cartesian product $R \times \{R^3,U\}$ in $\{R^3,U\}$ to the left and $\{R^3,U\} \times R$ in $\{R^3,U\}$ by the right, defined as a function of the external law «•» of R^3 or $\mathbf{V}^3$ over R. The vector space structure of R^3 guarantees that for the null and unit elements of R we have that it is $0 \bullet \overline{a} = \overline{a} \bullet 0 = \overline{0}$, being $\overline{0}$ the null vector of the vector addition in R^3, and $1 \bullet \overline{a} = \overline{a} \bullet 1 = \overline{a}$, where 1 is the unit element of the multiplication in R, so here too we have the same properties deduced for scalar dyads, according to the following schemes ilatives:

$$0 \circ (\overline{a}\ U) = (\overline{a}\ U) \circ 0 = (0 \bullet \overline{a})\ U = = (\overline{a} \bullet 0)\ U = \overline{0}\ U$$

$$1 \circ (\overline{a}\ U) = (\overline{a}\ U) \circ 1 = (1 \bullet \overline{a})\ U = (\overline{a} \bullet 1)\ U = \overline{a}\ U$$

Let us now use the dyadic notation with an inner comma for greater clarity. **The definition of multiplication by a scalar allows us to write $(a,U) = a \circ (1,U) = a \circ U$, then, $(a,U) = a \circ U$.** We must ask ourselves what happens to a dyad (a,U) when its unit is multiplied by a number p. Have:

$$(a, p \circ U) = a \circ (1, p \circ U) = a \circ (p \circ U) = a \circ (p,U) = (a \times p, U)$$

So, in the product of a dyad and a scalar, it is indifferent to multiply its primary or its secondary by said number.

Apartado IX

DEFINICIÓN DE MULTIPLICACIÓN
DE UN ESCALAR POR UNA DÍADA

Si una cantidad $a\ U$ o (a, U) de cierta magnitud escalar, del conjunto de los concretos uniformes $\{R, U\}$, se multiplica por un número real p, por definición, vamos a establecer que el resultado es una díada tal que su medida o primario es el producto real $a \times p$. De momento usaremos el mismo signo de multiplicación para la nueva operación, pero sabiendo que no es el producto de números reales, sino la multiplicación de un escalar por una cantidad de magnitud. Luego detallaremos una simbología propia para poner de manifiesto la diferencia entre estas dos operaciones. En términos analíticos la definición de este producto es:

$$(a\ U) \times p = (a \times p)\ U \qquad [9.1]$$

Si en el producto $(a\ U) \times p$ se conmutan sus factores para formar la multiplicación $p \times (a\ U)$, hemos de establecer por definición conveniente que ambas cantidades coinciden, por lo que se puede admitir axiomáticamente la **propiedad conmutativa** del producto de un número real por un concreto escalar, expresándola analíticamente mediante la expresión:

$$(a\ U) \times p = p \times (a\ U) \qquad [9.2]$$

En interés de la precisión matemática no está de más aclarar que la definición del producto de un escalar por un concreto, definida aquí analíticamente con las ecuaciones de definición [9.1] y [9.2], no representan sino una ley de composición externa o aplicación del producto cartesiano $R \times \{R, U\}$ en $\{R, U\}$ por la izquierda y la simétrica por la derecha de $\{R, U\} \times R$ en $\{R, U\}$.

Tendría que resultar ocioso ya recordar que el mismo signo de multiplicación, generalmente el aspa «×» o un espacio en blanco, se utilizan para simbolizar leyes de composición diferentes, según cuáles sean las parejas de elementos entre los que aparezcan. No obstante, precisémoslo: indicando «×» el producto de R, designando «•» el producto de un escalar por un vector, y señalando con el signo «∘» el producto de un número real por un elemento diádico escalar o vectorial, las definiciones [9.1] y [9.2] deben interpretarse conforme a las expresiones explícitas:

$$(a\ U)\circ p = p\circ(a\ U) = (a\times p)\ U = (p\times a)\ U$$

$$(\overline{a}\ U)\circ p = p\circ(\overline{a}\ U) = (\overline{a}\bullet p)\ U = (p\bullet\overline{a})\ U$$

Asimismo, se observa que las definiciones anteriores permiten manipular simbólicamente la unidad U como si fuese un elemento algebraico más, tal que formalmente en la escritura y junto a los otros elementos aparentan ser conmutativos y asociativos, aunque las operaciones en cada miembro sean distintas.

Multiplicando el elemento nulo de R o cero por cualquier concreto escalar $a\ U$ de $\{R,U\}$, se tendrá:

$$0\circ(a\ U) = (a\ U)\circ 0 = (0\times a)\ U = (a\times 0)\ U = 0\ U$$

Es decir, que cualquier concreto de $\{R,U\}$ multiplicado por el escalar cero, $0\in R$, elemento nulo de la adición en R, por la derecha y por la izquierda, es igual al elemento nulo $0\ U\in\{R,U\}$.

A su vez, el elemento unidad para la multiplicación de R, que usualmente se simboliza con el número 1, es tal que compuesto con esta nueva ley externa, deja inalterado cualquier concreto escalar, lo que podemos comprobar sin más que tomar el genérico $a\ U$ y componerlo con la unidad de R, de acuerdo con el razonamiento que se fundamenta en [9.1], [9.2] y en la condición del 1 como elemento unitario del producto en R, que es tal que $a\times 1 = 1\times a = a$, todo lo cual motiva el siguiente razonamiento:

$$1\circ(a\ U) = (a\ U)\circ 1 = (1\times a)\ U = (a\times 1)\ U = a\ U$$

De modo que, en efecto, la unidad $1\in R$ del grupo multiplicativo R, es tal que opera como escalar unitario para la ley externa «∘» de $R\times\{R,U\}$ en $\{R,U\}$ o $\{R,U\}\times R$ en $\{R,U\}$.

A su vez, para las díadas vectoriales uniformes de $\{R^3, U\}$, la multiplicación por un escalar «$\circ$» debe referirse a la ley externa «$\bullet$» del espacio vectorial R^3 o V^3 sobre R, con la definición que presenta la misma forma de [9.1] y [9.2], aunque con el significado propio de esta estructura algebraica:

$$(\overline{a}\ U)\circ p = (\overline{a} \bullet p)\ U \qquad\qquad [9.3]$$

$$(\overline{a}\ U)\circ p = p\circ(\overline{a}\ U) \qquad\qquad [9.4]$$

Estas ecuaciones de definición representan una ley de composición externa «$\circ$» o aplicación del producto cartesiano $R \times \{R^3, U\}$ en $\{R^3, U\}$ por la izquierda y de $\{R^3, U\} \times R$ en $\{R^3, U\}$ por la derecha, definida en función de la ley externa «$\bullet$» de R^3 o V^3 sobre R. La estructura de espacio vectorial de R^3 garantiza que para los elementos nulo y unidad de R se tenga que es $0 \bullet \overline{a} = \overline{a} \bullet 0 = \overline{0}$, siendo $\overline{0}$ el vector nulo de la adición vectorial en R^3, y $1 \bullet \overline{a} = \overline{a} \bullet 1 = \overline{a}$, siendo 1 el elemento unidad de la multiplicación en R, por lo que también aquí se tienen las mismas propiedades deducidas para las díadas escalares, de acuerdo con los siguientes esquemas ilativos:

$$0\circ(\overline{a}\ U) = (\overline{a}\ U)\circ 0 = (0\bullet\overline{a})\ U = = (\overline{a}\bullet 0)\ U = \overline{0}\ U$$

$$1\circ(\overline{a}\ U) = (\overline{a}\ U)\circ 1 = (1\bullet\overline{a})\ U = (\overline{a}\bullet 1)\ U = \overline{a}\ U$$

Utilicemos a continuación para mayor claridad la notación diádica con coma interior. **La definición de multiplicación por un escalar permite escribir** $(a, U) = a\circ(1, U) = a\circ U$, **luego,** $(a, U) = a\circ U$. Debemos preguntarnos qué le ocurre a una díada (a, U) cuando su unidad se multiplica por un número p. Tenemos:

$$(a, p\circ U) = a\circ(1, p\circ U) = a\circ(p\circ U) = a\circ(p, U) = (a\times p, U)$$

Es decir, que si el secundario de una díada (a, U) se multiplica por un número p, la díada resultante es $(a\times p, U)$, que es la misma que se obtiene al aplicar la definición de producto por un escalar. Conque podemos concluir la siguiente propiedad: **en el producto de una díada por un escalar resulta indiferente multiplicar su primario o su secundario por dicho número.**

Section X

DISTRIBUTIVE PROPERTIES OF THE DYADIC MULTIPLICATION BY A SCALAR

Given the scalar or real number p of R and the dyad scalars $a\,U$ and $b\,U$ of {R,U}, let's compose the dyad $p\circ(a\,U\oplus b\,U)$ from the left, the same reasoning would be had for the product from the right. Let us differentiate the signs of each operation. The definition of scalar concrete addition [5.3] will allow to write:

$$p\circ(a\,U\oplus b\,U)=p\circ[(a+b)\,U] \qquad\qquad [10.1]$$

The definition of the product of a scalar and a dyad, described by [9.3] and [9.4], allows us to transform the second member of the previous expression:

$$p\circ[(a+b)\,U]=[p\times(a+b)]\,U$$

The distributive property of the product with respect to the sum in R is $p\times(a+b)=(p\times a)+(p\times b)$, which allows the conversion:

$$[p\times(a+b)]\,U=[(p\times a)+(p\times b)]\,U$$

The definition of concrete addition [5.3] favors decomposing the second member into two addends:

$$[(p\times a)+(p\times b)]\,U=(p\times a)\,U\oplus(p\times b)\,U \qquad\qquad [10.2]$$

In conclusion, the first member of the equality [10.1] is equal to the second member of [10.2], resulting in the written form of the distributive property of the left product of a scalar with respect to the dyadic addition:

$$p\circ(a\,U\oplus b\,U)=(p\times a)\,U\oplus(p\times b)\,U$$

And by a completely similar reasoning, the right distributive property is also concluded:

$$(a \, U \oplus b \, U) \circ p = (a \times p) \, U \oplus (b \times p) \, U$$

Since in R they are $p \times a = a \times p$ and $p \times b = b \times p$, we verify the equality $p \circ (a \, U \oplus b \, U) = (a \, U \oplus b \, U) \circ p$, in coherence with the commutativity axiom [9.2].

To check that the reciprocal also holds, let's take two scalars p and q of R, and a dyad $a \, U$ of {R,U}. Let's compose the dyad $(p + q) \circ (a \, U)$. By definition of this external law with [9.1] and [9.2] or their explicit [9.3] and [9.4], we will have on the left and analogously on the right:

$$(p + q) \circ (a \, U) = [(p + q) \times a] \, U \qquad [10.3]$$

The distributive property in R, with $(p + q) \times a = (p \times a) + (q \times a)$, justifies taking the next logical step and writing:

$$[(p + q) \times a] \, U = [(p \times a) + (q \times a)] \, U$$

The definition of addition of concretes [5.3] motivates the passage to the following line of reasoning:

$$[(p \times a) + (q \times a)] \, U = (p \times a) \, U \oplus (q \times a) \, U \qquad [10.4]$$

Therefore, the first member of the expression [10.3] that initiates the reasoning is equal to the second of the last one [10.4], reflecting the reciprocal distributive property on the left:

$$(p + q) \circ (a \, U) = (p \times a) \, U \oplus (q \times a) \, U$$

The same scheme is presented for the reciprocal on the right, with the result:

$$(a \, U) \circ (p + q) = (a \times p) \, U \oplus (a \times q) \, U$$

Since in R they are $p \times a = a \times p$ and $q \times a = a \times q$, we check the equality $(p + q) \circ (a \, U) = (a \, U) \circ (p + q)$, in coherence with the axiom [9.2] of commutativity.

By means of a totally analogous reasoning scheme applied to the definition of concrete subtraction given by the simplified [8.1] or its explicit [8.3], we arrive at the distributive properties of the product by a scalar with respect to the dyadic and real subtractions, from the left and on the right, which can be expressed by the following four equations:

$$p \circ (a\ U \ominus b\ U) = (p \times a)\ U \ominus (p \times b)\ U$$

$$(p-q) \circ (a\ U) = (p \times a)\ U \ominus (q \times a)\ U$$

$$(a\ U \ominus b\ U) \circ p = (a \times p)\ U \ominus (b \times p)\ U$$

$$(a\ U) \circ (p-q) = (a \times p)\ U \ominus (a \times q)\ U$$

Even at the risk of making us heavy, it should be noted again that, for symbolic economy, it is usual to write expressions with the same multiplication, addition and subtraction symbols, to refer indistinctly to the respective composition laws in R and in {R,U} Therefore, depending on the positions of such signs between the composite pairs, the appropriate meanings will have to be attributed to them. With such a principle of symbolic economy, the equations derived earlier would be simplified as follows:

$$p \times (a\ U + b\ U) = (p \times a)\ U + (p \times b)\ U$$

$$(p+q) \times (a\ U) = (p \times a)\ U + (q \times a)\ U$$

$$(a\ U + b\ U) \times p = (a \times p)\ U + (b \times p)\ U$$

$$(a\ U) \times (p+q) = (a \times p)\ U + (a \times q)\ U$$

$$p \times (a\ U - b\ U) = (p \times a)\ U - (p \times b)\ U$$

$$(p-q) \times (a\ U) = (p \times a)\ U - (q \times a)\ U$$

$$(a\ U - b\ U) \times p = (a \times p)\ U - (b \times p)\ U$$

$$(a\ U) \times (p-q) = (a \times p)\ U - (a \times q)\ U$$

For the vector dyads of $\{R^3, U\}$, using an identical logical scheme, although based on the definition of addition of vector concretes [5.2] or its explicit [5.4] and on the distributive property, which guarantees the structure of the space itself R^3 vector, we will have the same formal equations as for the scalar dyads, on the left and on the right, although now referring to vector dyads. With all the explicit operations we will have:

$$p \circ (\overline{a} \, U \oplus \overline{b} \, U) = (p \bullet \overline{a}) \, U \oplus (p \bullet \overline{b}) \, U$$

$$(p+q) \circ (\overline{a} \, U) = (p \bullet \overline{a}) \, U \oplus (q \bullet \overline{a}) \, U$$

$$p \circ (\overline{a} \, U \ominus \overline{b} \, U) = (p \bullet \overline{a}) \, U \ominus (p \bullet \overline{b}) \, U$$

$$(p-q) \circ (\overline{a} \, U) = (p \bullet \overline{a}) \, U \ominus (q \bullet \overline{a}) \, U$$

$$(\overline{a} \, U \oplus \overline{b} \, U) \circ p = (\overline{a} \bullet p) \, U \oplus (\overline{b} \bullet p) \, U$$

$$(\overline{a} \, U) \circ (p+q) = (\overline{a} \bullet p) \, U \oplus (\overline{a} \bullet q) \, U$$

$$(\overline{a} \, U \ominus \overline{b} \, U) \circ p = (\overline{a} \bullet p) \, U \ominus (\overline{b} \bullet p) \, U$$

$$(p-q) \circ (\overline{a} \, U) = (p \bullet \overline{a}) \, U \ominus (q \bullet \overline{a}) \, U$$

And, applying the principle of symbolic economy, the same simplified expressions appear correlative:

$$p \times (\overline{a} \, U + \overline{b} \, U) = (p \times \overline{a}) \, U + (p \times \overline{b}) \, U$$

$$(p+q) \times (\overline{a} \, U) = (p \times \overline{a}) \, U + (q \times \overline{a}) \, U$$

$$p \times (\overline{a} \, U - \overline{b} \, U) = (p \times \overline{a}) \, U - (p \times \overline{b}) \, U$$

$$(p-q) \times (\overline{a} \, U) = (p \times \overline{a}) \, U - (q \times \overline{a}) \, U$$

$$(\overline{a} \, U + \overline{b} \, U) \times p = (\overline{a} \times p) \, U + (\overline{b} \times p) \, U$$

$$(\overline{a} \, U) \times (p+q) = (\overline{a} \times p) \, U + (\overline{a} \times q) \, U$$

$$(\overline{a} \, U - \overline{b} \, U) \times p = (\overline{a} \times p) \, U - (\overline{b} \times p) \, U$$

$$(p-q) \times (\overline{a} \, U) = (p \times \overline{a}) \, U - (q \times \overline{a}) \, U$$

In any case, although the operations of the first and second members do not coincide, it is justified to manipulate all the elements, including the unit U, with the fiction that they are all algebraic terms of R or R^3, even though U really is not, and this on the basis of the definitions and properties of these external laws, not because the illusion of the mere symbolism of the simplified formulas justifies it in any way.

Apartado X

PROPIEDADES DISTRIBUTIVAS DE LA
MULTIPLICACIÓN DIÁDICA POR UN ESCALAR

Dados el escalar o número real p de R y los concretos escalares $a\ U$ y $b\ U$ de $\{R, U\}$, compongamos la díada $p \circ (a\ U \oplus b\ U)$ por la izquierda, idéntico razonamiento se tendría para el producto por la derecha. Diferenciemos los signos de cada operación. La definición de adición concreta escalar [5.3] permitirá escribir:

$$p \circ (a\ U \oplus b\ U) = p \circ [(a + b)\ U] \qquad [10.1]$$

La definición del producto de un escalar por una díada, descrita por [9.3] y [9.4], permite transformar el segundo miembro de la anterior expresión:

$$p \circ [(a + b)\ U] = [p \times (a + b)]\ U$$

La propiedad distributiva del producto respecto de la suma en R es $p \times (a + b) = (p \times a) + (p \times b)$, lo que autoriza la conversión:

$$[p \times (a + b)]\ U = [(p \times a) + (p \times b)]\ U$$

La definición de adición concreta [5.3] propicia descomponer el segundo miembro en dos sumandos:

$$[(p \times a) + (p \times b)]\ U = (p \times a)\ U \oplus (p \times b)\ U \qquad [10.2]$$

En conclusión, el primer miembro de la igualdad [10.1] es igual al segundo miembro de [10.2], resultando la forma escrita de la propiedad distributiva del producto por la izquierda de un escalar respecto de la adición diádica:

$$p \circ (a\ U \oplus b\ U) = (p \times a)\ U \oplus (p \times b)\ U$$

Y mediante un razonamiento completamente similar, se concluye también la propiedad distributiva por la derecha:

$$(a\ U \oplus b\ U) \circ p = (a \times p)\ U \oplus (b \times p)\ U$$

Como en R son $p \times a = a \times p$ y $p \times b = b \times p$, verificamos la igualdad $p \circ (a\ U \oplus b\ U) = (a\ U \oplus b\ U) \circ p$, en coherencia con el axioma de conmutatividad [9.2].

Para comprobar que también se verifica la recíproca, tomemos dos escalares p y q de R, y una díada $a\ U$ de $\{R,U\}$. Compongamos la díada $(p+q) \circ (a\ U)$. Por definición de esta ley externa con [9.1] y [9.2] o sus explícitas [9.3] y [9.4], tendremos por la izquierda y análogamente resultará por la derecha:

$$(p+q) \circ (a\ U) = [(p+q) \times a]\ U \qquad [10.3]$$

La propiedad distributiva en R, con $(p+q) \times a = (p \times a) + (q \times a)$, justifica dar el siguiente paso lógico y escribir:

$$[(p+q) \times a]\ U = [(p \times a) + (q \times a)]\ U$$

La definición de adición de concretos [5.3] motiva el paso a la siguiente línea de razonamiento:

$$[(p \times a) + (q \times a)]\ U = (p \times a)\ U \oplus (q \times a)\ U \qquad [10.4]$$

Por tanto, el primer miembro de la expresión [10.3] que inicia el razonamiento es igual al segundo de la última [10.4], reflejando la propiedad distributiva recíproca por la izquierda:

$$(p+q) \circ (a\ U) = (p \times a)\ U \oplus (q \times a)\ U$$

Idéntico esquema se presenta para la recíproca por la derecha, con el resultado:

$$(a\ U) \circ (p+q) = (a \times p)\ U \oplus (a \times q)\ U$$

Como en R son $p \times a = a \times p$ y $q \times a = a \times q$, comprobamos la igualdad $(p+q) \circ (a\ U) = (a\ U) \circ (p+q)$, en coherencia con el axioma [9.2] de la conmutatividad.

Mediante un esquema de razonamiento totalmente análogo aplicado a la definición de sustracción concreta dada por la simplificada [8.1] o su explícita [8.3] se llega a las propiedades distributivas del producto por un escalar respecto a las restas diádica y real, por la izquierda y por la derecha, que se pueden expresar mediante las cuatro ecuaciones siguientes:

$$p \circ (a \: U \ominus b \: U) = (p \times a) \: U \ominus (p \times b) \: U$$

$$(p-q) \circ (a \: U) = (p \times a) \: U \ominus (q \times a) \: U$$

$$(a \: U \ominus b \: U) \circ p = (a \times p) \: U \ominus (b \times p) \: U$$

$$(a \: U) \circ (p-q) = (a \times p) \: U \ominus (a \times q) \: U$$

Aun a riesgo de hacernos pesados, debe observarse nuevamente que, por economía simbólica, es usual escribir las expresiones con los mismos símbolos de multiplicación, adición y sustracción, para referirse indistintamente a las respectivas leyes de composición en R y en $\{R, U\}$, por lo que, en función de las posiciones de tales signos entre los pares compuestos, habrá que atribuirles los significados adecuados. Con tal principio de economía simbólica, las ecuaciones deducidas antes quedarían simplificadas así:

$$p \times (a \: U + b \: U) = (p \times a) \: U + (p \times b) \: U$$

$$(p + q) \times (a \: U) = (p \times a) \: U + (q \times a) \: U$$

$$(a \: U + b \: U) \times p = (a \times p) \: U + (b \times p) \: U$$

$$(a \: U) \times (p + q) = (a \times p) \: U + (a \times q) \: U$$

$$p \times (a \: U - b \: U) = (p \times a) \: U - (p \times b) \: U$$

$$(p - q) \times (a \: U) = (p \times a) \: U - (q \times a) \: U$$

$$(a \: U - b \: U) \times p = (a \times p) \: U - (b \times p) \: U$$

$$(a \: U) \times (p - q) = (a \times p) \: U - (a \times q) \: U$$

Para las díadas vectoriales de $\{R^3, U\}$, mediante un esquema lógico idéntico, aunque con base en la definición de adición de concretos vectoriales [5.2] o su explícita [5.4] y en la propiedad distributiva, que garantiza la propia estructura del espacio vectorial R^3, tendremos las mismas ecuaciones formales que para los concretos escalares, por la izquierda y por la derecha, aunque ahora referidas a las díadas vectoriales. Con todas las operaciones explícitas tendremos:

$$p\circ(\overline{a}\ U\oplus\overline{b}\ U)=(p\bullet\overline{a})\ U\oplus(p\bullet\overline{b})\ U$$

$$(p+q)\circ(\overline{a}\ U)=(p\bullet\overline{a})\ U\oplus(q\bullet\overline{a})\ U$$

$$p\circ(\overline{a}\ U\ominus\overline{b}\ U)=(p\bullet\overline{a})\ U\ominus(p\bullet\overline{b})\ U$$

$$(p-q)\circ(\overline{a}\ U)=(p\bullet\overline{a})\ U\ominus(q\bullet\overline{a})\ U$$

$$(\overline{a}\ U\oplus\overline{b}\ U)\circ p=(\overline{a}\bullet p)\ U\oplus(\overline{b}\bullet p)\ U$$

$$(\overline{a}\ U)\circ(p+q)=(\overline{a}\bullet p)\ U\oplus(\overline{a}\bullet q)\ U$$

$$(\overline{a}\ U\ominus\overline{b}\ U)\circ p=(\overline{a}\bullet p)\ U\ominus(\overline{b}\bullet p)\ U$$

$$(p-q)\circ(\overline{a}\ U)=(p\bullet\overline{a})\ U\ominus(q\bullet\overline{a})\ U$$

Y, aplicando el principio de economía simbólica, aparecen correlativas las mismas expresiones simplificadas:

$$p\times(\overline{a}\ U+\overline{b}\ U)=(p\times\overline{a})\ U+(p\times\overline{b})\ U$$

$$(p+q)\times(\overline{a}\ U)=(p\times\overline{a})\ U+(q\times\overline{a})\ U$$

$$p\times(\overline{a}\ U-\overline{b}\ U)=(p\times\overline{a})\ U-(p\times\overline{b})\ U$$

$$(p-q)\times(\overline{a}\ U)=(p\times\overline{a})\ U-(q\times\overline{a})\ U$$

$$(\overline{a}\ U+\overline{b}\ U)\times p=(\overline{a}\times p)\ U+(\overline{b}\times p)\ U$$

$$(\overline{a}\ U)\times(p+q)=(\overline{a}\times p)\ U+(\overline{a}\times q)\ U$$

$$(\overline{a}\ U-\overline{b}\ U)\times p=(\overline{a}\times p)\ U-(\overline{b}\times p)\ U$$

$$(p-q)\times(\overline{a}\ U)=(p\times\overline{a})\ U-(q\times\overline{a})\ U$$

En todo caso, aunque las operaciones de los primeros y segundos miembros no coinciden, queda justificado manipular todos los elementos, incluido el unitario U, con la ficción de que todos sean términos algebraicos de R o de R^3, aunque realmente U no lo sea, y ello en base a las definiciones y propiedades de estas leyes externas, no porque la ilusión de la mera simbología de las fórmulas simplificadas lo justifique en modo alguno.

Section XI

DEFINITION OF DIVISION BETWEEN HOMOGENOUS DYADS

Let us first deal with the quotient between two homogeneous units, and remember that the units are always scalar concretes. For this, if the homogeneous units U_1 and U_2 are of the same magnitude, the continuity axiom [4.3] makes it possible to ensure that the real number k exists such that $U_2 = k\ U_1$. The symbolic definition [3.1] of scalar concrete determines that this expression means the same as this: $(1\ U_2) = k \times (1\ U_1)$. Note that this multiplication is actually the dyadic symbolized by «∘», given by $(1\ U_2) = k \circ (1\ U_1)$. The common concept of division allows us to consider in the abstract that $(1\ U_2)$ is associated with a dividend, that $(1\ U_1)$ is a divisor and that k is a quotient, and thus it can be considered that the division between $(1\ U_2)$ and $(1\ U_1)$, which nothing prevents symbolizing with the form of a quotient with a double bar $(1\ U_2)/\!/(1\ U_1)$, or what would mean the same $U_2/\!/U_1$, will be equal to a quotient k, which is a real number. From this we can conclude that the ratio or division between two homogeneous units must always be equal to a real number. If the units were the same, we would have that k is the unit of R, that is, the quotient of every unit between itself will be the real unit, which justifies the way of operating by simplifying the symbols of the units that appear in the numerator and denominator of physical equations. We thus observe that it is in the axiom of continuity [4.3] that the germ of the definition of division of homogeneous units is found.

To practice, let's reason with symbolic economy and now take two homogeneous scalar concretes a1 $a_1\ U_1$ and $a_2\ U_2$, we know that there exists k such that $U_2 = k\ U_1$, so that the concrete $a_2\ U_2$, given the definition [3.1] of scalar dyad, is you can write like this:

$$a_2 \, U_2 = a_2 \times (1 \, U_2) = a_2 \times (k \, U_1) = (a_2 \times k) \, U_1$$

Multiplying the equality by a_1 and, operating with the product [9.1] and [9.2] by a scalar and the laws of R, we will have:

$$a_1 \times (a_2 \, U_2) = a_1 \times [(a_2 \times k) \, U_1] = (a_1 \times a_2 \times k) \, U_1 = (a_2 \times k)(a_1 \, U_1)$$

Now multiplying by a_1^{-1}, knowing that $a_1 \times a_1^{-1} = a_1^{-1} \times a_1 = 1$, because these entities are elements of R, we will have:

$$(a_2 \, U_2) = (a_1^{-1} \times a_2 \times k)(a_1 \, U_1)$$

Assuming that $(a_2 \, U_2)$ is a dividend and $(a_1 \, U_1)$ a divisor, the quotient between the two, which can be symbolized with the common notation $(a_2 \, U_2)/(a_1 \, U_1)$, will be given by the real number that results from the operation of the second member of [11.1]:

$$\frac{a_2 \, U_2}{a_1 \, U_1} = a_1^{-1} \times a_2 \times k = \frac{a_2}{a_1} \times k \qquad [11.1]$$

This means that the quotient between any two homogeneous scalar dyads is the real number given by the last member of the definition equation [11.1].

For vector concretes, we must note that vector algebra is such that multiplication by a scalar relates collinear vectors, so that the concrete division will only be possible when the dividend and divisor primaries are in turn collinear. So that the vector concretes $\overline{a}_1 \, U_1$ and $\overline{a}_2 \, U_2$ are now, such that the vectors $\overline{a}_1$ and $\overline{a}_2$ are collinear and that the units U_1 and U_2 are homogeneous. As in the previous case of scalar concretes, homogeneity assumes that there exists k such that $U_2 = k \, U_1$. In turn, vector algebra ensures that there exists a scalar λ such that $\overline{a}_2 = \lambda \times \overline{a}_1$. Operating with vector algebra and using the definition [9.3] and [9.4] of a concrete product by a scalar, we can write with full justification the following reasoning:

$$\overline{a}_2 \, U_2 = (\lambda \times \overline{a}_1)(k \, U_1) = [(\lambda \times \overline{a}_1) \times k] \, U_1 =$$

$$=[(\lambda \times k) \times \overline{a}_1] U_1 = (\lambda \times k) \times (\overline{a}_1 U_1)$$

Looking at the first and last members, according to the usual division criteria, we can assume that $(\overline{a}_2 U_2)$ is a dividend, that $(\overline{a}_1 U_1)$ is a divisor and that $(\lambda \times k)$ is a quotient. With this, we will arrive at the division formulation between collinear and homogeneous vector concretes, which with the usual symbology will be established by the following definition equation:

$$\frac{\overline{a}_2 U_2}{\overline{a}_1 U_1} = \lambda \times k \qquad [11.2]$$

So, by definition, the quotient of two collinear vector dyads is the scalar $(\lambda \times k) \in R$.

This section has been developed and reasoned with intention, taking into account the principle of symbolic economy, without explicitly differentiating the homonymous operations of multiplication and division, for which, strictly speaking, divisions [11.1] and [11.2] should be written with their own dyadic quotient sign, for which at the beginning we had chosen the double bar, with which these equations should explicitly understand as follows:

$$\frac{a_2 U_2}{a_1 U_1} = \frac{a_2}{a_1} \times k$$

$$\frac{\overline{a}_2 U_2}{\overline{a}_1 U_1} = \lambda \times k$$

These expressions justify the simplification of the homogeneous units in numerator and denominator, so that the dyadic ratio of two homogeneous quantities will always result in a determined real number. On the other hand, it is possible to conceive of two other dyadic divisions, which we will also indicate with a double line, although they are different composition laws: those that correspond to a dyadic, scalar or vector dividend, divided by a divisor that is a real number. It is enough to observe the previous equations to deduce these other two, which relate all the elements of the division of a scalar dyad by a real number and the division of a vector dyad by a real number, according to the following two equations:

$$\frac{a_2\,U_2}{\dfrac{a_2}{a_1} \times k} = a_1\,U_1 \qquad [11.3]$$

$$\frac{\overline{a_2\,U_2}}{\lambda \times k} = \overline{a_1\,U_1} \qquad [11.4]$$

Therefore, the division of a scalar dyad by a real number produces another scalar dyad, but not just any one, but the one established by equation [11.3]. In turn, the division of a vector dyad by a real number will give rise to another collinear vector dyad that verifies equation [11.4].

Continuing with the completion of the dyadic equality criterion, to indicate that two quantities of the same magnitude $(a_1\,U_1)$ and $(a_2\,U_2)$ are equal, we will write $(a_1\,U_1)=(a_2\,U_2)$. The parentheses are superfluous, but we specify them to mark the dyadic pairs well. By the scalar multiplication defined above we can write $(a_1\,U_1)=1\circ(a_2\,U_2)$, with which $(a_1\,U_1)=1\circ(a_2\,U_2)$. And in this equality we can define $(a_1\,U_1)$ as a dividend, $(a_2\,U_2)$ as a divisor and 1 as a quotient, with respect to the multiplication «○», which we can notice $(a_1\,U_1)/\!\!/(a_2\,U_2)=1$. Since we have called this operation homogeneous dyadic division, we will have that the dyadic quotient of two equal dyads, representative of the same quantity of a certain magnitude, is the unit of real numbers. In analytical terms, the following important property of equality results, which is the basis of physical equations:

$$\left(a_1\,U_1\right)=\left(a_2\,U_2\right) \;\Rightarrow\; \frac{\left(a_1\,U_1\right)}{\left(a_2\,U_2\right)}=1$$

Apartado XI
DEFINICIÓN DE DIVISIÓN ENTRE
DÍADAS HOMOGÉNEAS

Abordemos en primer lugar el cociente entre dos unidades homogéneas, y recordemos que las unidades son siempre concretos escalares. Para ello, sean las unidades homogéneas U_1 y U_2 de una misma magnitud, el axioma de continuidad [4.3] permite asegurar que exista el número real k tal que $U_2 = k \ U_1$. La definición simbólica [3.1] de concreto escalar determina que esta expresión signifique lo mismo que esta: $(1 \ U_2) = k \times (1 \ U_1)$. Obsérvese que realmente dicha multiplicación es la diádica simbolizada con «°», dada por $(1 \ U_2) = k \circ (1 \ U_1)$. El concepto común de división permite considerar en abstracto que $(1 \ U_2)$ se asocie a un dividendo, que $(1 \ U_1)$ sea un divisor y que k sea un cociente, y así se podrá considerar que la división entre $(1 \ U_2)$ y $(1 \ U_1)$, que nada impide simbolizar con la forma de un cociente con doble barra $(1 \ U_2) /\!/ (1 \ U_1)$, o lo que significaría lo mismo $U_2 /\!/ U_1$, resultará igual a un cociente k, que es un número real. De ello podemos concluir que la razón o división entre dos unidades homogéneas siempre ha de ser igual a un número real. Si las unidades fuesen la misma, se tendría que k es la unidad de R, es decir, el cociente de toda unidad entre sí misma será la unidad real, lo que justifica la manera de operar simplificando los símbolos de las unidades que aparecen en el numerador y el denominador de las ecuaciones físicas. Observamos así que es en el axioma de continuidad [4.3] donde se halla el germen de la definición de división de unidades homogéneas.

Para practicar, razonemos con economía simbólica y tomemos ahora dos concretos escalares homogéneos $a_1 \ U_1$ y $a_2 \ U_2$, sabemos que existe k tal que $U_2 = k \ U_1$, de modo que el concreto $a_2 \ U_2$, dada la definición [3.1] de concreto escalar, se podrá escribir así:

$$a_2\ U_2 = a_2 \times (1\ U_2) = a_2 \times (k\ U_1) = (a_2 \times k)\ U_1$$

Multiplicando la igualdad por a_1 y, operando con el producto [9.1] y [9.2] por un escalar y las leyes de R, se tendrá:

$$a_1 \times (a_2\ U_2) = a_1 \times [(a_2 \times k)\ U_1] = (a_1 \times a_2 \times k)\ U_1 = (a_2 \times k)\ (a_1\ U_1)$$

Multiplicando ahora por a_1^{-1}, sabiendo que $a_1 \times a_1^{-1} = a_1^{-1} \times a_1 = 1$, porque estos entes son elementos de R, tendremos:

$$(a_2\ U_2) = (a_1^{-1} \times a_2 \times k)\ (a_1\ U_1)$$

Imaginando que $(a_2\ U_2)$ sea un dividendo y $(a_1\ U_1)$ un divisor, el cociente entre ambos, que se puede simbolizar con la notación común $(a_2\ U_2)/(a_1\ U_1)$, vendrá dado por el número real que resulte de la operación del segundo miembro de [11.1]:

$$\frac{a_2\ U_2}{a_1\ U_1} = a_1^{-1} \times a_2 \times k = \frac{a_2}{a_1} \times k \qquad [11.1]$$

Lo que significa que el cociente entre dos díadas escalares homogéneas cualesquiera es el número real dado por el último miembro de la ecuación de definición [11.1].

Para concretos vectoriales hemos de advertir que el álgebra vectorial es tal que la multiplicación por un escalar relaciona vectores colineales, por lo que la división concreta, solo será posible cuando los primarios de dividendo y divisor sean a su vez colineales. De modo que sean ahora los concretos vectoriales $\overline{a}_1\ U_1$ y $\overline{a}_2\ U_2$, tales que los vectores $\overline{a}_1$ y $\overline{a}_2$ sean colineales y que las unidades U_1 y U_2 sean homogéneas. Como en el caso anterior de concretos escalares, la homogeneidad supone que exista k tal que $U_2 = k\ U_1$. A su vez, el álgebra vectorial asegura que exista un escalar λ tal que $\overline{a}_2 = \lambda \times \overline{a}_1$. Operando con el álgebra vectorial y sirviéndonos de la definición [9.3] y [9.4] de producto concreto por un escalar, podremos escribir con plena justificación el siguiente razonamiento:

$$\overline{a}_2\ U_2 = (\lambda \times \overline{a}_1)\ (k\ U_1) = [(\lambda \times \overline{a}_1) \times k]\ U_1 =$$

$$=[(\lambda \times k) \times \overline{a}_1] \; U_1 = (\lambda \times k) \times (\overline{a}_1 \; U_1)$$

Observando el primer y el último miembro, de acuerdo con el criterio usual de división, podemos asumir que $(\overline{a}_2 \; U_2)$ sea un dividendo, que $(\overline{a}_1 \; U_1)$ sea un divisor y que $(\lambda \times k)$ sea un cociente. Con ello, llegaremos a la formulación de división entre concretos vectoriales colineales y homogéneos, que con la simbología usual quedará establecida mediante la siguiente ecuación de definición:

$$\frac{\overline{a}_2 \; U_2}{\overline{a}_1 \; U_1} = \lambda \times k \qquad [11.2]$$

De modo que, por definición, el cociente de dos concretos vectoriales colineales es el escalar $(\lambda \times k) \in R$.

Este apartado lo hemos desarrollado y razonado con intención teniendo en cuenta el principio de economía simbólica, sin diferenciar explícitamente las operaciones homónimas de multiplicación y división, por lo que, en rigor, las divisiones [11.1] y [11.2] deberían escribirse con su propio signo de cociente diádico, para el que al principio habíamos elegido la doble barra, con lo que explícitamente dichas ecuaciones deberían entender así:

$$\frac{a_2 \; U_2}{a_1 \; U_1} = \frac{a_2}{a_1} \times k$$

$$\frac{\overline{a}_2 \; U_2}{\overline{a}_1 \; U_1} = \lambda \times k$$

Estas expresiones justifican la simplificación de las unidades en numerador y denominador, de modo que la razón diádica de dos cantidades homogéneas dará como resultado siempre un número real determinado. Por otra parte, es posible concebir otras dos divisiones diádicas, que señalaremos también con doble raya, aunque se trate de leyes de composición diferentes: las que corresponden a un dividendo diádico, escalar o vectorial, dividido entre un divisor que sea un número real. Basta observar las ecuaciones anteriores para deducir estas otras dos, que relacionan todos los elementos de la división de una díada escalar por un número real y la división de una díada vectorial por un número real, de acuerdo con las dos ecuaciones siguientes:

$$\frac{a_2\,U_2}{\dfrac{a_2}{a_1}\times k} = a_1\,U_1 \qquad [11.3]$$

$$\frac{\overline{a}_2\,U_2}{\lambda\times k} = \overline{a}_1\,U_1 \qquad [11.4]$$

Por tanto, la división de una díada escalar por un número real produce otra díada escalar, pero no una cualquiera, sino la establecida por la ecuación [11.3]. A su vez, la división de una díada vectorial entre un número real dará lugar a otra díada vectorial colineal tal que verifique la ecuación [11.4].

Siguiendo con la compleción del **criterio de igualdad diádica**, para indicar que dos cantidades de la misma magnitud ($a_1\,U_1$) y ($a_2\,U_2$) son iguales, escribiremos ($a_1\,U_1$)=($a_2\,U_2$). Los paréntesis son superfluos, pero los especificamos para marcar bien los pares diádicos. Por la multiplicación escalar antes definida podremos escribir ($a_2\,U_2$)=1∘($a_2\,U_2$), con lo cual ($a_1\,U_1$)=1∘($a_2\,U_2$). Y en esta igualdad podemos definir ($a_1\,U_1$) como dividendo, ($a_2\,U_2$) como divisor y 1 como cociente, respecto de la multiplicación «∘», lo que podemos notar ($a_1\,U_1$)⫽($a_2\,U_2$)=1. Puesto que esta operación la hemos llamado división diádica homogénea, tendremos que el cociente diádico de dos díadas iguales, representativas de la misma cantidad de cierta magnitud, es la unidad de los números reales. En términos analíticos resulta la siguiente importante propiedad de la igualdad, que es la base de las ecuaciones físicas:

$$\left(a_1\,U_1\right)=\left(a_2\,U_2\right) \;\Rightarrow\; \frac{\left(a_1\,U_1\right)}{\left(a_2\,U_2\right)}=1$$

Section XII

DEFINITION OF MULTIPLICATION
DYADIC OF SCALAR DYADS

Just as the addition of dyads is defined as an internal law, we will see that geometry teaches us that multiplication must be conceived as an external law. To support this conception, we will begin by referring to the geometric experiment of areas and volumes, described in figure 1, which reveals how the multiplication of lengths defined by metric geometry gives rise to two new magnitudes, the surface and the volume, according to let the multiplied factors be two lengths or three, respectively [7]. We observe that, if we take two lengths $a\,U_1$ and $b\,U_2$, where U_1 and U_2 are any two units of length, and whose measures a and b are integers, that is, a and $b \in Z$, with both lengths forming an abstract rectangle without base scale $a\,U_1$ and height $b\,U_2$, the magnitude that we call the area or surface of the rectangle thus formed would be expressed as a concrete scalar equal to $a{\times}b$ times the area of an elementary rectangle with a base unit U_1 and a height unit U_2, which is would indicate with the dyadic form $(a{\times}b)\,(U_1{\times}U_2)$. This geometric operation is called length multiplication, and we immediately observe that it does not correspond to the classical algebraic notion of as many times the multiplicand as indicated by the multiplier, which reducing to an addition would require homogeneity in both factors, that is, they should be referred to the same unit, and in turn the product would also be given in that same unit. By contrast, geometric

---

[7] If the reader is not familiar with the geometric multiplication of segments, or what is the same, of lengths, you can consult the detailed development of this topic in «Lesson 32» of Mathematize 1 and in «Lesson 3» of Mathematize 3, both publications of the same author's agenda.

Geometric experiment of the areas

Given two lengths expressed in the same unit U, if an **abstract rectangle without scale** is formed with its numerical parts, it is observed that, dividing it into ideal squares of unit side, the number of these is equal to the product of the measures of the lengths given with respect to unity. This observation of the geometry allows defining the product of two lengths $a\,U$ and $b\,U$, or two dyad quantities with the same unit, interpreting it as an area that is symbolized:

$$a\,U \times b\,U = (a \times b)\,(U \times U) = (a \times b)\,U^2$$

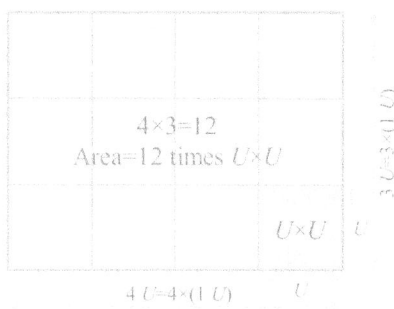

On the left, the case in which the lengths or dyads are not expressed in the same unit as $a\,U_1$ and $b\,U_2$, in the abstract rectangle built with them it is observed that their product can be associated with the quantity called area, which is measured by means of rectangles equal to the symbolized area unit $U_1 \times U_2$, justifying the same definition of product:

$$a\,U_1 \times b\,U_2 = (a \times b)\,(U_1 \times U_2)$$

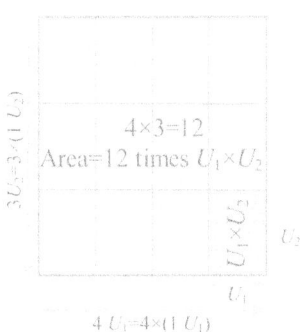

On the right the product of two lengths with fractional measure $(3/5)\,U_1 \cdot (2/3)\,U_2$. Dividing one of the dimensions into five equal segments and the other into three, results in a set of equal rectangles whose sides measure $1/5$ of U_1 and $1/3$ of U_2, the number of these equal elements that make up the unit is equal to $5 \times 3 = 15$, which coincides with the product of the denominators, and the number of equal elements that fit in the assumed fractional measure is $3 \times 2 = 6$, which coincides with the product of the numerators; the fractional area will be $3 \cdot 2$ elements of the 5×3 total rectangles, which is the fraction $(2 \times 3)/(3 \times 5)$, which is equal to the product $(3/5) \times (2/3) = 6/15$, so here also the form of the definition of the dyad multiplication is fulfilled.

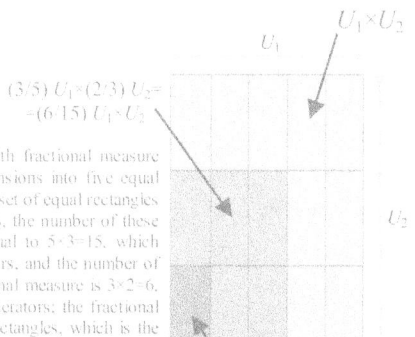

Figure 1

multiplication allows lengths to be expressed in different units, provided that the unit in which the product is measured is an abstract rectangle with dimensions precisely equal to the units in which the multiplied factors or segments are expressed.

If the measures of the segments were given, instead of with integers, with rational numbers, the dyads $(p/q)\,U_1$ and $(r/s)\,U_2$, it is also observed that the area of the abstract rectangle formed with both measures also It can be expressed as the scalar concrete with a numerical part equal to the product of the initial rationals, according to the definition of multiplication of fractions, which corresponds to the rational number that has the product of the numerators as the numerator and the product of the denominators as the denominator , and this as a multiplier of the area of the unit rectangle whose base and height are the given units U_1 and U_2, which would be expressed analytically with the written form $[(p \times r)/(q \times s)]\,(U_1 \times U_2)$.

In sum, given any two segments or lengths $a\,U_1$ and $b\,U_2$, forming an abstract rectangle with them and taking as a unit of surface the unit rectangle with dimensions U_1 and U_2, which are the units of the multiplied lengths, whose area will be written By definition, in the form of the product $U_1 \times U_2$, it turns out that the area of the rectangle is measured with the scalar dyad $(a \times b)\,(U_1 \times U_2)$.

Well, this geometric fact serves as the basis for defining the multiplication of lengths, and by axiomatic generalization, it also conceptualizes the **multiplication of quantities of scalar magnitudes** or **dyads** $a\,U_1$ and $b\,U_2$, no matter what the units are associated with them, from according to the analytic expression:

$$a\,U_1 * b\,U_2 = (a \times b)\,(U_1 * U_2) \qquad\qquad [12.1]$$

The definition equation above symbolizes an external composition law of the Cartesian product $\{R,U_1\} \times \{R,U_2\}$ in $\{R,U_1 * U_2\}$. So that the multiplication of dyads does not operate on the same set, as was had with the addition, which is an internal law, but the three sets involved in the multiplication, in general, can be different. If the units of the multiplied concretes coincide, designating them with the letter U, the generating external law of multiplication would apply the Cartesian product $\{R,U\} \times \{R,U\}$ in $\{R,U * U\}$. The product $U * U$, for symbolic convenience, is by definition symbolized by the power form U^2, so that the set $\{R,U * U\}$ can also be written $\{R,U^2\}$. And here we see clearly how the notation of square meters m^2, or seconds squared s^2, or any other unit squared U^2 arises; with this, the unknown that eminent mathematical philosophers of the stature of Fourier or Maxwell wondered, as we explained in the exordium, about what could be the meaning or motivation of multiplication and of the powers of physical units or quantities of magnitudes. In turn, the division between quantities of any scalar magnitudes and, therefore, between non-homogeneous units, will also be described and justified later, as soon as it is conceived as an operation derived from multiplication.

As with the previous definitions on laws of composition, it should be noted here that the product of $a\,U_1 * b\,U_2$ means the newly defined dyadic multiplication, that the product of $(a \times b)$ points to the multiplication of the real numbers, and that the asterisk of $(U_1 * U_2)$ refers again to the multiplication of dyads, in this case as a particular case that multiplies the two units, with the meaning of $(U_1 * U_2) = (1\,U_1) * (1\,U_2)$.

If instead of two segments or lengths we think of composing three, $a\,U_1$, $b\,U_2$ and $c\,U_3$, where U_1, U_2 and U_3 are three units of length, which can be different, the geometric multiplication of segments is defined as the composition of a abstract straight parallelepiped with the dimensions of the three segments. We observe that a certain volume thus results, which can be measured as a function of the volume of the unit parallelepipeds of dimensions U_1, U_2 and U_3, which can be indicated with the form of the product $U_1 \times U_2 \times U_3$, with the meaning of the new unit of volume magnitude created in this way; well, the total volume will be described by the real product of the numerical part of the multiplied lengths, which is $a \times b \times c$, or the number of times that the total volume comprises the abstract or numerically unspecified unit volume $U_1 \times U_2 \times U_3$. A numerical example with lengths is shown in figure 2 to illustrate the reasoning. With which, all that has been said can be included in an analytical formulation, which defines the geometric multiplication of three quantities of lengths or segments and, generalized in the

Experimental significance of the product geometric of three lengths, and by abstraction of any three magnitudes

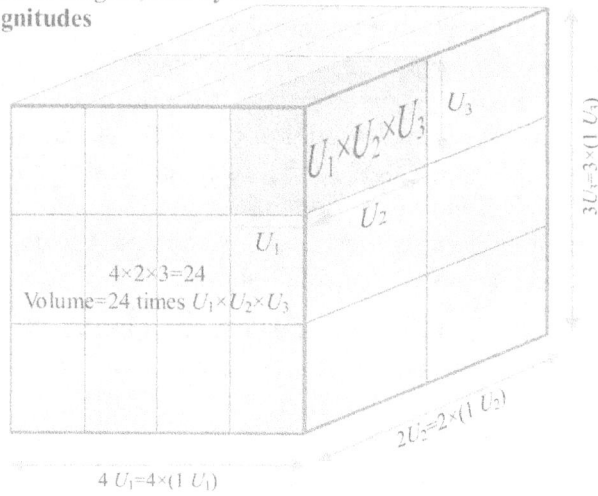

$4 U_1 = 4 \times (1\ U_1)$

Given three lengths $4\ U_1$, $2\ U_2$ and $3\ U_3$, an abstract straight parallelepiped without scale can be formed with them and ideally decomposed by delimiting the corresponding symbolic length on each edge. Thus, they result in a series of parallelepipeds with the same ideal unit measurements, which is why they are congruent and equal. The new magnitude that results from composing three lengths is called volume, and the fact that the number of elementary parallelepipeds is equal to 24 makes it possible to refer to the amount of volume indicating that one of these elements measures 24 times, which nothing prevents symbolizing with the Notation similar to the algebraic $U_1 \times U_2 \times U_3$, writing this result $24\ (U_1 \times U_2 \times U_3)$. With this, the operation of composing three lengths consisting of forming a straight parallelepiped with them can be called multiplication of the initial dyads given by three lengths, symbolized $(4\ U_1) \times (2\ U_2) \times (3\ U_3) = (4 \times 2 \times 3)\ (U_1 \times U_2 \times U_3)$, resulting in that the numerical part is given by $4 \times 2 \times 3 = 24$. So, in general, it can be defined that multiplying dyads is obtaining another dyad whose numerical element is the usual product of the numerical parts of the factors and whose dimension is expressed as the product of the units of the factors. As the unit elements are composed in the same way regardless of the order in which the factor units are composed, the commutative and associative properties of concrete multiplication must be axiomatized.

Figure 2

abstract, of any three quantities of magnitudes or any specific scalars, according to the definition equation following, explicit for dyadic algebra:

$$a\,U_1 * b\,U_2 * c\,U_3 = (a \times b \times c)\,(U_1 * U_2 * U_3) \qquad [12.2]$$

Equation [12.2] is the analytical form of the external composition law of the Cartesian product $\{R,U_1\}\times\{R,U_2\}\times\{R,U_3\}$ in $\{R,U_1*U_2*U_3\}$. So that the multiplication of three dyads, as when it operates on two, does not act on the same set, but the sets related by it, in general, can be different. If the units of the multiplied concretes coincide, designating them with the letter U, **the generating external law of the new magnitude** would apply the Cartesian product set $\{R,U\}\times\{R,U\}\times\{R,U\}$ in $\{R,U*U*U\}$. The product $U*U*U$, for symbolic convenience, is designated by definition with the form of the power U^3, so that the set $\{R,U*U*U\}$ can also be written $\{R,U^3\}$. And we once again clearly appreciate how the notation of the new generated units arises with mathematical coherence, cubic meters m^3, seconds cubed s^3, or any other unit raised to the cube U^3.

To explicitly describe dyadic multiplication without symbolic economy, let us differentiate this operation with the asterisk sign «*» in the definition equations [12.1] and [12.2]. We avoid the character «⊗» so as not to confuse it with the tensor product. Such definitions provide the logical motivation to generalize the definition of the product of scalar concretes with any number n of factors and units U_1, U_2, up to U_n; by applying the multiple Cartesian product $\{R,U_1\}\times\{R,U_2\}\times \dots \times\{R,U_n\}$ in $\{R,U_1*U_2* \dots *U_n\}$, so that, given the dyads $a_1 U_1$, $a_2 U_2$, up to $a_n U_n$, the dyadic product will be defined by the generic equation:

$$a_1 U_1 * a_2 U_2 * \dots * a_n U_n =$$

$$= (a_1 \times a_2 \times \dots \times a_n)(U_1 * U_2 * \dots * U_n) \qquad [12.3]$$

The multiplication of the first member «*» is the dyadic, as well as the one that appears between the units U_i of the second member; while the multiplication marked with the cross sign «×» between the numbers a_i symbolizes the product of the real numbers:

$$a_1 U * a_2 U * \dots * a_n U = (a_1 \times a_2 \times \dots \times a_n)(U * U * \dots * U) =$$

$$= (a_1 \times a_2 \times \dots \times a_n) U^n \text{ con } U^n = U * U * \dots * U \text{ y } n \text{ factores}$$

Apartado XII

DEFINICIÓN DE MULTIPLICACIÓN
DIÁDICA DE CONCRETOS ESCALARES

Así como la adición de díadas se define como una ley interna, veremos que la geometría nos enseña que la multiplicación debe concebirse como ley externa. Para fundamentar esta concepción comenzaremos refiriéndonos al experimento geométrico de las áreas y de los volúmenes, descrito en la figura 1, que nos revela cómo la multiplicación de longitudes definida por la geometría métrica da lugar a dos nuevas magnitudes, la superficie y el volumen, según que los factores multiplicados sean dos longitudes o tres, respectivamente[8].

Observamos que, si se toman dos longitudes $a\,U_1$ y $b\,U_2$, donde U_1 y U_2 sean dos unidades cualesquiera de longitud, y cuyas medidas a y b sean números enteros, es decir, a y $b \subset Z$, formando con ambas longitudes un rectángulo abstracto sin escala de base $a\,U_1$ y altura $b\,U_2$, la magnitud que llamamos área o superficie del rectángulo así formado quedaría expresada como un concreto escalar igual a $a{\times}b$ veces el área de un rectángulo elemental de base la unidad U_1 y de altura la unidad U_2, lo que se indicaría con la forma diádica $(a{\times}b)\,(U_1{\times}U_2)$. Esta operación geométrica se denomina multiplicación de longitudes, y observamos de inmediato que no se corresponde con la noción algebraica clásica de tantas veces el multiplicando como indique el multiplicador, que reduciéndose a una adición, exigiría homogeneidad en ambos factores, es decir, deberían estar referidos a la misma unidad, y a

---

[8]   Si el lector no estuviera familiarizado con la multiplicación geométrica de segmentos, o lo que es igual, de longitudes, puede consultar el desarrollo detallado de este tema en la «Lección 32» de *Matematizar 1* y en la «Lección 3» de *Matematizar 3*, ambas publicaciones del temario del mismo autor.

Experimento geométrico de las áreas

Dadas dos longitudes expresadas en la misma unidad U, si se forma un **rectángulo abstracto sin escala** con sus partes numéricas, se observa que, dividiéndolo en cuadrados ideales de lado la unidad, el número de estos resulta igual al producto de las medidas de las longitudes dadas respecto de la unidad. Esta observación de la geometría permite definir el producto de dos longitudes $a \cdot U$ y $b \cdot U$ o dos números concretos con la misma unidad, interpretándola como un área que se simboliza:

$$a \cdot U \cdot b \cdot U = (a \times b)(U \times U) = (a \times b) U^2$$

A la izquierda el caso en que las longitudes o concretos no se expresan en la misma unidad $a \cdot U_1$ y $b \cdot U_2$, en el rectángulo abstracto construido con ellas, se observa que su producto se puede asociar a la magnitud denominada área, que queda medida por medio de rectángulos iguales a la unidad de área simbolizada $U_1 \cdot U_2$, justificándose la misma definición de producto.

$$a \cdot U_1 \cdot b \cdot U_2 = (a \times b)(U_1 \times U_2)$$

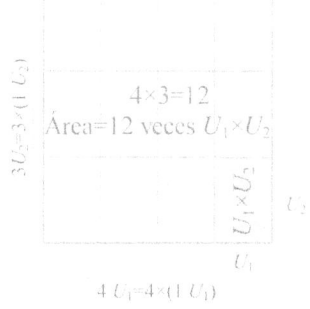

A la derecha el producto de dos longitudes con medida fraccionaria $(3/5) U_1 \cdot (2/3) U_2$. Dividiendo una de las dimensiones en cinco segmentos iguales y en tres la otra, resulta un conjunto de rectángulos iguales cuyos lados miden 1/5 de U_1 y 1/3 de U_2, el número de estos elementos iguales que componen la unidad es igual a $5 \times 3 = 15$, que coincide con el producto de los denominadores, y el número de elementos iguales que caben en la medida fraccionaria supuesta es de $3 \times 2 = 6$, que coincide con el producto de los numeradores; el área fraccionaria será 3×2 elementos de los 5×3 rectángulos totales, que es la fracción $(2 \times 3)/(3 \times 5)$, que resulta igual al producto $(3/5) \cdot (2/3) = 6/15$, conque aquí también se cumple la forma de la definición de la multiplicación concreta.

Figura 1

su vez el producto vendría dado también en esa misma unidad. Por el contrario, la multiplicación geométrica permite que las longitudes se expresen en unidades diferentes, a condición de que la unidad en que se mida el producto sea un rectángulo abstracto de dimensiones precisamente iguales a las unidades en que vengan expresados los factores o segmentos multiplicados.

Si las medidas de los segmentos viniesen dadas, en vez de con números enteros, con números racionales, las díadas (p/q) U_1 y (r/s) U_2, se observa igualmente que el área del rectángulo abstracto formado con ambas medidas también se puede expresar como el concreto escalar con parte numérica igual al producto de los racionales iniciales, según la definición de multiplicación de fracciones, que se corresponde con el número racional que tenga por numerador el producto de los numeradores y por denominador el producto de los denominadores, y ello como multiplicador del área del rectángulo unitario que tenga por base y altura las unidades dadas U_1 y U_2, lo que se expresaría analíticamente con la forma escrita $[(p \times r)/(q \times s)]$ $(U_1 \times U_2)$.

En suma, dados dos segmentos o longitudes cualesquiera a U_1 y b U_2, formando con ellas un rectángulo abstracto y tomando como unidad de superficie el rectángulo unitario que tenga por dimensiones U_1 y U_2, que son las unidades de las longitudes multiplicadas, cuya área se escribirá, por definición, en forma del producto $U_1 \times U_2$, resulta que el área del rectángulo queda medida con el concreto escalar $(a \times b)$ $(U_1 \times U_2)$.

Pues bien, este hecho geométrico sirve de fundamento para definir la multiplicación de longitudes, y por generalización axiomática, conceptúa también la **multiplicación de cantidades de magnitudes escalares** o **díadas** a U_1 y b U_2, no importa cuáles sean las unidades que a ellos estén asociadas, de acuerdo con la expresión analítica:

$$a\ U_1 * b\ U_2 = (a \times b)\ (U_1 * U_2) \qquad [12.1]$$

La ecuación de definición anterior simboliza una ley de composición externa del producto cartesiano $\{R, U_1\} \times \{R, U_2\}$ en $\{R, U_1 * U_2\}$. De modo que la multiplicación de concretos no opera sobre el mismo conjunto, como se tenía con la adición, que es una ley interna, sino que los tres conjuntos implicados en la multiplicación, en general, pueden ser distintos. Si las unidades de los concretos multiplicados coincidiesen, designándolas con la letra U, la ley externa generatriz de la multiplicación aplicaría el producto cartesiano $\{R, U\} \times \{R, U\}$ en $\{R, U * U\}$. El producto

$U * U$, por conveniencia simbólica, se representa por definición con la forma de potencia U^2, de modo que el conjunto $\{R, U*U\}$ se puede escribir también $\{R, U^2\}$. Y aquí ya vemos con claridad cómo surge la notación de los metros cuadrados m^2, o de los segundos al cuadrado s^2, o de cualquier otra unidad elevada al cuadrado U^2; con lo cual, queda explicada la incógnita que eminentes filósofos matemáticos de la talla de Fourier o Maxwell se preguntaban, tal como expusimos en el exordio, sobre cuál podría ser el sentido o motivación de la multiplicación y de las potencias de unidades o magnitudes físicas. A su vez, las división entre cantidades de magnitudes escalares cualesquiera y, por ende, entre unidades no homogéneas, también quedará descrita y justificada más adelante, en cuanto se la conciba como operación derivada de la multiplicación.

Al igual que con las anteriores definiciones sobre leyes de composición, hay que advertir aquí que el producto de $a\, U_1 * b\, U_2$ significa la multiplicación diádica recién definida, que el producto de $(a \times b)$ señala a la multiplicación de los números reales, y que el asterisco de $(U_1 * U_2)$ se refiere nuevamente a la multiplicación de díadas, en este supuesto como caso particular que multiplica las dos unidades, con el significado de $(U_1 * U_2) = (1\ U_1) * (1\ U_2)$.

Si en lugar de dos segmentos o longitudes pensamos en componer tres, $a\, U_1$, $b\, U_2$ y $c\, U_3$, donde U_1, U_2 y U_3 sean tres unidades de longitud, que pueden ser distintas, la multiplicación geométrica de segmentos se define como la composición de un paralelepípedo recto abstracto con las dimensiones de los tres segmentos. Observamos que resulta así un volumen determinado, que puede medirse en función del volumen de los paralelepípedos unitarios de dimensiones U_1, U_2 y U_3, que puede indicarse con la forma del producto $U_1 \times U_2 \times U_3$, con el significado de la nueva unidad de la magnitud volumen creada de este modo; pues bien, el volumen total quedará descrito por el producto real de la parte numérica de las longitudes multiplicadas, que es $a \times b \times c$, o número de veces que el volumen total comprende al volumen unitario abstracto o no especificado numéricamente $U_1 \times U_2 \times U_3$. En la figura 2 se representa un ejemplo numérico con longitudes que ilustra el razonamiento. Con lo cual, todo lo dicho se puede incluir en una formulación analítica, que define la multiplicación geométrica de tres cantidades de longitudes o segmentos y, generalizada en abstracto, de tres cantidades de magnitudes o concretos escalares cualesquiera, de acuerdo con la ecuación de definición siguiente:

Significado experimental del producto
geométrico de tres longitudes, y por
abstracción de tres magnitudes
cualesquiera

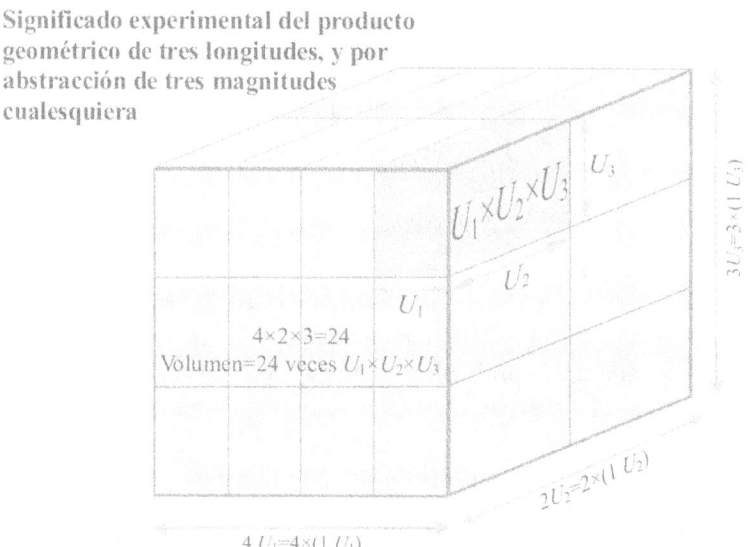

Dadas tres longitudes $4\,U_1$, $2\,U_2$ y $3\,U_3$, se puede formar con ellas un paralelepípedo recto abstracto sin escala y descomponerlo idealmente delimitando en cada arista la longitud simbólica que corresponda. Resultan así una serie de paralelepípedos con las mismas medidas unitarias ideales, por lo que son congruentes e iguales. La nueva magnitud que resulta de componer tres longitudes se denomina volumen, y el hecho de que el número de paralelepípedos elementales resulte igual a 24 permite referirse a la cantidad de volumen indicando que mide 24 veces uno de esos elementos, que nada impide simbolizar con la notación semejante a la algebraica $U_1 \times U_2 \times U_3$, escribiendo este resultado $24\,(U_1 \times U_2 \times U_3)$. Con ello, la operación de componer tres longitudes consistente en formar con ellas un paralelepípedo recto se puede denominar multiplicación de los números concretos iniciales dados por tres longitudes, y esta operación se simboliza $(4\,U_1) \times (2\,U_2) \times (3\,U_3) = (4 \times 2 \times 3)\,(U_1 \times U_2 \times U_3)$, resultando que la parte numérica es dada por $4 \times 2 \times 3 = 24$. De modo que, en general, se puede definir que multiplicar concretos es obtener otro concreto cuyo elemento numérico es el producto usual de las partes numéricas de los factores y cuya dimensión se expresa como producto de las unidades de los factores. Como los elementos unitarios quedan compuestos de la misma manera con independencia del orden en que se compongan las unidades de los factores, deben axiomatizarse las propiedades conmutativa y asociativa de la multiplicación concreta.

Figura 2

122

$$a\ U_1 *b\ U_2 *c\ U_3 = (a \times b \times c)\ (U_1 * U_2 * U_3) \qquad [12.2]$$

La ecuación [12.2] es la forma analítica de la ley de composición externa del producto cartesiano $\{R, U_1\} \times \{R, U_2\} \times \{R, U_3\}$ en $\{R, U_1 * U_2 * U_3\}$. De modo que la multiplicación de tres díadas, como cuando opera sobre dos, no actúa sobre el mismo conjunto, sino que los conjuntos relacionados por ella, en general, pueden ser distintos. Si las unidades de los concretos multiplicados coincidiesen, designándolas con la letra U, la **ley externa generatriz de la nueva magnitud** aplicaría el conjunto producto cartesiano $\{R, U\} \times \{R, U\} \times \{R, U\}$ en $\{R, U * U * U\}$. El término $U * U * U$, por conveniencia simbólica, se designa por definición con la forma de la potencia U^3, de modo que el conjunto $\{R, U * U * U\}$ se puede escribir también $\{R, U^3\}$. Y volvemos a apreciar con claridad cómo surge con coherencia matemática la notación de las nuevas unidades generadas, metros cúbicos m^3, segundos al cubo s^3, o cualquier otra unidad elevada al cubo U^3.

Para describir explícitamente sin economía simbólica la multiplicación diádica, diferenciamos con el signo asterisco «$*$» esta operación en las ecuaciones de definición [12.1] y [12.2]. Evitamos el carácter «$\otimes$» para no confundirla con el producto tensorial. Tales definiciones proporcionan la motivación lógica para generalizar la definición del producto de concretos escalares con cualquier número n de factores y unidades U_1, U_2, hasta U_n; mediante una aplicación del producto cartesiano múltiple $\{R, U_1\} \times \{R, U_2\} \times ... \times \{R, U_n\}$ en $\{R, U_1 * U_2 * ... * U_n\}$, de modo que, dadas las díadas $a_1\ U_1$, $a_2\ U_2$, hasta $a_n\ U_n$, el producto diádico quedará definido por la ecuación genérica:

$$a_1\ U_1 * a_2\ U_2 * ... * a_n\ U_n =$$
$$= (a_1 \times a_2 \times ... \times a_n)\ (U_1 * U_2 * ... * U_n) \qquad [12.3]$$

La multiplicación del primer miembro «$*$» es la diádica, así como la que aparece entre las unidades U_i del segundo miembro; mientras que la multiplicación señalada con el signo del aspa «$\times$» entre los números a_i simboliza el producto de los números reales. Si todas las unidades fuesen la misma, con $U_i = U$ para todo i, con la misma notación de potencias en R, se tendrá en este caso la siguiente expresión:

$$a_1\,U*a_2\,U*\,...\,*a_n\,U=(a_1\times a_2\times\,...\,\times a_n)\,(U*U*\,...\,*U)=$$

$$=(a_1\times a_2\times\,...\,\times a_n)\,U^n\text{ con }U^n=U*U*\,...\,*U\text{ y }n\text{ factores}$$

Section XIII

COMMUTATIVE AND ASSOCIATIVE PROPERTIES
OF THE MULTIPLICATION OF SCALAR DYADS

Taking into account that the abstract rectangle that integrates the unit of the multiplication of two dyads has to geometrically present the same quantity of area whether it is taken as base U_1 and height U_2, or if the base is equal to U_2 and the height to U_1, because both rectangles will be geometrically congruent and, therefore, of equal area, it is justified **to axiomatize the commutativity of the multiplication of units whatever they are:**

$$U_1 * U_2 = U_2 * U_1 \qquad\qquad [13.1]$$

And this unavoidable axiom results in the product of any two scalar dyads being commutative. Indeed, given the dyads $a\,U_1$ and $b\,U_2$, equation [12.1] for the definition of the explicit product is:

$$a\,U_1 * b\,U_2 = (a \times b)\,(U_1 * U_2) \qquad\qquad [13.2]$$

The commutative axiom [13.1] of units and the commutative property of multiplication in R of real numbers allow us to write:

$$(a \times b)\,(U_1 * U_2) = (b \times a)\,(U_2 * U_1)$$

The same equation [12.1] for the definition of multiplication justifies the following logical step:

$$(b \times a)\,(U_2 * U_1) = b\,U_2 * a\,U_1 \qquad\qquad [13.3]$$

Resulting in conclusion that the first member of [13.2] is equal to the second member of [13.3], whose meaning is that of the commutative property of the multiplication of diads, which will be written as follows:

$$a\,U_1 * b\,U_2 = b\,U_2 * a\,U_1$$

If instead of two factors there were three, it would be necessary to consider that the abstract unitary right parallelepiped formed with the units U_1, U_2 and U_3 should include the same amount of volume regardless of how its edges are ordered, because all of them will be geometrically congruent and equals; so that here too we are obliged to axiomatize that the dyad product of three units is commutative, which will be written:

$$U_1*U_2*U_3=U_1*U_3*U_2=U_2*U_1*U_3=$$
$$=U_2*U_3*U_1=U_3*U_1*U_2=U_3*U_2*U_1$$

This means that the units can be multiplied in any order, because the dyadic product will always represent the same quantity of the compound magnitude. This axiom, associated with the commutative property of the multiplication of real numbers for any number of factors, allows us to conclude, with the same ease as for the product of two dyads, that in the case of three the commutativity of the dyadic product can also be affirmed , regardless of the order of the multiplied dyads. And, generalized in the abstract for any number of factors, this geometric foundation authorizes us to **postulate that the dyadic product is commutative whatever the multiplied dyadic factors are.**

As usual, understanding the parentheses as a priority command in the operations that group, with the same geometric foundations indicated, **the multiplication of units must be axiomatized that it is associative,** which means that the units can be associated in any way in the multiplication without varying your product:

$$U_1*(U_2*U_3)=(U_1*U_2)*U_3=U_1*U_2*U_3 \qquad [13.4]$$

So the general associative property is immediate. Indeed, let the product $a\,U_1*(b\,U_2*c\,U_3)$, by the definition described in [12.3], we will have:

126

$$a\,U_1*(b\,U_2*c\,U_3)=$$
$$=a\,U_1*[(b\times c)\,(U_2*U_3)]=$$
$$=[a\times(b\times c)]\,[U_1*(U_2*U_3)] \qquad [13.5]$$

The associative property of the product in R is $a\times(b\times c)=(a\times b)\times c$, with which, substituting in [13.5] and taking into account [13.4]:

$$[a\times(b\times c)]\,[U_1*(U_2*U_3)]=[(a\times b)\times c]\,[(U_1*U_2)*U_3]$$

The definition [12.3] of the multiplication of scalar dyad entities allows us to write:

$$[(a\times b)\times c]\,[(U_1*U_2)*U_3]=$$
$$=[(a\times b)\,(U_1*U_2)]*c\,U_3=$$
$$=(a\,U_1*b\,U_2)*c\,U_3 \qquad [13.6]$$

Since the first member of [13.5] is equal to the last of [13.6], the associative property of the dyadic multiplication of scalar dyads finally results:

$$a\,U_1*(b\,U_2*c\,U_3)=(a\,U_1*b\,U_2)*c\,U_3$$

Therefore, scalar dyads behave with the associative form because of the definition of dyadic multiplication and its properties, not because the validity of traditional symbolic logic, whose lack of prior foundation is patently and radically unacceptable, must be admitted outright. since it clearly consists of an invalid scheme of reasoning.

Apartado XIII

PROPIEDADES CONMUTATIVA Y ASOCIATIVA DE LA MULTIPLICACIÓN DE DÍADAS ESCALARES

Teniendo en cuenta que el rectángulo abstracto que integra la unidad de la multiplicación de dos concretos ha de presentar geométricamente la misma cantidad de área tanto si se toma por base U_1 y altura U_2, como si la base se hace igual a U_2 y la altura a U_1, porque ambos rectángulos serán geométricamente congruentes y, por tanto, de igual área, está justificado **axiomatizar la conmutatividad de la multiplicación de unidades sean las que sean:**

$$U_1 * U_2 = U_2 * U_1 \qquad [13.1]$$

Y este axioma inevitable trae como consecuencia que el producto de dos díadas escalares cualesquiera resulte conmutativo. En efecto, dadas las díadas $a\ U_1$ y $b\ U_2$, la ecuación [12.1] de definición del producto explicitado es:

$$a\ U_1 * b\ U_2 = (a \times b)\ (U_1 * U_2) \qquad [13.2]$$

El axioma conmutativo [13.1] de las unidades y la propiedad conmutativa de la multiplicación en R de números reales permiten escribir:

$$(a \times b)\ (U_1 * U_2) = (b \times a)\ (U_2 * U_1)$$

La misma ecuación [12.1] de definición de la multiplicación justifica el siguiente paso lógico:

$$(b \times a)\ (U_2 * U_1) = b\ U_2 * a\ U_1 \qquad [13.3]$$

Resultando como conclusión que el primer miembro de [13.2] es igual al segundo miembro de [13.3], cuyo significado es el de la propiedad conmutativa de la multiplicación de concretos:

$$a\ U_1 * b\ U_2 = b\ U_2 * a\ U_1$$

Si en lugar de dos factores se tuvieran tres, habría que considerar que el paralelepípedo recto unitario abstracto formado con las unidades U_1, U_2 y U_3 deberá incluir la misma cantidad de volumen con independencia de cómo se ordenen sus aristas, porque todos ellos resultarán geométricamente congruentes e iguales; de modo que también aquí estamos obligados a axiomatizar que el producto concreto de tres unidades sea conmutativo, lo que se escribirá:

$$U_1 * U_2 * U_3 = U_1 * U_3 * U_2 = U_2 * U_1 * U_3 =$$
$$= U_2 * U_3 * U_1 = U_3 * U_1 * U_2 = U_3 * U_2 * U_1$$

Ello significa que las unidades puedan multiplicarse en cualquier orden, porque el producto diádico siempre representará la misma cantidad de la magnitud compuesta. Este axioma, asociado a la propiedad conmutativa de la multiplicación de números reales para cualquier número de factores, permite concluir, con la misma facilidad que para el producto de dos díadas, que en el caso de tres también se pueda afirmar la conmutatividad del producto diádico, sin importar el orden de las díadas multiplicadas. Y, generalizado en abstracto para cualquier número de factores, este fundamento geométrico nos autoriza a **postular que el producto diádico sea conmutativo cualesquiera que sean los factores diádicos multiplicados.**

Como es usual, entendiendo los paréntesis como un mandato de prioridad en las operaciones que agrupen, con los mismos fundamentos geométricos indicados, **la multiplicación de unidades debe axiomatizarse que sea asociativa**, lo que significa que las unidades se puedan asociar de cualquier modo en la multiplicación sin variar su producto:

$$U_1 * (U_2 * U_3) = (U_1 * U_2) * U_3 = U_1 * U_2 * U_3 \qquad [13.4]$$

Con lo que la propiedad asociativa general es inmediata. En efecto, sea el producto $a\ U_1 * (b\ U_2 * c\ U_3)$, por la definición descrita en [12.3], tendremos:

129

$$a\ U_1 * (b\ U_2 * c\ U_3) =$$
$$= a\ U_1 * [(b \times c)\ (U_2 * U_3)] =$$
$$= [a \times (b \times c)]\ [U_1 * (U_2 * U_3)] \qquad [13.5]$$

La propiedad asociativa del producto en R es $a \times (b \times c) = (a \times b) \times c$, con lo que, sustituyendo en [13.5] y teniendo en cuenta [13.4]:

$$[a \times (b \times c)]\ [U_1 * (U_2 * U_3)] = [(a \times b) \times c]\ [(U_1 * U_2) * U_3]$$

La definición [12.3] de la multiplicación de entes concretos escalares permite escribir:

$$[(a \times b) \times c]\ [(U_1 * U_2) * U_3] =$$
$$= [(a \times b)\ (U_1 * U_2)] * c\ U_3 =$$
$$= (a\ U_1 * b\ U_2) * c\ U_3 \qquad [13.6]$$

Puesto que el primer miembro de [13.5] es igual al último de [13.6], resulta finalmente la propiedad asociativa de la multiplicación diádica de concretos escalares:

$$a\ U_1 * (b\ U_2 * c\ U_3) = (a\ U_1 * b\ U_2) * c\ U_3$$

Por tanto, las díadas escalares se comportan con la forma asociativa en razón de la definición de multiplicación diádica y de sus propiedades, no porque deba admitirse de plano la validez de la lógica simbólica tradicional, cuya falta de fundamento previo es patente y radicalmente inaceptable, pues a todas luces consiste en un esquema no válido de razonamiento.

Section XIV

INEXISTENCE OF UNIT OR REVERSE ELEMENTS
FOR THE MULTIPLICATION OF SCALAR DYADS

Given a set of scalar dyads associated to the unit U, which we have symbolically indicated with the notation $\{R,U\}$, let 1 be the multiplicative unit in R, whose existence is guaranteed by the body structure of the set of real numbers; so it will always be possible to form the dyad $1\ U$ of $\{R,\ U\}$. It could be intuited that the dyad $1\ U$ would have to be the unit element of the multiplication of homogeneous dyads. To check this, let's take any other element $a\ U$ of $\{R,U\}$, being any number of R. The product between the two will give $a\ U^2$, because the dyadic multiplication $a\ U * 1\ U$, by the definition equation [12.1] and the condition of unit element in R, they allow to write the following:

$$a\ U * 1\ U = (a \times 1)\ (U * U) = a\ U^2$$

The commutative property of dyadic multiplication, followed by the same two previous foundations, lead us to string together this reasoning:

$$a\ U * 1\ U = 1\ U * a\ U = (1 \times a)\ (U * U) = a\ U^2$$

Therefore, in any case $1\ U$ is such that multiplied dyadically by any dyad $a\ U$ gives another dyad $a\ U^2$, with the same primary a, but with different secondary, U in one case and U^2 in the other. Observing that U^2 is a unit of another magnitude different from the magnitude that corresponds to the unit U, the dyads $a\ U$ and $a\ U^2$ are different and it can be stated that $1\ U$ is not the neutral element or unit of the concrete multiplication defined on $\{R,U\}$. Furthermore, as this would happen with any element of $\{R,U\}$, because multiplying it by $a\ U$ would always result in another different measurable magnitude in the unit U^2, it can be concluded that it is not possible to find any element of {R,U} that behaves like a typical unit element. We insist, this is because concrete or dyadic multiplication is an external generating law that applies the Cartesian product {R,U}×{R,U} in {R,U²}, and it turns out that the elements of {R,U²} they are different from those of {R,U}, because their unit parts are units of different magnitudes, so both sets are different and this prevents the algebraic behavior of the concept of unit element from being verified.

With the inverse element something similar happens. For all non-zero concrete $a\,U$, that is, with $a \neq 0$, given the field structure of R, there exists $a^{-1}\,U$, a^{-1} being the inverse of a for the product of the reals. One could suspect that the $a^{-1}\,U$ dyad is the inverse of $a\,U$, but this dyad is such that multiplied dyadically with the other produces the $a\,U * a^{-1}\,U$ dyad. Thus, considering definition [12.1], knowing that in R it is $a \times a^{-1} = 1$, and taking into account the commutative property in the second chain of equalities, we easily have the reasoning scheme of the following two lines:

$$a\,U * a^{-1}\,U = (a \times a^{-1})\,(U * U) = 1\,U^2$$

$$a\,U * a^{-1}\,U = a^{-1}\,U * a\,U = (a^{-1} \times a)\,(U * U) = 1\,U^2$$

As $1\,U$ is a dyad of different magnitude than $1\,U^2$, they cannot be equal and this result allows us to ensure that $a^{-1}\,U$ is not the inverse for the multiplication of dyad $a\,U$, because U^2 is a unit of certain magnitude always different from the magnitude associated with U.

And this happens not only for the dyad $a^{-1}\,U$, but for any other, since multiplying it with $a\,U$ will produce in any case a quantity of another magnitude, to which the unit $U^2 \neq U$, so the secondaries will always be different and the criterion of dyadic equality can never be applied.

The foregoing is reinforced by the fact that, as for this composition law there is no unitary element, because it is an external generating law, neither can there be an inverse element in the strict sense of ordinary algebra, leaving possible innovations safe, because no one can be multiplied. dyad by another so that an element that does not exist results.

This same result is quickly reached as follows: since the set $\{R,U\}$ is different from $\{R,U^2\}$, for the dyadic multiplication or application of $\{R,U\}\times\{R,U\}$ in $\{R,U^2\}$, the unit and inverse elements of any dyad of $\{R,U\}$ cannot be found in $\{R,U\}$, because the product of any element of $\{R,U\}$ by another element of the same set it produces elements from another set, that is, the factors belong to a different magnitude than their multiplication, so the criterion of dyadic equality cannot be applied to them. In particular, the inverse elements U^{-1} of the unit $U=(1,U)$ cannot be found. So unit notations with negative exponents, such as m^{-1}, kg^{-1}, or s^{-1} are absurd and non-existent, unless they are given the proper and consistent meaning as divisors of the dyadic division to be defined later. As will be duly verified, these symbols, which cannot be rationally associated with any quantity of magnitude, must have as their only coherent algebraic meaning that of mere indications that the quantities indicated with negative exponents are part of the divisor of a certain unit composed in the form of quotient, resulting from the composition laws of dyadic algebra.

If this is the case for the multiplication of homogeneous dyads, all the more so for the heterogeneous products or applications of $\{R,U_1\}\times\{R,U_2\}$ in $\{R,U_1 *U_2\}$, since in these the three related sets are different, so they are associated with measurements of different magnitudes between which the equality relationship cannot be established.

So that the multiplication of physical measurements or scalar dyads can never satisfy the algebraic conditions of existence of unit and inverse elements, departing in this aspect from isomorphism with the field of real numbers.

It could be judged that an unnecessary detour has been taken to demonstrate the obvious non-existence of unitary and inverse elements with respect to dyadic multiplications, but we have wanted to underpin this fact to the maximum, which due to its negative nature always presents greater evidentiary resistance, to refute its current presumed existence, since currently the International System of Units admits by mere conventionalism that the set of physical quantities has an abelian multiplicative group structure in which every magnitud can be expressed as a function of the integer powers of a certain number of magnitudes called base.

The algebra of magnitudes established here shows us that there cannot be such an abelian multiplicative group, because the operations defined are external laws and, therefore, lacking unitary and inverse elements, two qualities that cannot be lacking in every group. We are, therefore, before a crass error of the International System of Units, caused by the omission of definition of the multiplicative laws of composition for physical magnitudes. Omission that is saved with full physical and mathematical coherence in this First Algebra of Magnitudes.

Apartado XIV

INEXISTENCIA DE ELEMENTOS UNIDAD NI INVERSO PARA LA MULTIPLICACIÓN DE DÍADAS ESCALARES

Dado un conjunto de díadas escalares asociadas a la unidad U, que simbólicamente hemos indicado con la notación $\{R,U\}$, sea 1 la unidad multiplicativa en R, cuya existencia está garantizada por la estructura de cuerpo del conjunto de los números reales; así que siempre será posible formar el concreto 1 U de $\{R,U\}$. Podría intuirse que la díada 1 U habría de ser el elemento unitario de la multiplicación de díadas homogéneas. Para comprobarlo tomemos otro elemento cualquiera a U de $\{R,U\}$, siendo a cualquier número de R. El producto entre ambos dará a U^2, porque la multiplicación a $U*1$ U, por la ecuación de definición [12.1] y la condición de elemento unitario en R, permiten escribir lo siguiente:

$$a\ U*1\ U=(a\times1)\ (U*U)=a\ U^2$$

La propiedad conmutativa de la multiplicación concreta, seguida de los mismos dos fundamentos anteriores, nos llevan a hilar este razonamiento:

$$a\ U*1\ U=1\ U*a\ U=(1\times a)\ (U*U)=a\ U^2$$

Luego, en todo caso 1 U es tal que multiplicado diádicamente por cualquier díada a U da otra díada a U^2, con el mismo primario a, pero con distinto secundario, U en un caso y U^2 en el otro. Observando que U^2 es una unidad de otra magnitud diferente de la magnitud que corresponde a la unidad U, las díadas a U y a U^2 son diferentes y se puede afirmar que 1 U no es el elemento neutro o unidad de la multiplicación concreta definida sobre $\{R,U\}$. Es más, como esto sucedería con cualquier elemento de $\{R,U\}$, porque al multiplicarlo por a U siempre resultaría otra magnitud distinta medible en la unidad U^2, se puede concluir que no es posible encontrar ningún elemento de $\{R,U\}$ que se comporte como un típico elemento unitario. Insistimos, es así porque la multiplicación concreta o diádica es una ley externa generatriz que aplica el producto cartesiano $\{R,U\}\times\{R,U\}$ en $\{R,U^2\}$, y resulta que los elementos de $\{R,U^2\}$ son diferentes de los de $\{R,U\}$, porque sus partes unitarias son unidades de magnitudes diversas, conque ambos conjuntos son distintos y ello impide que se pueda verificar el comportamiento algebraico propio del concepto de elemento unitario.

Con el elemento inverso sucede algo análogo. Para todo concreto $a\ U$ no nulo, es decir, con $a\neq 0$, dada la estructura de cuerpo de R, existe el $a^{-1}\ U$, siendo a^{-1} el inverso de a para el producto de los reales. Se podría sospechar que la díada $a^{-1}\ U$ sea la inversa de $a\ U$, pero esa díada es tal que multiplicada diádicamente con la otra produce la díada $a\ U * a^{-1}\ U$. Conque, considerando la definición [12.1], sabiendo que en R es $a\times a^{-1}=1$, y teniendo en cuenta la propiedad conmutativa en la segunda cadena de igualdades, tenemos con facilidad el esquema de razonamiento de las dos líneas siguientes:

$$a\ U * a^{-1}\ U = (a\times a^{-1})\ (U * U) = 1\ U^2$$

$$a\ U * a^{-1}\ U = a^{-1}\ U * a\ U = (a^{-1}\times a)\ (U * U) = 1\ U^2$$

Como $1\ U$ es una díada de diferente magnitud que la $1\ U^2$, no pueden ser iguales y este resultado nos permite asegurar que $a^{-1}\ U$ no es el inverso para la multiplicación del concreto $a\ U$, porque U^2 es una unidad de cierta magnitud siempre distinta de la magnitud asociada a U. Y esto sucede no solo para la díada $a^{-1}\ U$, sino para cualquier otra, ya que al multiplicarla con $a\ U$ producirá en todo caso una cantidad de otra magnitud, a la que pertenece la unidad $U^2 \neq U$, por lo que los secundarios siempre serán diferentes y nunca podrá aplicarse el criterio de igualdad diádica.

Lo anterior se refuerza con el hecho de que, como para esta ley de composición no existe elemento unitario, porque es ley externa generatriz, tampoco puede existir elemento inverso en el sentido estricto del álgebra ordinaria, dejando a salvo posibles innovaciones, porque no podrá multiplicarse cualquier díada por otra de modo que resulte un elemento que no existe.

A este mismo resultado se llega rápidamente de la forma siguiente: puesto que el conjunto $\{R, U\}$ es distinto del $\{R, U^2\}$, para la multiplicación diádica o aplicación de $\{R, U\} \times \{R, U\}$ en $\{R, U^2\}$, no se pueden encontrar en $\{R, U\}$ los elementos unidad ni inversos de cualquier díada de $\{R, U\}$, porque el producto de cualquier elemento de $\{R, U\}$ por otro elemento del mismo conjunto produce elementos de otro conjunto, es decir, que los factores pertenecen a una magnitud diferente de su multiplicación, por lo que no se les puede aplicar el criterio de igualdad diádica. En particular, no se pueden encontrar los elementos inversos U^{-1} de la unidad $U = (1, U)$. Así que las notaciones de unidades con exponentes negativos, tales como m^{-1}, kg^{-1} o s^{-1} son absurdas e inexistentes, salvo que se les dé el significado adecuado y coherente como divisores de la división diádica que se definirá más adelante. Como se comprobará oportunamente, esos símbolos, que no pueden asociarse racionalmente a ninguna cantidad de magnitud, han de tener como único significado algebraico coherente el de meras indicaciones de que las magnitudes indicadas con exponentes negativos forman parte del divisor de cierta unidad compuesta en forma de cociente, resultante de las leyes de composición del álgebra diádica.

Si esto es así para la multiplicación de díadas homogéneas, tanto más para los productos heterogéneos o aplicaciones de $\{R, U_1\} \times \{R, U_2\}$ en $\{R, U_1 * U_2\}$, ya que en estos los tres conjuntos relacionados son distintos, por lo que están asociados a mediciones de magnitudes diferentes entre las que no se puede establecer la relación de igualdad.

De modo que la multiplicación de mediciones físicas o díadas escalares no puede satisfacer nunca las condiciones algebraicas de existencia de elementos unidad ni inverso, apartándose en este aspecto del isomorfismo con el cuerpo de los números reales.

Podría juzgarse que se ha dado un rodeo innecesario para demostrar la obvia inexistencia de elementos unitarios e inversos respecto de las multiplicaciones diádicas, pero hemos querido apuntalar al máximo este hecho, que por su carácter negativo presenta siempre una mayor resistencia probatoria, para refutar su vigente presumida existencia, dado que actualmente el Sistema Internacional de Unidades admite por mero convencionalismo que el conjunto de las magnitudes físicas tiene estructura de grupo multiplicativo abeliano en el que toda magnitud se puede expresar en función de las potencias enteras de un número determinado de magnitudes llamadas de base.

El álgebra de magnitudes aquí establecida nos manifiesta que no puede haber tal grupo multiplicativo abeliano, porque las operaciones definidas son leyes externas y, por tanto, carentes de elementos unitarios e inversos, dos cualidades que no pueden faltar en todo grupo. Estamos, pues, ante un craso error del Sistema Internacional de Unidades, propiciado por la omisión de definición de las leyes de composición multiplicativas para las magnitudes físicas. Omisión que queda salvada con plena coherencia física y matemática en esta *Primera álgebra de magnitudes*.

Section XV

DISTRIBUTIVE OWNERSHIP OF THE MULTIPLICATION
ON THE ADDITION OF THE DYADIC ALGEBRA

Let's check if the distributive behavior of dyadic multiplication with respect to the addition of measurements holds. To do this, let's take the dyads $a\,U_1$, $b\,U_2$ and $c\,U_2$. The second and third dyad with the same unit have been chosen so that they are uniform and can be added together. Let us form the compound dyad $a\,U_1 * (b\,U_2 \oplus c\,U_2)$. Equation [5.3] for the definition of addition supports the first logical step:

$$a\,U_1 * (b\,U_2 \oplus c\,U_2) = a\,U_1 * [(b+c)\,U_2)] \qquad [15.1]$$

The equation [12.1] for the definition of the dyadic product allows us to write the equality:

$$a\,U_1 * [(b+c)\,U_2)] = [a \times (b+c)]\,(U_1 * U_2)$$

The distributive property of the product with respect to the multiplication in the field R of the reals is $a \times (b+c) = (a \times b) + (a \times c)$ and allows us to write:

$$[a \times (b+c)]\,(U_1 * U_2) = [(a \times b) + (a \times c)]\,(U_1 * U_2)$$

The definition [5.3] of addition of dyads, makes it easier for us to advance again in the reasoning, by simply doubling the sum of the second member:

$$[(a \times b) + (a \times c)]\,(U_1 * U_2) = (a \times b)\,(U_1 * U_2) \oplus (a \times c)\,(U_1 * U_2)$$

The definition [12.1] of the product leads us directly to the expression:

$$(a \times b)\,(U_1 * U_2) \oplus (a \times c)\,(U_1 * U_2) =$$

$$= (a\,U_1 * b\,U_2) \oplus (a\,U_1 * c\,U_2) \qquad [15.2]$$

The conclusion arises from the equality between the first member of equation [15.1] and the second of [15.2]:

$$a\,U_1 * (b\,U_2 \oplus c\,U_2) = (a\,U_1 * b\,U_2) \oplus (a\,U_1 * c\,U_2)$$

This formula reproduces the well-known distributive form, which here reflects this property of the product with respect to the sum of scalar dyads and for equal units it is reduced to:

$$a\,U * (b\,U \oplus c\,U) = (a\,U * b\,U) \oplus (a\,U * c\,U)$$

In any case, the unit symbols behave formally like any other algebraic element; but, as for the other properties analyzed, this is not an immediate consequence of the symbolic operation, but due to the definitions and properties of dyadic algebra.

As a consequence of everything explained so far in the previous sections, any set of scalar measurements referring to the same unit $\{R,U\}$, endowed with the internal laws that we have called dyadic addition and multiplication, which respond to the definition equations [5.3] and [12.1], would verify all the conditions that configure the algebraic structure of the commutative field, if it were not for the fact that dyadic multiplication is an external law instead of an internal one, making the existence of the unit and inverse elements impossible. However, the algebraic structure of scalar dyadic elements is isomorphic with that of the field of real numbers. On the other hand, we have verified the fact that every set of homogeneous scalar $\{R,U\}$ or vector $\{R^3,U\}$ concretes, endowed with the internal composition law of addition, defined in section V, and with the law external of the multiplication by a scalar, defined in section IX, satisfies all the conditions of a vector space over the field R of the real numbers[9].

---

[9]   The vector space structure can be found in «Lesson 32» of Mathematize 1, on the fundamentals of algebraic structures, and more extensively in «Lesson 2» of Mathematize 2, which studies vector spaces.

Apartado XV

PROPIEDAD DISTRIBUTIVA DE LA MULTIPLICACIÓN SOBRE LA ADICIÓN DEL ÁLGEBRA DIÁDICA

Comprobemos si se cumple el comportamiento distributivo de la multiplicación diádica respecto de la adición de mediciones. Para ello, tomemos los concretos $a\ U_1$, $b\ U_2$ y $c\ U_2$. Se han elegido el segundo y el tercer concreto con la misma unidad para que sean uniformes y puedan sumarse. Formemos la díada compuesta $a\ U_1*(b\ U_2 \oplus c\ U_2)$. La ecuación [5.3] de definición de la adición fundamenta el primer paso lógico:

$$a\ U_1*(b\ U_2 \oplus c\ U_2) = a\ U_1*[(b+c)\ U_2)] \qquad [15.1]$$

La ecuación [12.1] de definición del producto diádico permite escribir la igualdad:

$$a\ U_1*[(b+c)\ U_2)] = [a \times (b+c)]\ (U_1*U_2)$$

La propiedad distributiva del producto respecto de la multiplicación en el cuerpo R de los reales es $a \times (b+c) = (a \times b) + (a \times c)$ y autoriza a escribir:

$$[a \times (b+c)]\ (U_1*U_2) = [(a \times b) + (a \times c)]\ (U_1*U_2)$$

La definición [5.3] de adición de díadas, nos facilita avanzar nuevamente en el razonamiento, bastando desdoblar la suma del segundo miembro:

$$[(a \times b) + (a \times c)]\ (U_1*U_2) = (a \times b)\ (U_1*U_2) \oplus (a \times c)\ (U_1*U_2)$$

La definición [12.1] del producto nos conduce directamente a la expresión:

$$(a \times b)\ (U_1*U_2) \oplus (a \times c)\ (U_1*U_2) =$$
$$= (a\ U_1*b\ U_2) \oplus (a\ U_1*c\ U_2) \qquad [15.2]$$

La conclusión brota de la igualdad entre el primer miembro de la ecuación [15.1] y el segundo de [15.2]:

$$a \ U_1 * (b \ U_2 \oplus c \ U_2) = (a \ U_1 * b \ U_2) \oplus (a \ U_1 * c \ U_2)$$

Esta fórmula reproduce la conocida forma distributiva, que aquí refleja esa propiedad del producto respecto de la suma de díadas escalares y para unidades iguales se reduce a:

$$a \ U * (b \ U \oplus c \ U) = (a \ U * b \ U) \oplus (a \ U * c \ U)$$

En todo caso, los símbolos de las unidades se comportan formalmente como cualquier otro elemento algebraico; pero, como para las demás propiedades analizadas, ello no es consecuencia inmediata de la operativa simbólica, sino debido a las definiciones y propiedades del álgebra diádica.

Como consecuencia de todo lo expuesto hasta aquí en los apartados anteriores, cualquier conjunto de mediciones escalares referidas a la misma unidad $\{R, U\}$, dotado de las leyes internas que hemos denominado adición y multiplicación diádicas, que responden a las ecuaciones de definición [5.3] y [12.1], verificaría todas las condiciones que configuran la estructura algebraica de cuerpo conmutativo, si no fuera porque la multiplicación concreta es una ley externa en vez de interna, haciendo imposible la existencia de los elementos unidad e inverso. No obstante, la estructura algebraica de los elementos diádicos escalares resulta isomorfa con la del cuerpo de los números reales. Por otra parte, hemos constatado el hecho de que todo conjunto de concretos homogéneos escalares $\{R, U\}$ o vectoriales $\{R^3, U\}$, dotado de la ley de composición interna de la adición, definida en el apartado V, y con la ley externa de la multiplicación por un escalar, definida en el apartado IX, satisface todas las condiciones de un espacio vectorial sobre el cuerpo R de los números reales[10].

---

[10] La estructura de espacio vectorial se puede encontrar en la «Lección 32» de *Matematizar 1*, sobre los fundamentos de las estructuras algebraicas, y con mayor extensión en la «Lección 2» de *Matematizar 2*, que estudia los espacios vectoriales.

Section XVI

DEFINITION OF DYADIC DIVISION
BETWEEN SCALAR MEASUREMENTS

In section XI we define the division between dyads homogeneous scalars as a function of dyadic multiplication by a scalar. In this section we will define the division based on the multiplication of dyads defined by the definition equation [12.1]. With it, the dyadic multiplication with two factors establishes the relationship between the multiplicand, the multiplier and the product, by means of an abstract rectangle in which the base is the multiplicand, the height the multiplier and the area of the rectangle the product. Well, changing the symbology and identifying said area with a dividend, one of its dimensions with a divisor and the other with the resulting quotient, we will have the notion of division as an operation such that the quotient multiplied by the divisor equals the dividend, and all this through the dyadic algebra of measurements.

Let's start by shaping the division between any two units U_1 and U_2. Nothing prevents us from establishing as a symbol of the quotient between the two a notation similar to that of abstract algebraic elements, for example, separating them with two inclined or horizontal bars $U_1/\!/U_2$. We must pay attention to the fact that the unit U_1, dividend, will be associated here with the area of the abstract rectangle of dimensions U_2, divisor, and $U_1/\!/U_2$, quotient. Therefore, the units related by the dyadic product must satisfy the following equation:

$$(U_1/\!/U_2) * U_2 = U_1 \qquad [16.1]$$

It is vital to observe that [16.1] justifies the simplification rule of the factors U_2, as it would happen with the algebra of R, but not because it applies, but because the dyadic definition of the product by means of abstract rectangles creates a specific algebra that thus determines it .

Under these conditions, the writing of the division between the concretes or dyads $a\,U_1$ and $b\,U_2$ must have the symbolic form of this definition equation:

$$\frac{a\,U_1}{b\,U_2} = \frac{a}{b}\frac{U_1}{U_2} \qquad [16.2]$$

The epistemic equation [16.2] will be fully justified when verifying that it satisfies the condition that the dyadic product of the quotient and the divisor is equal to the dividend. To do this, let us write this dyadic product and operate with the properties of R and taking into account equation [16.1] for the definition of the quotient of units, resulting in:

$$\left(\frac{a}{b}\frac{U_1}{U_2}\right) * \left(b\,U_2\right) = \left(\frac{a}{b} \times b\right)\left(\frac{U_1}{U_2} * U_2\right) = a\,U_1$$

Therefore, definition [16.2] is motivated and sufficiently grounded, because it describes with the general criterion of division the quotient between any two scalar dyads or concretes.

It is possible to deduce the definition [16.2] without more than attending to the generic concept of division. To do this, just imagine an abstract rectangle whose surface is identified with a dyadic dividend $a\,U_1$, one of its dimensions with $b\,U_2$ and the other with the concrete quotient $c\,(U_1//U_2)$. The unit associated with c must be identified with the dyadic quotient of units $U_1//U_2$, because the unit rectangle must have the unit U_1 by area and by dimensions U_2 and $U_1//U_2$, as seen in [16.1]. In the same way, the three indicated dyads cannot be independent, but must satisfy the division condition, that is, the quotient multiplied by the divisor must give the dividend; or, in other words, the dyadic product of the two dimensions of the abstract rectangle must be equal to its surface; and it will be written analytically like this:

$$a\,U_1 = b\,U_2 * c\,\frac{U_1}{U_2} \qquad [16.3]$$

The equality [16.3] can be interpreted as a dyadic division, for which it is enough to consider the factor $c\,(U_1 /\!/ U_2)$ as the quotient between the total surface of the abstract rectangle $a\,U_1$ and the other of its two dimensions $b\,U_2$. And this analytically can be described in this way:

$$c\,\frac{U_1}{U_2} = \frac{a\,U_1}{b\,U_2}$$

The geometry of the abstract rectangle is such that $a=b\times c$, so $c=a/b$ with the algebra of R. So, substituting $c=a/b$ in the first member of the last equality, we will finally have this other:

$$c\,\frac{U_1}{U_2} = \frac{a\,U_1}{b\,U_2} = \frac{a}{b}\,\frac{U_1}{U_2}$$

And the same equality of the definition [16.2] is already observed between the second and third terms, to establish the dyadic division between the dyads $a\,U_1$ and $b\,U_2$, which was postulated there and has been deduced here by means of the preceding reasoning. So it can be concluded that the quotient of two dyads is equal to a concrete whose primary is the ordinary quotient of the primaries of the factors and whose secondary is the dyadic division of the units of the dividend and the divisor. Expressed analytically:

$$\frac{a\,U_1}{b\,U_2} = \frac{a}{b}\,\frac{U_1}{U_2} \qquad\qquad [16.4]$$

We check in this way, as for the rest of the operations previously analyzed, that the symbols of the units behave ideally like the other elements of R, but this consequence is not due to the traditional symbolic logic, and we insist, it would be a crass error and it is inadmissible to consider it this way, because we have irrefutably justified that this formal behavior is due to the concept of dyadic multiplication by means of abstract rectangles.

On the other hand, let us note that the division of scalar concretes analyzed in this chapter has been symbolized with the double bar, a different operation from the quotient of homogeneous dyads in section XI, which we have agreed to represent with the same sign. And it is that the diversity of algebraic laws is such that, although symbolic exhaustiveness is sought for didactic clarity, it is inevitable and even at times convenient to resort to a certain degree to the principle of symbolic economy, if one does not want to fall into a kind of confused batahola operational.

To better visualize the operation of the dyadic division, let's analyze the case of the magnitude that is known by the name of density. The analysis begins with the observation that bodies have two proper magnitudes, the volume they occupy and the mass that corresponds to that volume. The geometry of the case would be the one described in figure 3. We can start by defining an abstract rectangle such that its area is identified with the amount of mass of the considered body, which we will assume equal to M kg, and one of its dimensions with the amount of volume that occupies that same mass, indicated by $V\,m^3$. It is not incongruous that a volume is represented by a length, because in dyadic or physical algebra the quantity of any magnitude can be indistinctly similar, by definition, to a segment, an area, a volume or a hypervolume. In turn, in this case, these two quantities, mass

146

Dyadic analysis of the composite magnitude called DENSITY

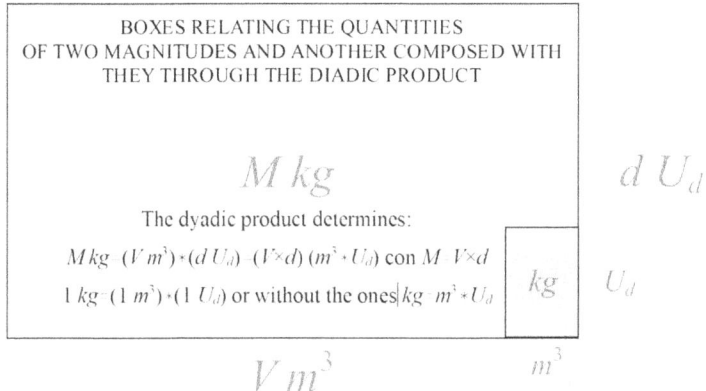

BOXES RELATING THE QUANTITIES
OF TWO MAGNITUDES AND ANOTHER COMPOSED WITH
THEY THROUGH THE DIADIC PRODUCT

$M\ kg$

The dyadic product determines:

$M\ kg\ (V\ m^3) \cdot (d\ U_d)\ (V \times d)\ (m^3 \cdot U_d)$ con $M\ V \times d$

$1\ kg\ (1\ m^3) \cdot (1\ U_d)$ or without the ones $kg\ m^3 \cdot U_d$

$V\ m^3$

$d\ U_d$

Given a body of mass $M\ kg$ and volume $V\ m^3$, the dyadic product allows us to relate these quantities to that of a third magnitude derived from the first ones and called density, such that $M\ kg$ corresponds to the area of the rectangle with dimensions equal to the others two, $V\ m^3$ and $d\ U_d$. Thus, the density measure is given by the dyadic quotient $M\ kg / V\ m^3$. In turn, in the unit rectangle the three units of the related quantities must be such that the unit of density U_d is the dyadic quotient between a kg and a m^3, that is, $1\ U_d = (1\ kg) / (1\ m^3)$ or written for short $U_d = kg / m^3$.

Figure 3

and volume, are related to each other through the dyadic product and through another magnitude represented by the second dimension of the abstract rectangle. Let us symbolize the quantity of this third magnitude with $d\ U_d$ and to understand ourselves let's call it density. The three quantities thus related, mass, volume and density, it is clear that they are not independent, but must satisfy the condition imposed by the dyadic product, that is, the following concrete equation:

$$M\ kg = (V\ m^3) * (d\ U_d) \qquad [16.5]$$

In the unit rectangle of dimensions 1 m^3 and 1 U_d, whose area must be identified with the quantity of mass to 1 kg, we will have:

$$1\,kg=(1\,m^3)*(1\,U_d),\text{ abbreviated } kg=m^3*U_d \qquad [16.6]$$

The definition of the dyadic product [12.1] transforms the expression [16.5] into this one:

$$M\,kg=(V\times d)\,(m^3*U_d) \qquad [16.7]$$

Equation [16.7] means that the quantity of mass M kg or area of the rectangle is equal to V×d times the area of the unit rectangle of dimensions 1 m^3 and 1 U_d, which is symbolized by the dyadic product m^3*U_d and that has to be equal to the unit of mass or kg, given the equation of the product operation [16.6].

The dyadic product is such that M=V×d, multiplication that is that of R, with the justification of the geometric experiment of the areas exposed in section XII. Expressing it as a quotient in R, we will have:

$$d = \frac{M}{V} \qquad [16.8]$$

In turn, the general division criterion that arises from multiplication, applied to the dyadic products [16.5] and [16.6], allows them to be symbolized in this other way:

$$d\,U_d = \frac{M\,kg}{V\,m^3} \;;\; U_d = \frac{kg}{m^3}$$

Taking into account [16.8] and the second of the last two equations, substituting d and U_d in the first member of the first of these, it results:

$$\frac{M\,kg}{V\,m^3} = \frac{M}{V}\,\frac{kg}{m^3} \qquad [16.9]$$

It is observed that [16.9] indicates the same result as [16.4] or generic dyadic quotient, in the present assumption applied to the case of density and its other two magnitudes related by the dyadic product. The reading of [16.9] should be understood as follows: the dyadic quotient between a quantity of mass M kg and a quantity of volume V m^3 is a quantity of a compound magnitude called density, which is equal to the physical or concrete dyad whose primary is the real quotient of the primaries M and V, and whose secondary is the dyadic quotient of the secondaries, that is, in this case the unit of mass or kilogram kg divided by the unit of volume or cubic meter m^3. Even at the risk of being repetitive, the concept of «quantity» is emphasized so that the dyadic nature of the entities that are composed is not forgotten.

The physical meaning of density can be assessed simply by multiplying it by the unit of volume to determine its corresponding mass, according to the following reasoning from physical algebra:

$$\left(1 \ m^3\right) * \left(d \ \frac{kg}{m^3}\right) = \left(1 \times d\right) \left(m^3 * \frac{kg}{m^3}\right)$$

The compound unit of the second member indicates the dyadic product of m^3 by the dyadic quotient $kg/\!/m^3$, which is U_d, so it is the unit rectangle whose abstract surface is one kg; so that the symbols of m^3 that appear both as a multiplier and as a divisor can be simplified as in R, but not by the properties of R, but by the definition of a dyadic product with abstract rectangles, as defined in section XII . Under these conditions, it is concluded that the first member must be equal to d kg, with which it is observed that the mass of the unit of volume called m^3 is precisely d kg; and this must be the meaning of density: mass of each unit of volume of material bodies.

Another significant case and analogous to that of density is the composite magnitud called velocity. His analysis starts from the material observation that every mobile takes a certain time to travel each specific distance. So suppose that the distance L m, which is a length measured in meters, is covered in an interval t s, time expressed in seconds. Nothing prevents us from assembling the abstract rectangle in figure 4, which has an area L m and such that one of its sides is t s, so that the other dimension will be univocally established by both measurements. It is not incongruous that the area represents the length L m, although geometrically this does not seem to make sense, because we are operating in the abstract with the dyadic algebra of magnitudes. The argument for the velocity magnitude is completely analogous to that seen above for density; however, it is developed again step by step for greater clarity.

In this assumption, the three related quantities are length, time and velocity, which are not independent, but must satisfy the condition imposed by the dyadic product, that is:

Dyadic analysis of the compound magnitude called VELOCITY

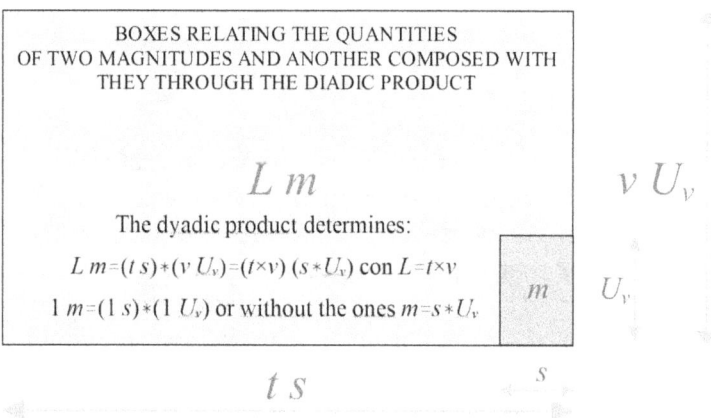

Given a mobile that travels the distance $L\ m$ in the time of $t\ s$, the dyadic product allows us to relate these quantities to that of a third magnitude derived from the first ones and called velocity, such that $L\ m$ corresponds to the area of the rectangle of equal dimensions to the other two, $t\ s$ and $v\ U_v$. Thus, the measure of velocity is given by the dyadic quotient $L\ m//t\ s$. In turn, in the unit rectangle the three units of the related magnitudes must be such that the unit of speed U_v is the dyadic quotient between one m and one s, that is, $1\ U_v = (1\ m)//(1\ s)$ or abbreviated $U_v = m//s$.

Figure 4

$$L\,m = (t\,s) * (v\,U_v)$$ [16.10]

In the unit rectangle of dimensions 1 s and 1 U_v, whose area must be identified with the amount of length equal to 1 m, we will have:

$$1\,m = (1\,s) * (1\,U_v), \text{ abbreviated } m = s * U_v$$ [16.11]

The definition of dyadic product [12.1] transforms the expression [16.10] into this one:

$$L\,m = (t \times v)\,(s * U_v)$$ [16.12]

Equation [16.12] means that the amount of length L m o area of the rectangle is equal to $t \times v$ times the area of the unit rectangle of dimensions 1 s and 1 U_v, which is symbolized by the dyadic product $s * U_v$ and must be equal to the unit of length om, given the equation of the product operation [16.11].

The dyadic product is such that $L = t \times v$, multiplication that is that of R, with the justification of the geometric experiment of the areas exposed in section XII. Expressing it as a quotient in R, we will have:

$$v = \frac{L}{t}$$ [16.13]

In turn, the general division criterion that arises from multiplication, applied to the dyadic products [16.10] and [16.11], allows them to be symbolized in this other way:

$$v\,U_v = \frac{L\,m}{t\,s} \quad ; \quad U_v = \frac{m}{s}$$

Taking into account [16.13] and the second of the last two equations, substituting v and Uv in the first member of the first one, it results:

$$\frac{L\,m}{t\,s} = \frac{L}{t}\,\frac{m}{s}$$ [16.14]

It is observed that [16.14] indicates the same result as [16.4] or generic dyadic quotient, in the present assumption applied to the case of speed and its other two magnitudes related by the dyadic product. The reading of [16.14] should be understood as follows: the dyadic quotient between a quantity of length L m and a quantity of time t s is the quantity of a compound magnitude called velocity, which is equal to dyad whose primary is the real quotient of the primaries L and v, and whose secondary is the dyadic quotient of the secondaries, that is, in this case the unit of length or meter m divided by the unit of time or second s.

The physical meaning of velocity can be assessed simply by multiplying it by the unit of time to determine the corresponding length, according to the following reasoning from dyadic algebra:

$$\left(1s\right)* \left(v \, \frac{m}{s} \right) = \left(1 \times v\right) \left(s* \frac{m}{s} \right)$$

The compound unit of the second member indicates the dyadic product of a second s by the dyadic quotient m//s, which is U_v, so it is the unit rectangle whose abstract surface is one meter; so that the symbols of s that appear both as a multiplier and as a divisor can be simplified as in R, but not by the properties of R, but by the definition of a dyadic product with abstract rectangles, as defined in section XII. Under these conditions, it is concluded that the first member must be equal to v m, with which it is observed that the length traveled in the unit of time called second s is precisely v m; and this must be the meaning of speed: length or distance traveled in each unit of time.

Any other magnitude derived from two others by means of dyadic division will show an analysis completely analogous to those of density and speed, without more than taking into account the units that correspond to the dividend and the divisor, which will determine the compound unit in which it should be measured the derived magnitude or quotient.

Finally, we must now give the meaning in dyadic algebra of negative exponents, which is nothing more than a simple notation to write the division. Let's take definition [14.4], it is obvious that nothing prevents us from writing it with the product notation, following the steps of ordinary algebra, according to the following notation:

$$\frac{aU_1}{bU_2} = \frac{a}{b}\frac{U_1}{U_2} = \frac{a}{b}\left(U_1 * U_2^{-1}\right)$$

But here it should be noted that the inverse notation U_2^{-1} does not mean that there exists an entity such that multiplied by U_2 gives unit of real numbers, because we have shown in section XIV that such a thing does not exist. There we conclude that U_2^{-1} cannot be a quantity of the same magnitude U_2, so the dyadic notation $U_1 * U_2^{-1}$ should be considered equivalent to the dyadic quotient $U_1 /\!/ U_2$, for symbolic purposes only.

Therefore, in dyadic algebra of magnitudes, unlike what occurs in arithmetic algebraic structures, inverse notations with units or physical quantities have their own meaning and do not correspond to the ghost of inverse elements isolated from internal operations, but to divisors of some ratio that cannot be missing. Thus, in general, an expression such as $U_1 * U_2^{-n}$ does not mean the product of U_1 by the figurative inverse element of U_2^n (see the following section on exponentiation), but rather represents the dyadic quotient $U_1 /\!/ U_2^n$, resulting from the operation of dividing the dividend U_1 by the power U_2^n. We insist, $U_1 * U_2^{-n}$ does not denote the product of the quantity U_1 by the inverse quantity U_2^{-n}, which has no physical meaning. In affine geometry, it symbolizes a rectangle with abstract surface U_1 and sides U_2^n and $U_1 /\!/ U_2^n$. In turn, with algebraic operation nomenclature, the expression indicated by $U_1 /\!/ U_2^n = U_1 * U_2^{-n}$ intervenes in a generating external composition law that applies the Cartesian product $\{R,U_1\} \times \{R,U_2^n\}$ in the dyadic set $\{R,U_1 /\!/ U_2^n\}$, which is a dyadic division.

Apartado XVI

DEFINICIÓN DE DIVISIÓN DIÁDICA
ENTRE MEDICIONES ESCALARES

En el apartado XI definimos la división entre concretos escalares homogéneos en función de la multiplicación diádica por un escalar. En este apartado definiremos la división en base a la multiplicación de díadas definida mediante la ecuación de definición [12.1]. Con ella la multiplicación diádica con dos factores establece la relación entre el multiplicando, el multiplicador y el producto, mediante un rectángulo abstracto en que la base sea el multiplicando, la altura el multiplicador y el área del rectángulo el producto. Pues bien, cambiando la simbología e identificando dicha área con un dividendo, una de sus dimensiones con un divisor y la otra con el cociente resultante, se tendrá la noción de división como operación tal que el cociente multiplicado por el divisor sea igual al dividendo, y todo ello mediante el álgebra diádica de las mediciones.

Comencemos dando forma a la división entre dos unidades cualesquiera U_1 y U_2. Nada nos impide establecer como símbolo del cociente entre ambas una notación similar a la de elementos algebraicos abstractos, por ejemplo, separándolos con dos barras inclinadas u horizontales $U_1 /\!/ U_2$. Hay que atender al hecho de que la unidad U_1, dividendo, quedará aquí asociada al área del rectángulo abstracto de dimensiones U_2, divisor, y $U_1 /\!/ U_2$, cociente. Por tanto, las unidades relacionadas por el producto diádico deberán satisfacer la ecuación siguiente:

$$(U_1 /\!/ U_2) * U_2 = U_1 \qquad [16.1]$$

Es vital observar que [16.1] justifica la regla de simplificación de los factores U_2, como ocurriría con el álgebra de R, pero no porque se aplique esta, sino porque la definición diádica del

154

producto mediante rectángulos abstractos crea un álgebra específica que así lo determina.

En estas condiciones, la escritura de la división entre los concretos o díadas $a\,U_1$ y $b\,U_2$ debe tener la forma simbólica de esta ecuación de definición:

$$\frac{a\,U_1}{b\,U_2} = \frac{a}{b}\frac{U_1}{U_2} \qquad [16.2]$$

La ecuación epistémica [16.2] quedará plenamente justificada al comprobar que satisface la condición de que el producto concreto del cociente por el divisor sea igual al dividendo. Para ello, escribamos dicho producto diádico y operemos con las propiedades de R y teniendo en cuenta la ecuación [16.1] de definición del cociente de unidades, resultando:

$$\left(\frac{a}{b}\frac{U_1}{U_2}\right) * \left(b\,U_2\right) = \left(\frac{a}{b}\times b\right)\left(\frac{U_1}{U_2}*U_2\right) = a\,U_1$$

Por lo que la definición [16.2] está motivada y suficientemente fundamentada, porque describe con el criterio general de la división el cociente entre dos díadas o concretos escalares cualesquiera.

Es posible deducir la definición [16.2] sin más que atender al concepto genérico de división. Para ello, basta con imaginar un rectángulo abstracto cuya superficie quede identificada con un dividendo diádico $a\,U_1$, una de sus dimensiones con $b\,U_2$ y la otra con el cociente concreto $c\,(U_1/\!/U_2)$. La unidad asociada a c debe identificarse con el cociente diádico de unidades $U_1/\!/U_2$, porque el rectángulo unitario ha de tener por área la unidad U_1 y por dimensiones U_2 y $U_1/\!/U_2$, tal como se vio para [16.1]. De la misma manera, las tres díadas indicadas no pueden ser independientes, sino que deben satisfacer la condición de la división, es decir, que el cociente multiplicado por el divisor debe dar el dividendo; o, dicho de otro modo, el producto diádico de las dos dimensiones del rectángulo abstracto debe ser igual a su superficie; y ello se escribirá analíticamente así:

$$a\, U_1 = b\, U_2 * c\frac{U_1}{U_2} \qquad [16.3]$$

La igualdad [16.3] se puede interpretar como una división diádica, para lo cual basta considerar el factor $c\,(U_1 /\!/ U_2)$ como el cociente entre la superficie total del rectángulo abstracto $a\, U_1$ y la otra de sus dos dimensiones $b\, U_2$. Y ello analíticamente se podrá describir de esta manera:

$$c\frac{U_1}{U_2} = \frac{a\, U_1}{b\, U_2}$$

La geometría del rectángulo abstracto es tal que $a=b \times c$, por lo que $c=a/b$ con el álgebra de R. De modo que, sustituyendo $c=a/b$ en el primer miembro de la última igualdad, tendremos finalmente esta otra:

$$c\frac{U_1}{U_2} = \frac{a\, U_1}{b\, U_2} = \frac{a}{b}\frac{U_1}{U_2}$$

Y ya se observa entre los términos segundo y tercero la misma igualdad de la definición [16.2], para establecer la división diádica entre las díadas $a\, U_1$ y $b\, U_2$, que allí se postuló y aquí se ha deducido mediante el razonamiento precedente. Conque se puede concluir que el cociente de dos díadas es igual a un concreto cuyo primario es el cociente ordinario de los primarios de los factores y cuyo secundario es la división diádica de las unidades del dividendo y del divisor. Expresado analíticamente:

$$\frac{a\, U_1}{b\, U_2} = \frac{a}{b}\frac{U_1}{U_2} \qquad [16.4]$$

Comprobamos de este modo, como para el resto de las operaciones anteriormente analizadas, que los símbolos de las unidades se comportan idealmente como los demás elementos de R, pero esta consecuencia no es debida a la lógica simbólica tradicional, e insistimos, sería un error craso e inadmisible considerarlo así, porque hemos justificado irrefutablemente que ese comportamiento formal se debe al concepto de multiplicación diádica mediante rectángulos abstractos.

Por otra parte, advirtamos que con la doble barra se ha simbolizado la división de concretos escalares analizada en este capítulo, operación distinta del cociente de díadas homogéneas del apartado XI, que hemos convenido en representar con ese mismo signo. Y es que la diversidad de leyes algebraicas es tal que, aunque se busque la exhaustividad simbólica por claridad didáctica, resulta inevitable y hasta a veces conveniente recurrir en cierto grado al principio de economía simbólica, si no se quisiera caer en una especie de confusa batahola operacional.

Para visualizar mejor el funcionamiento de la división diádica, analicemos el caso de la magnitud que se conoce con el nombre de densidad. El análisis comienza con la observación de que los cuerpos presentan dos magnitudes propias, el volumen que ocupan y la masa que corresponde a ese volumen. La geometría del caso sería la descrita en la figura 3. Se puede empezar definiendo un rectángulo abstracto tal que su área se identifique con la cantidad de masa del cuerpo considerado, que supondremos igual a $M\,kg$, y una de sus dimensiones con la cantidad de volumen que ocupa esa misma masa, indicada por $V\,m^3$. No es incongruente que un volumen quede representado por una longitud, porque en álgebra diádica o física la cantidad de cualquier magnitud se puede semejar indistintamente, por definición, con un segmento, con un área, con un volumen o con un hipervolumen. A su vez, en este caso, estas dos magnitudes, masa y volumen, quedan relacionadas entre sí mediante el producto diádico y a través de otra magnitud representada por la segunda dimensión del rectángulo abstracto. Simbolicemos la cantidad de esta tercera magnitud con $d\,U_d$ y para entendernos llamémosla densidad. Las tres magnitudes así relacionadas, la masa, el volumen y la densidad, es claro que no son independientes, sino que deben satisfacer la condición

Análisis diádico de la magnitud compuesta llamada DENSIDAD

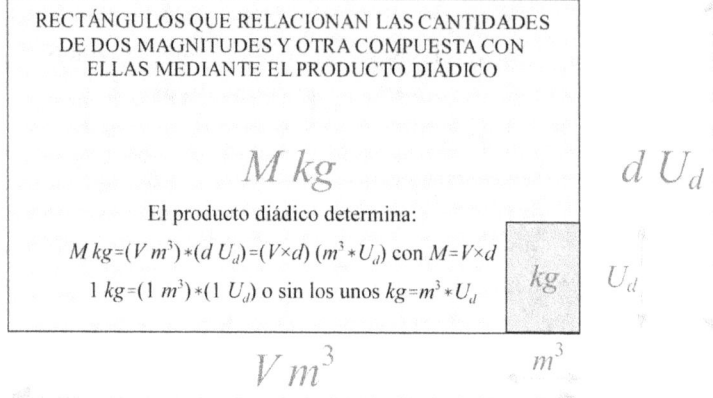

RECTÁNGULOS QUE RELACIONAN LAS CANTIDADES
DE DOS MAGNITUDES Y OTRA COMPUESTA CON
ELLAS MEDIANTE EL PRODUCTO DIÁDICO

$M \ kg$

El producto diádico determina:

$M \ kg = (V \ m^3) * (d \ U_d) = (V \times d)(m^3 * U_d)$ con $M = V \times d$

$1 \ kg = (1 \ m^3) * (1 \ U_d)$ o sin los unos $kg = m^3 * U_d$

$V \ m^3$

$d \ U_d$

Dado un cuerpo de masa $M \ kg$ y volumen $V \ m^3$, el producto diádico permite relacionar estas cantidades con la de una tercera magnitud derivada de las primeras y llamada densidad, tal que $M \ kg$ se corresponda con el área del rectángulo de dimensiones iguales a las otras dos, $V \ m^3$ y $d \ U_d$. De este modo, la medida de la densidad es dada por el cociente diádico $M \ kg // V \ m^3$. A su vez, en el rectángulo unitario las tres unidades de las magnitudes relacionadas han de ser tales que la unidad de densidad U_d sea el cociente diádico entre un kg y un m^3, es decir, $1 \ U_d = (1 \ kg) // (1 \ m^3)$ o escrito abreviadamente $U_d = kg // m^3$.

Figura 3

impuesta por el producto diádico, esto es la ecuación concreta siguiente:

$$M \ kg = (V \ m^3) * (d \ U_d) \qquad [16.5]$$

En el rectángulo unitario de dimensiones $1 \ m^3$ y $1 \ U_d$, cuya área debe identificarse con la cantidad de masa igual a $1 \ kg$, se tendrá:

$$1 \ kg = (1 \ m^3) * (1 \ U_d), \text{ abreviadamente } kg = m^3 * U_d \quad [16.6]$$

La definición de producto diádico [12.1] transforma la expresión [16.5] en esta otra:

$$M \ kg = (V \times d)(m^3 * U_d) \qquad [16.7]$$

158

La ecuación [16.7] significa que la cantidad de masa M kg o área del rectángulo es igual a $V \times d$ veces el área del rectángulo unitario de dimensiones 1 m^3 y 1 U_d, que se simboliza con el producto diádico $m^3 * U_d$ y que ha de ser igual a la unidad de masa o kg, dada la ecuación de la operación producto [16.6].

El producto diádico es tal que $M = V \times d$, multiplicación que es la de R, con la justificación del experimento geométrico de las áreas expuesto en el apartado XII. Expresándolo en forma de cociente en R, se tendrá:

$$d = \frac{M}{V} \qquad [16.8]$$

A su vez, el criterio general de división que nace de la multiplicación, aplicado a los productos diádicos [16.5] y [16.6], permite simbolizarlos de esta otra manera:

$$d\,U_d = \frac{M\,kg}{V\,m^3} \;\; ; \;\; U_d = \frac{kg}{m^3}$$

Teniendo en cuenta [16.8] y la segunda de las dos últimas ecuaciones, sustituyendo d y U_d en el primer miembro de la primera de estas, resulta:

$$\frac{M\,kg}{V\,m^3} = \frac{M}{V}\,\frac{kg}{m^3} \qquad [16.9]$$

Se observa que [16.9] indica el mismo resultado que [16.4] o cociente diádico genérico, en el supuesto presente aplicado al caso de la densidad y sus otras dos magnitudes relacionadas por el producto diádico. La lectura de [16.9] debe entenderse así: el cociente diádico entre una cantidad de masa $M\,kg$ y una cantidad de volumen $V\,m^3$ es la cantidad de una magnitud compuesta denominada densidad, que resulta igual a la díada física o concreto cuyo primario sea el cociente real de los primarios M y V, y cuyo secundario sea el cociente diádico de los secundarios, es decir, en este caso la unidad de masa o kilogramo kg entre la unidad de volumen o metro cubico m^3. Aún a riesgo de resultar reiterativos, se hace hincapié en el concepto «cantidad» para que no se olvide la naturaleza diádica de los entes que se componen.

El significado físico de la densidad puede valorarse sin más que multiplicarla por la unidad de volumen para determinar la masa que le corresponda, de acuerdo con el siguiente razonamiento de álgebra física:

$$\left(1\,m^3\right)*\left(d\,\frac{kg}{m^3}\right)=\left(1\times d\right)\left(m^3*\frac{kg}{m^3}\right)$$

La unidad compuesta del segundo miembro indica el producto diádico del m^3 por el cociente diádico $kg/\!/m^3$, que es U_d, por lo que se trata del rectángulo unitario que tenga por superficie abstracta un kg; de modo que pueden simplificarse como en R los símbolos de m^3 que aparecen a la vez como multiplicador y como divisor, pero no por las propiedades de R, sino por la propia definición de producto diádico con rectángulos abstractos, conforme se define en el apartado XII. En estas condiciones, se concluye que el primer miembro ha de ser igual a $d\ kg$, con lo que se observa que la masa de la unidad de volumen denominada m^3 es precisamente $d\ kg$; y este ha de ser el significado de la densidad: masa de cada unidad de volumen de los cuerpos materiales.

Otro caso significativo y análogo al de la densidad es la magnitud compuesta llamada velocidad. Su análisis parte de la observación material de que todo móvil emplea un cierto tiempo en recorrer cada distancia específica. Así que supongamos que la distancia $L\ m$, que es una longitud medida en metros, se cubra en un intervalo $t\ s$, tiempo expresado en segundos. Nada nos impide montar el rectángulo abstracto de la figura 4, que tiene por área $L\ m$ y tal que uno de sus lados sea $t\ s$, de modo que la otra dimensión quedará establecida unívocamente por ambas medidas. No es incongruente que el área represente la longitud $L\ m$, aunque geométricamente esto no parezca tener sentido, porque estamos operando en abstracto con el álgebra diádica de magnitudes. El argumento para la magnitud velocidad es completamente análogo al visto anteriormente para la densidad; no obstante, se desarrolla nuevamente paso a paso para mayor claridad expositiva.

En este supuesto, las tres magnitudes relacionadas son la longitud, el tiempo y la velocidad, que no son independientes, sino que deben satisfacer la condición impuesta por el producto diádico, esto es:

$$L\ m = (t\ s) * (v\ U_v) \qquad [16.10]$$

En el rectángulo unitario de dimensiones 1 s y 1 U_v, cuya área debe identificarse con la cantidad de longitud igual a 1 m, se tendrá:

$$1\ m = (1\ s) * (1\ U_v), \text{ abreviadamente } m = s * U_v \qquad [16.11]$$

Análisis diádico de la magnitud compuesta llamada VELOCIDAD

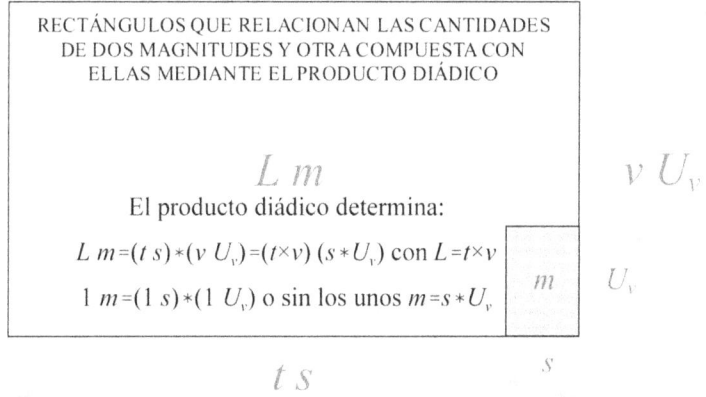

RECTÁNGULOS QUE RELACIONAN LAS CANTIDADES
DE DOS MAGNITUDES Y OTRA COMPUESTA CON
ELLAS MEDIANTE EL PRODUCTO DIÁDICO

$L\ m$

El producto diádico determina:

$L\ m = (t\ s) * (v\ U_v) = (t \times v)\ (s * U_v)$ con $L = t \times v$

$1\ m = (1\ s) * (1\ U_v)$ o sin los unos $m = s * U_v$

$v\ U_v$

m U_v

$t\ s$ s

Dado un móvil que recorra la distancia $L\ m$ en el tiempo de $t\ s$, el producto diádico permite relacionar estas cantidades con la de una tercera magnitud derivada de las primeras y llamada velocidad, tal que $L\ m$ se corresponda con el área del rectángulo de dimensiones iguales a las otras dos, $t\ s$ y $v\ U_v$. De este modo, la medida de la velocidad es dada por el cociente diádico $L\ m/t\ s$. A su vez, en el rectángulo unitario las tres unidades de las magnitudes relacionadas han de ser tales que la unidad de velocidad U_v sea el cociente diádico entre un m y un s, es decir, $1\ U_v = (1\ m)/(1\ s)$ o escrito abreviadamente $U_v = m/s$.

Figura 4

161

La definición de producto diádico [12.1] transforma la expresión [16.10] en esta otra:

$$L\,m = (t \times v)\,(s * U_v) \qquad [16.12]$$

La ecuación [16.12] significa que la cantidad de longitud $L\,m$ o área del rectángulo es igual a $t \times v$ veces el área del rectángulo unitario de dimensiones 1 s y 1 U_v, que se simboliza con el producto diádico $s * U_v$ y que ha de ser igual a la unidad de longitud o m, dada la ecuación de la operación producto [16.11].

El producto diádico es tal que $L = t \times v$, multiplicación que es la de R, con la justificación del experimento geométrico de las áreas expuesto en el apartado XII. Expresándolo en forma de cociente en R, se tendrá:

$$v = \frac{L}{t} \qquad [16.13]$$

A su vez, el criterio general de división que nace de la multiplicación, aplicado a los productos diádicos [16.10] y [16.11], permite simbolizarlos de esta otra manera:

$$v\,U_v = \frac{L\,m}{t\,s} \;\; ; \;\; U_v = \frac{m}{s}$$

Teniendo en cuenta [16.13] y la segunda de las dos últimas ecuaciones, sustituyendo v y U_v en el primer miembro de la primera de estas, resulta:

$$\frac{L\,m}{t\,s} = \frac{L}{t}\,\frac{m}{s} \qquad [16.14]$$

Se observa que [16.14] indica el mismo resultado que [16.4] o cociente diádico genérico, en el supuesto presente aplicado al caso de la velocidad y sus otras dos magnitudes relacionadas por el producto diádico. La lectura de [16.14] debe entenderse así: el cociente diádico entre una cantidad de longitud $L\,m$ y una cantidad de tiempo $t\,s$ es la cantidad de una magnitud compuesta denominada velocidad, que resulta igual al concreto cuyo primario sea el cociente real de los primarios L y v, y cuyo secundario sea el cociente diádico de los secundarios, es decir, en este caso la unidad de longitud o metro m entre la unidad de tiempo o segundo s.

El significado físico de la velocidad puede valorarse sin más que multiplicarla por la unidad de tiempo para determinar la longitud que le corresponda, de acuerdo con el siguiente razonamiento de álgebra física:

$$\left(1\,s\right) * \left(v\ \frac{m}{s}\right) = \left(1 \times v\right) \left(s * \frac{m}{s}\right)$$

La unidad compuesta del segundo miembro indica el producto diádico de un segundo s por el cociente diádico $m\,/\!/s$, que es U_v, por lo que se trata del rectángulo unitario que tenga por superficie abstracta un metro; de modo que pueden simplificarse como en R los símbolos de s que aparecen a la vez como multiplicador y como divisor, pero no por las propiedades de R, sino por la propia definición de producto diádico con rectángulos abstractos, conforme se define en el apartado XII. En estas condiciones, se concluye que el primer miembro ha de ser igual a $v\,m$, con lo que se observa que la longitud recorrida en la unidad de tiempo denominada segundo s es precisamente $v\,m$; y este ha de ser el significado de la velocidad: longitud o distancia recorrida en cada unidad de tiempo.

Cualquier otra magnitud derivada de otras dos mediante la división diádica mostrará un análisis completamente análogo a los de la densidad y la velocidad, sin más que tener en cuenta las unidades que correspondan al dividendo y al divisor, que determinarán la unidad compuesta en que deba medirse la magnitud derivada o cociente.

Finalmente, debemos dar a continuación el significado en álgebra diádica de los exponentes negativos, que no es sino una simple notación para escribir la división. Tomemos la definición [14.4], es obvio que nada nos impide escribirla con la notación de producto, siguiendo los pasos del álgebra ordinaria, de acuerdo con la siguiente notación:

$$\frac{a\,U_1}{b\,U_2} = \frac{a}{b}\frac{U_1}{U_2} = \frac{a}{b}\left(U_1 * U_2^{-1}\right)$$

Pero aquí hay que advertir que la notación del inverso U_2^{-1} no significa que exista un ente tal que multiplicado por U_2 dé la unidad de los números reales, porque hemos demostrado en el apartado XIV que tal cosa no existe. Allí concluimos que U_2^{-1} no puede ser una cantidad de la misma magnitud U_2, por lo que la notación diádica $U_1 * U_2^{-1}$ debe considerarse equivalente al cociente diádico $U_1/\!/U_2$, únicamente a efectos simbólicos.

Por tanto, en álgebra diádica de magnitudes, a diferencia de lo que ocurre en las estructuras algebraicas aritméticas, las notaciones inversas con unidades o cantidades físicas tienen significado propio y no se corresponden con el fantasma de elementos inversos aislados de las operaciones internas, sino con divisores de alguna razón que no puede faltar. Así, en general, una expresión como $U_1 * U_2^{-n}$ no significa el producto de U_1 por el figurado elemento inverso de U_2^n (ver apartado siguiente sobre potenciación), sino que representa el cociente diádico $U_1/\!/U_2^n$, resultante de la operación de dividir el dividendo U_1 entre la potencia U_2^n. Insistimos, $U_1 * U_2^{-n}$ no denota el producto de la cantidad U_1 por la cantidad inversa U_2^{-n}, que no tiene sentido físico. En geometría afín simboliza un rectángulo de superficie abstracta U_1 y lados U_2^n y $U_1/\!/U_2^n$. A su vez, con nomenclatura de operación algebraica, la expresión indicada por $U_1/\!/U_2^n = U_1 * U_2^{-n}$ interviene en una ley de composición externa generatriz que aplica el producto cartesiano $\{R, U_1\} \times \{R, U_2^n\}$ en el conjunto diádico $\{R, U_1/\!/U_2^n\}$, que es una división diádica.

Section XVII

DEFINITION OF POWER AND
RADICATION OF SCALAR DYADS

The definition equation [12.3] of the dyadic product allows us to define potentiation without controversy. Let us consider an element aU of the set $\{R,U\}$ and let n be any natural number of N, we will call the n power of aU the scalar dyad $(aU)^n$ defined by the following equation:

$$(aU)^n = (a^n)(U^n) \qquad [17.1]$$

It is required that n be natural, because the definition of the concrete multiplication law operates on whole units, not on fractions of units, a concept that would be meaningless by the definition of unit itself.

Let's analyze the meaning of equation [17.1] that defines the power of a scalar dyad, because it is not strange to find opinions that judge it as obvious, since these understandings suppose that it obeys the most elementary algebra of real numbers. Well, such a prejudice would mean not having understood anything about the algebra of magnitudes, let's see why: equation [17.1] does not relate real numbers but scalar physical dyads; if its elements were considered real numbers, it would mean that the power n of the product aU would be equal to the product of the powers of the factors $(a^n) \times (U^n)$; however, since aU is not a product of real numbers, but a quantity of some magnitude, with the meaning of the quantity of it equal to times that of the unit U, its power n, given by [17.1], means the measure a^n of another magnitude in the unit indicated by $U^n = U * U * \ldots * U$, with n factors, and this product is not the real but the dyadic of scalar measurements that arises from the abstraction of the multiplication of the geometric segments. Thus, the definition equation [17.1] is not obvious from the algebra of R, but is a consequence motivated by two causes: the first, due to the geometric fact derived from the definition of segment multiplication and its subsequent abstraction generic; the second, because when developing the algebra of the quantities of magnitudes we lead the symbology along the path of formal similarity with the common of real numbers, so that the resulting notation is isomorphic with it. But, of course, dyadic algebra, like any mathematical or scientific entity, cannot be considered valid or obvious without having adequately defined it in advance, as we are doing here or by means of any other scheme of appropriate definitions. Hence the scruples shown by eminent authors such as Planck or Maxwell, among many others, in relation to operations with magnitudes and the meanings of the expressions constructed with them, as we outlined in the exordium, concerns that can only be overcome by means of the definition of an epistemic algebra that supports operations with magnitudes, since the surprising scientific gap in this matter is not admissible, and solving it is the object of this humble and well-intentioned work.

Once the empowerment is defined, the formulation of the filing is immediate. To do this, the first step must be to fix the natural root of any unit U, which can be established analytically with the following definition equation:

$$\sqrt[n]{U} = U_n \text{ such that } U_n * U_n * \ldots * U_n = U \text{ with } n \text{ factors}$$

This definition equation allows to define analytically the natural dyadic root n of any concrete to U:

$$\sqrt[n]{a\ U} = \sqrt[n]{a}\ U_n \qquad\qquad [17.2]$$

Definition [17.2] could also be struck off as obvious on the wrong basis of assimilating it to an algebraic expression of R, which can be refuted with the same reasons given for potentiation.

Apartado XVII

DEFINICIÓN DE POTENCIACIÓN Y
RADICACIÓN DE DÍADAS ESCALARES

La ecuación de definición [12.3] del producto concreto nos permite definir sin controversia la potenciación. Consideremos un elemento $a\ U$ del conjunto $\{R, U\}$ y sea n un número natural cualquiera de N, llamaremos potencia n de $a\ U$ al concreto escalar $(a\ U)^n$ definido por la ecuación siguiente:

$$(a\ U)^n = (a^n)\ (U^n) \qquad\qquad [17.1]$$

Se exige que n sea natural, porque la definición de la ley de multiplicación concreta opera sobre unidades íntegras, no sobre fracciones de unidades, concepto este que carecería de sentido por la propia definición de unidad.

Analicemos el significado de la ecuación [17.1] que define la potencia de un concreto escalar, porque no es extraño encontrar opiniones que la juzguen de obvia, pues suponen esos entendimientos que obedezca al álgebra más elemental de los números reales. Pues bien, tal prejuicio significaría no haber comprendido nada del álgebra de magnitudes, veamos por qué: la ecuación [17.1] no relaciona números reales sino díadas físicas escalares; si sus elementos fuesen considerados números reales, significaría que la potencia n del producto $a\ U$ sería igual al producto de las potencias de los factores $(a^n) \times (U^n)$; sin embargo, puesto que $a\ U$ no es un producto de números reales, sino una cantidad de alguna magnitud, con el significado de la cantidad de ella igual a a veces la de la unidad U, su potencia n, dada por [17.1], significa la medida a^n de otra magnitud en la unidad indicada por $U^n = U * U * \ldots * U$, con n factores, y este producto no es el real sino el diádico de mediciones escalares que nace por la abstracción de la multiplicación de los segmentos geométricos.

Así, pues, la ecuación de definición [17.1] no es obvia por el álgebra de R, sino que es una consecuencia motivada en dos causas: la primera, debida al hecho geométrico derivado de la definición de multiplicación de segmentos y de su posterior abstracción genérica; la segunda, porque al desarrollar el álgebra de las cantidades de magnitudes conducimos la simbología por la senda de la semejanza formal con la común de los números reales, para que la notación resultante sea isomorfa con esta. Pero, desde luego, el álgebra diádica, como todo ente matemático o científico, no se puede considerar válida ni obvia sin haberla definido adecuadamente con carácter previo, tal como estamos haciendo aquí o por medio de cualquier otro esquema de definiciones apropiadas. De ahí los escrúpulos que muestran eminentes autores como Planck o Maxwell, entre otros muchos, en relación con las operaciones con magnitudes y con los significados de las expresiones construidas con ellas, tal como esbozamos en el exordio, inquietudes que solo pueden salvarse mediante la definición de un álgebra epistémica que dé soporte a las operaciones con magnitudes, pues la sorprendente laguna científica existente en esta materia no es admisible, y solventarla es el objeto de este humilde y bien intencionado trabajo.

Definida la potenciación, es inmediata la formulación de la radicación. Para ello, el primer paso habrá de ser fijar la raíz natural de una unidad cualquiera U, que puede establecerse analíticamente con la ecuación de definición siguiente:

$$\sqrt[n]{U} = U_n \text{ tal que } U_n * U_n * \ldots * U_n = U \text{ con } n \text{ factores}$$

Esta ecuación de definición permite definir analíticamente la raíz diádica natural n de cualquier concreto $a\,U$:

$$\sqrt[n]{a\,U} = \sqrt[n]{a}\,U_n \qquad\qquad [17.2]$$

La definición [17.2] podría también tacharse de obvia con la base errónea de asimilarla a una expresión algebraica de R, lo cual se puede refutar con los mismos motivos dados para la potenciación.

Section XVIII

DEFINITION OF DYADIC LOGARITHMATION
SCALAR AND THE LEGENDARY CALCULATION RULE

Equation [17.1] for the definition of the empowerment of measurements, in combination with the dyadic multiplication and division of the operations [12.3] and [16.2] allow the notion of logarithmation to be defined in an isomorphic way with R. It is enough to consider that in [17.1] the number n represents the logarithm at the base $(a\,U)$ of the given scalar concrete and equal to $[(a^n)(U^n)]$. Thus, the dyadic logarithm is established as follows: given a dyad $(a\,U^n)$ called antilogarithm, and another $(b\,U)$ called base, it will be said that the real number n is the dyadic logarithm in the considered base of the indicated antilogarithm if and only if the condition that $(b\,U)^n = (a\,U^n)$ is verified, which will be written with the following form:

$$\mathscr{L}og_{(b\,U)}(a\,U^n) = n \qquad\qquad [18.1]$$

The intention of this definition is none other than to maintain the isomorphism with the definition of logarithm in R, hence, according to the criterion of dyadic equality, it must be $b^n = a$, which means that in R it is also n the logarithm base b of a, that is:

$$\log_b(a) = n$$

Remember that, for the logarithmic definition to be coherent and reproduce a one-to-one and continuous function between every logarithm n and its corresponding antilogarithm associated to, the base b must be positive and different from unity ($b>0$ y $b \neq 1$). Remember also the equivalent notations:

$$b^n = a \Leftrightarrow \log_b(a) = n; \; b^{\log_b(a)} = a; \; antilog_b(n) = a; \; antilog_b(n) = b^n$$

169

Let's check if the dyadic logarithm satisfies the important properties that are given in R, insofar as the logarithm of a product is equal to the sum of the logarithms, and that of a quotient is the difference of logarithms. To do this, take a base $(b\,U)$ and two dyadic antilogarithms $(a_1\,U^m)$ and $(a_2\,U^n)$. It is clear that m and n are, by definition, the dyadic logarithms of these two antilogarithms, so they are verified:

$$(b\,U)^m = (a_1\,U^m) \text{ y } (b\,U)^m = (a_2\,U^n) \qquad [18.2]$$

By multiplying these two equations dyadically, we arrive at the formulation:

$$(b\,U)^m * (b\,U)^n = (a_1\,U^m) * (a_2\,U^n)$$

The definition $[12.3]$ of the scalar dyadic product allows transforming the first member of this formula into this other:

$$(b\,U)^{m+n} = (a_1\,U^m) * (a_2\,U^n)$$

Reading this equation according to the definition of dyadic logarithm $[18.1]$, we have that $m+n$ is the base logarithm $(b\,U)$ of the dyadic product $(a_1\,U^m) * (a_2\,U^n)$; therefore, with m being the logarithm of the first factor and n being the logarithm of the second, we have that the logarithm of a dyadic product is the sum of the logarithms of the factors.

In a totally analogous way, simply dividing equations $[18.2]$ member by member, it is concluded that the logarithm of a dyadic quotient is the difference of the logarithms, by virtue of the following reasoning:

$$\frac{\left(b\,U^m\right)}{\left(b\,U^n\right)} = \left(b\,U^{m-n}\right) = \frac{\left(a_1\,U^m\right)}{\left(a_2\,U^n\right)}$$

Indeed, it is observed that $m-n$ is the logarithm of the dyadic quotient and, in turn, $m-n$ is the difference of the logarithms of the numerator and denominator.

It has already been established that dyadic composition laws are based on the algebra of geometric segments. Hence one of the most spectacular applications of the fascinating relationship between segment addition, logarithmation, and arithmetic operations cannot be omitted. This is the **slide rule**, which facilitated the development of technology for centuries. Suffice it to say that bridges like the Golden Gate and so many great engineering works or that NASA's Apollo missions were possible thanks to this ingenious calculation instrument, until modern electronics removed it from circulation, not without some loss of training in mental calculation and mathematical foundations, undermining above all the quality of education, since the slide rule infused those values in those who understood it. Well, **the slide rule is based on dyadic algebra and the properties of logarithms**, which make it possible to convert numerical multiplication into a geometric sum of segments and numerical division into a geometric subtraction of segments, in addition to other operations such as the numerical empowerment, which is reduced to the geometric addition of segments through the double logarithm. In any case, the principle of the slide rule is always the same: additive geometric algebra of segments or lengths. Let's see below the foundations of this principle.

A slide rule is a device that consists of three elements: a fixed part or ruler; another mobile rule that moves over the other with a rectilinear movement; and a cursor, which serves to improve the accuracy of the readings and indicate the vertical lines that relate the readings on the ruler and the mobile ruler. The fact of being able to slide the mobile ruler over the ruler is clear that it allows the geometric addition and subtraction of segments to be easily reproduced. To do this, it is enough to arrange a certain quantity of length or segment S_1 on the ruler, juxtapose another segment S_2, this one located on the mobile ruler, and make the final end of S_1 coincide with the initial of S_2. In this way, the sum segment will be given by the dyadic addition $S_1 \oplus S_2$. In turn, the subtraction of segments is reproduced with the mobile ruler placing the end of S_2 on the end of S_1, resulting in the geometric difference $S_1 \ominus S_2$. The schematic in figure 5 clarifies the above and what follows. If both the ruler and the mobile ruler are graduated

SLIDE RULE
Linear scales for design unit U in ruler and mobile ruler
Configurations for arithmetic addition and subtraction

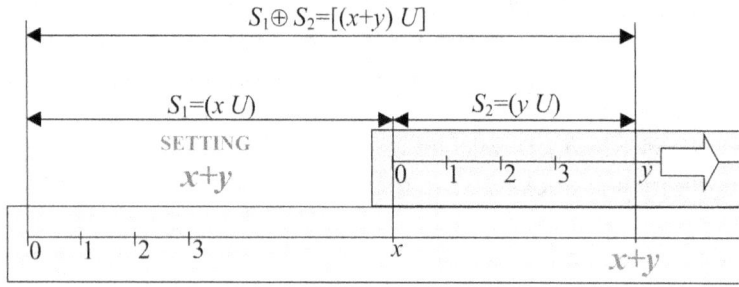

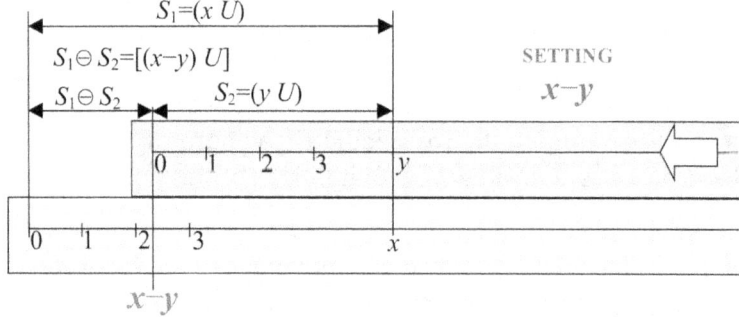

With linear scales in the design length unit U, the distances to the origin are lengths or dyads ($x\,U$) or ($y\,U$), so segment addition and subtraction serves to materialize arithmetic addition and subtraction.

Figure 5

in a certain unit of length U and proportionally, forming linear scales in ruler and mobile ruler, the slide rule will allow to reproduce the addition and subtraction of numbers, representative of the measurements of the segments in the unit of length U set by the design.

And this is where the great contribution of logarithms to mathematical calculations comes: if the ruler and the mobile ruler, instead of linear graduations, are arranged with a logarithmic scale at any base b and always with reference to the design length unit U, adding segments will reproduce arithmetic multiplication and subtracting segments will materialize arithmetic division. Let us see why with the help of figure 6: given two dyads ($M U$), supposed by multiplying, and the multiplier ($m U$), let's look on the rule for the number that indicates the measure M, indicative of the multiplicand. This is equivalent to establishing the segment S_1 such that its measure in the design unit U is the logarithm in the base b of M, that is, $log_b M$. Let us place the number m on the grid, which will indicate the multiplier or measure of segment S_2, which is $log_b m$. **It should be noted that the measures of the logarithms are not labeled in the ruler and the mobile ruler on the grid, but the antilogarithms M and m themselves**, as is usual in logarithmic scales, so that the multiplicand and multiplier can be read directly and easily. This is the reason why on the logarithmic scale the graduation starts at one, which is the antilogarithm base b of zero, since $b^0 = 1$. So, if on a linear scale the graduation with the unit of length U one by one is 0, 1, 2, 3, etc., on a logarithmic scale with base b, the correlative graduation from unit to unit is marked with the antilogarithms of the previous sequence, thus resulting in the series b^0, b^1, b^2, b^3, etc. Under these conditions, the geometric addition of the segments S_1 and S_2, symbolized $S_1 \oplus S_2$, will have as its measure the sum of the logarithms:

$$log_b M + log_b m = log_b (M \times m)$$

Therefore, looking in the rule for the end of the segment $S_1 \oplus S_2$, which must be in the vertical of m, the result of the arithmetic product $M \times m$ will be read.

The configuration for the division is the inverse of the previous one and is based on the geometric subtraction of segments with logarithmic scale. On the rule we look for the dividend D and the divisor d is placed on the mobile ruler. With this configuration, the difference segment $S_1 \ominus S_2$ is given by the dyadic expression:

SLIDE RULE
Logarithmic scales for design unit *U* in ruler and mobile ruler
Settings for arithmetic multiplication and division

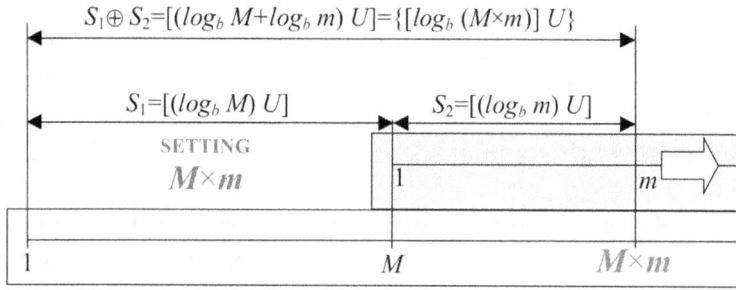

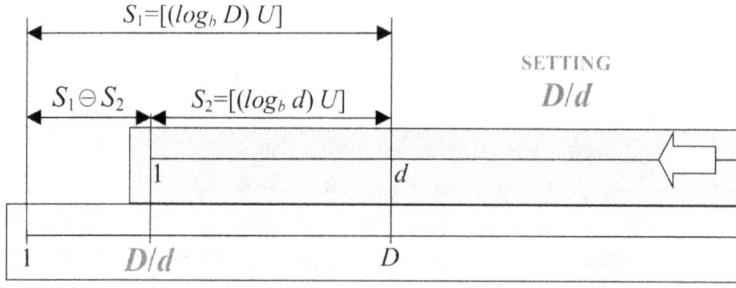

$$S_1 \ominus S_2 = [(log_b D - log_b d)\ U] = \{[(log_b (D/d)]\ U\}$$

With logarithmic scales the geometric sum of segments reproduces the arithmetic multiplication and the difference of segments materializes the arithmetic division.

Figure 6

$$S_1 \ominus S_2 = [(log_b D - log_b d)\ U] = \{[(log_b (D/d)]\ U\}$$

Therefore, the measure of the segment $S_1 \ominus S_2$ in unit U is $log_b (D/d)$, resulting in that the arithmetic quotient between D and d is indicated in the rule by the vertical of the origin of the

174

mobile ruler, marked with one. Configuration that obviously coincides with the multiplication of the quotient D/d by d to obtain D, and which could be considered the geometric proof of the division.

The last operation to be analyzed is potentiation, with the help of figure 7. On a doubly logarithmic scale, which is symbolized LL, the reference point of the rule must be the antilogarithm of the antilogarithm base b of zero, which is b. With this arrangement, take the arithmetic power $B^P=x$ and apply the logarithm in any base b, resulting in the equality indicated below:

$$log_b B^P = p \times log_b B = log_b x$$

SLIDE RULE
Logarithmic scale L on the mobile ruler for design unit U
and double logarithmic scale LL on the ruler
Settings for arithmetic empowerment

SETTING
B^P

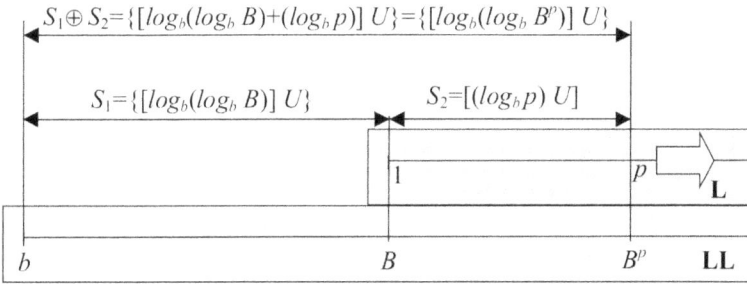

With scales doubly logarithmic LL on the ruler and logarithmic L on the ruler, the configuration of the power B^P results as that which would correspond to the multiplication $B \times p$, but here with the meaning of power, instead of product.

Figure 7

In the previous identity, let's calculate the logarithm base b again, we will have:

$$log_b(p \times log_b B) = log_b p + log_b(log_b B) = log_b(log_b x)$$

Let the segments $S_1 = \{[log_b(log_b B)] U\}$ and $S_2 = [(log_b p) U]$. The geometric addition of these two segments is represented by the analytical form and the reasoning presented in the following logical sequence:

$$S_1 \oplus S_2 = \{[log_b(log_b B)] U\} \oplus [(log_b p) U] = \{[log_b(log_b B) + log_b p] U\} =$$

$$= \{[log_b(log_b x)] U\} = \{[log_b(log_b B^p)] U\}$$

In conclusion, the resulting configuration indicates that, taking the base B of the power on the double-logarithmic scale rule LL, moving the scale on logarithmic scale L until its origin is located on B and reading the quantity p, the power $x = B^p$ will lie on the rule LL in the vertical of p.

The slide rule became possible after the invention of logarithms in 1614 by the Scottish mathematician John Napier. In 1622 the englishman William Aughtred, the first to use the Greek letter π to symbolize the constant quotient between the length of every circumference and its diameter, was the one who knew how to apply the properties of logarithms to operate with numbers and is considered the inventor of the slide rule. Until 1972, when the first electronic pocket calculators appeared, the slide rule was an emblematic and essential instrument for technicians of all fields, it was widely used in Europe and the US. Its manufacture stopped in 1975 due to the advancement of modern computer equipment available to everyone.

To operate with the slide rule, it is necessary to understand the mathematical foundations on which it is based, as well as skill in mental calculation, in order to be able to establish the integer and decimal parts of the results for the different operations and configurations. Hence, it is an instrument of great pedagogical potential. Although its practical benefits have been left behind by advances in electronics, it is no less true that its close relationship with Mathematics and, especially, with additive dyadic algebra, make it a first-rate teaching tool. And this without forgetting that it is an iconic instrument and a very powerful assistant for technicians and scientists until the seventies of the 20th century, with whose contribution the vigorous social development of recent modernity was possible.

Apartado XVIII

DEFINICIÓN DE LOGARITMACIÓN DIÁDICA
ESCALAR Y LA LEGENDARIA REGLA DE CÁLCULO

La ecuación [17.1] de definición de la potenciación de mediciones, en combinación con la multiplicación y la división diádicas de las operaciones [12.3] y [16.2] permiten definir de manera isomorfa con R la noción de logaritmación. Basta considerar que en [17.1] el número n represente el logaritmo en la base $(a\ U)$ del concreto escalar dado e igual a $[(a^n)\ (U^n)]$. Así, el logaritmo diádico queda establecido como se hace a continuación: dada una díada $(a\ U^n)$ llamada antilogaritmo, y otra $(b\ U)$ llamada base, se dirá que el número real n es el logaritmo diádico en la base considerada del antilogaritmo indicado si y solo si se verifica la condición que $(b\ U)^n = (a\ U^n)$, lo cual se escribirá con la forma siguiente:

$$\mathscr{L}\!og_{(b\ U)}(a\ U^n) = n \qquad\qquad [18.1]$$

La intención de esta definición no es otra que mantener el isomorfismo con la definición de logaritmo en R, de ahí que, a tenor del criterio de igualdad diádica, haya de ser $b^n = a$, lo que supone como consecuencia que en R también sea n el logaritmo base b de a, es decir:

$$\log_b(a) = n$$

Recuérdese que, para que la definición logarítmica sea coherente y reproduzca una función biunívoca y continua entre todo logaritmo n y su correspondiente antilogaritmo asociado a, ha de ser la base b positiva y distinta de la unidad ($b>0$ y $b\neq1$). Recuérdense también las notaciones equivalentes:

$$b^n = a \Leftrightarrow \log_b(a) = n;\ b^{\log_b(a)} = a;\ antilog_b(n) = a;\ antilog_b(n) = b^n$$

Comprobemos si el logaritmo diádico satisface las importantes propiedades que se dan en R, en cuanto a que el logaritmo de un producto sea igual a la suma de los logaritmos, y el de un cociente sea la diferencia de logaritmos. Para ello, tómese una base $(b\ U)$ y dos antilogaritmos diádicos $(a_1\ U^m)$ y $(a_2\ U^n)$. Está claro que m y n son, por definición, los logaritmos diádicos de estos dos antilogaritmos, por lo que se verifican:

$$(b\ U)^m = (a_1\ U^m) \text{ y } (b\ U)^m = (a_2\ U^n) \qquad [18.2]$$

Multiplicando diádicamente estas dos ecuaciones, se llega a la formulación:

$$(b\ U)^m * (b\ U)^n = (a_1\ U^m) * (a_2\ U^n)$$

La definición [12.3] del producto diádico escalar permite transformar el primer miembro de esta fórmula en este otro:

$$(b\ U)^{m+n} = (a_1\ U^m) * (a_2\ U^n)$$

Leyendo esta ecuación con arreglo a la definición de logaritmo diádico [18.1] se tiene que $m+n$ es el logaritmo base $(b\ U)$ del producto diádico $(a_1\ U^m) * (a_2\ U^n)$; por lo que, siendo m el logaritmo del primer factor y n el del segundo, se tiene que el logaritmo de un producto diádico es la suma de los logaritmos de los factores.

De manera totalmente análoga, simplemente dividiendo diádicamente miembro a miembro las ecuaciones [18.2], se concluye que el logaritmo de un cociente diádico es la diferencia de los logaritmos, en virtud del siguiente razonamiento:

$$\frac{\left(b\,U^m\right)}{\left(b\,U^n\right)} = \left(b\ U^{m-n}\right) = \frac{\left(a_1\,U^m\right)}{\left(a_2\,U^n\right)}$$

En efecto, se observa que $m-n$ es el logaritmo del cociente diádico y, a su vez, $m-n$ es la diferencia de los logaritmos de numerador y denominador.

Ya se ha establecido que las leyes de composición diádicas tienen como fundamento el álgebra de los segmentos geométricos.

De ahí que no se pueda omitir una de las aplicaciones más espectaculares de la fascinante relación existente entre la adición de segmentos, la logaritmación y las operaciones aritméticas. Se trata de la **regla de cálculo**, que facilitó el desarrollo de la tecnología durante siglos. Baste decir que puentes como el Golden Gate y tantas grandes obras de ingeniería o que las misiones Apolo de la NASA fueron posibles gracias a este ingenioso instrumento de cálculo, hasta que la electrónica moderna lo retiró de la circulación, no sin cierta pérdida de adiestramiento en cálculo mental y fundamentos matemáticos, menoscabando sobre todo la calidad de la educación, puesto que la regla de cálculo insuflaba esos valores en quienes la entendieran. Pues bien, **la regla de cálculo se basa en el álgebra diádica y en las propiedades de los logaritmos**, que permiten convertir la multiplicación numérica en una suma geométrica de segmentos y la división numérica en una sustracción geométrica de segmentos, además de otras operaciones tales como la potenciación numérica, que se reduce a la adición geométrica de segmentos a través del doble logaritmo. En todo caso, el principio de la regla de cálculo es siempre el mismo: el álgebra geométrica aditiva de segmentos o longitudes. Veamos a continuación los fundamentos de dicho principio.

Una regla de cálculo es un dispositivo que consta de tres elementos: una parte fija o regla; otra regla móvil que se traslada sobre la otra con movimiento rectilíneo, llamada reglilla; y un cursor, que sirve para mejorar la precisión de las lecturas e indicar las líneas verticales que relacionen las lecturas sobre la regla y la reglilla. El hecho de poder deslizar la reglilla sobre la regla es claro que permite reproducir con facilidad la adición y la sustracción geométricas de segmentos. Para ello, basta disponer una cierta cantidad de longitud o segmento S_1 sobre la regla, yuxtaponer otro segmento S_2, este situado sobre la reglilla, y hacer coincidir el extremo final de S_1 con el inicial de S_2. De este modo el segmento suma vendrá dado por la adición diádica $S_1 \oplus S_2$. A su vez, la sustracción de segmentos se reproduce con la reglilla situando el extremo final de S_2 sobre el final de S_1, resultando la diferencia geométrica $S_1 \ominus S_2$. El esquema de la figura 5 aclara lo

REGLA DE CÁLCULO
Escalas lineales para la unidad de diseño *U* en regla y reglilla
Configuraciones para la adición y sustracción aritméticas

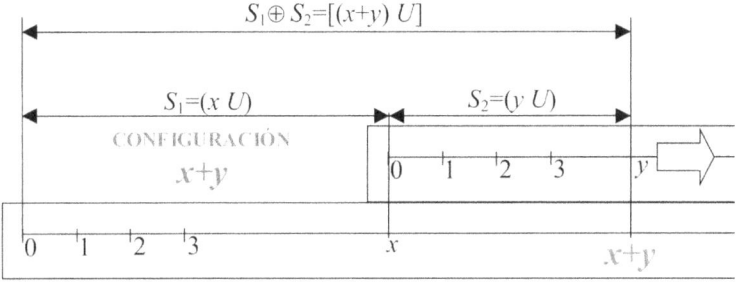

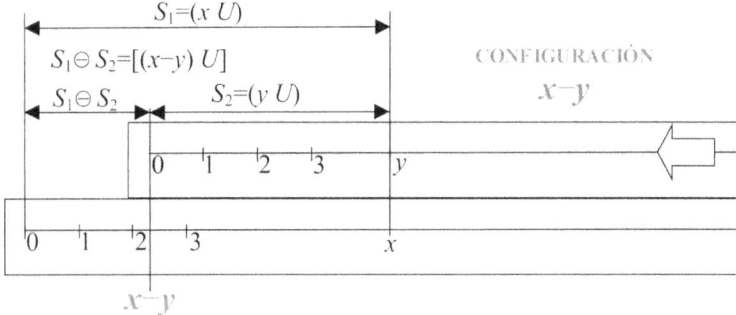

Con escalas lineales en la unidad de longitud de diseño *U*, las distancias al origen son longitudes o diadas ($x\,U$) o ($y\,U$), de modo que la adición y la sustracción de segmentos sirve para materializar la adición y la sustracción aritméticas.

Figura 5

anterior y lo que viene a continuación. Si tanto la regla como la reglilla se gradúan en una cierta unidad de longitud *U* y de forma proporcional, formando escalas lineales en regla y reglilla, la regla de cálculo permitirá reproducir la adición y la sustracción de números, representativos de las medidas de los segmentos en la unidad de longitud *U* establecida por el diseño.

181

Y a continuación es donde viene la gran aportación de los logaritmos a los cálculos matemáticos: si la regla y la reglilla, en vez de graduaciones lineales, se disponen con escala logarítmica en cualquier base *b* y siempre con referencia a la unidad de longitud de diseño *U*, la adición de segmentos reproducirá la multiplicación aritmética y la sustracción de segmentos materializará la división aritmética. Veamos por qué con ayuda de la figura 6: dadas dos díadas (*M U*), supuesto multiplicando, y el multiplicador (*m U*),

REGLA DE CÁLCULO
Escalas logarítmicas para la unidad de diseño *U* en regla y reglilla
Configuraciones para la multiplicación y división aritméticas

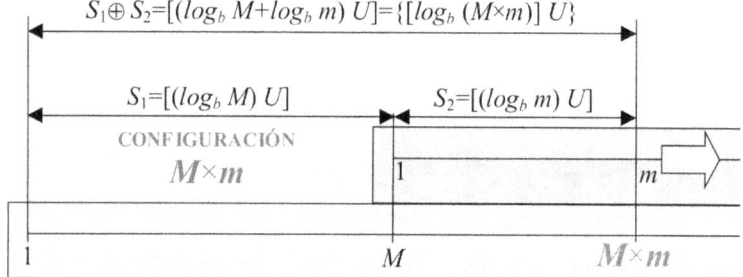

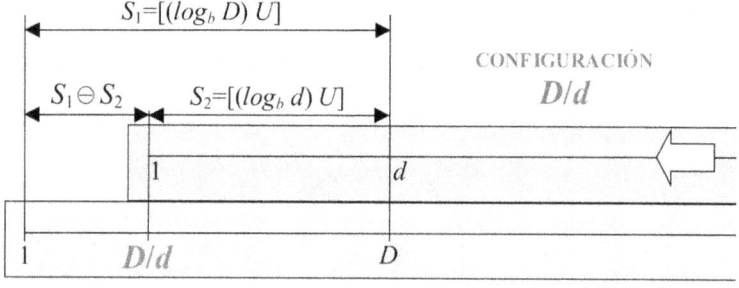

$$S_1 \ominus S_2 = [(log_b D - log_b d)\, U] = \{[(log_b (D/d)]\, U\}$$

Con escalas logarítmicas la suma geométrica de segmentos reproduce la multiplicación aritmética y la diferencia de segmentos materializa la división aritmética.

Figura 6

182

busquemos sobre la regla el número que indica la medida M, indicativa del multiplicando. Ello equivale a establecer el segmento S_1 tal que su medida en la unidad de diseño U sea el logaritmo en la base b de M, es decir, $log_b\, M$. Situemos sobre la reglilla el número m, que indicará el multiplicador o medida del segmento S_2, que es $log_b\, m$. Hay que notar que **en la regla y en la reglilla no se etiquetan las medidas de los logaritmos, sino los propios antilogaritmos M y m**, como es usual en las escalas logarítmicas, para que se puedan leer directa y cómodamente el multiplicando y el multiplicador. Esta es la razón por la que en escala logarítmica la graduación empieza en uno, que es el antilogaritmo base b de cero, pues $b^0 = 1$. De modo que, si en escala lineal la graduación con la unidad de longitud U de una en una es 0, 1, 2, 3, etc., en escala logarítmica con base b la graduación correlativa de unidad en unidad se marca con los antilogaritmos de la sucesión anterior, resultando así la serie b^0, b^1, b^2, b^3, etc. En estas condiciones, la adición geométrica de los segmentos S_1 y S_2, simbolizada $S_1 \oplus S_2$, tendrá por medida la suma de los logaritmos:

$$log_b\, M + log_b\, m = log_b\, (M \times m)$$

Por tanto, buscando en la regla el extremo del segmento $S_1 \oplus S_2$, que ha de estar en la vertical de m, se leerá el resultado del producto aritmético $M \times m$.

La configuración para la división es la inversa de la anterior y se basa en la sustracción geométrica de segmentos con escala logarítmica. Sobre la regla se busca el dividendo D y en la reglilla se sitúa encima el divisor d. Con esta configuración el segmento diferencia $S_1 \ominus S_2$ viene dado por la expresión diádica:

$$S_1 \ominus S_2 = [(log_b\, D - log_b\, d)\; U] = \{[(log_b\, (D/d)]\; U\}$$

Por tanto, la medida del segmento $S_1 \ominus S_2$ en la unidad U es $log_b\, (D/d)$, resultando que el cociente aritmético entre D y d está señalado en la regla por la vertical del origen de la reglilla, marcado con el uno. Configuración que obviamente coincide con la multiplicación del cociente D/d por d para obtener D, y que podría considerarse la prueba geométrica de la división.

REGLA DE CÁLCULO
Escala logarítmica L en la reglilla para la unidad de diseño *U*
y escala doblemente logarítmica LL en la regla
Configuración para la potenciación aritmética

CONFIGURACIÓN
$$B^p$$

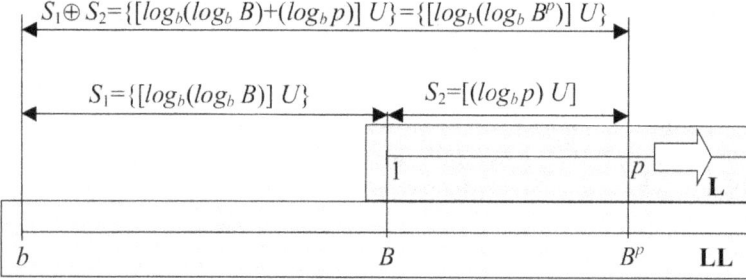

Con escalas doblemente logarítimica LL en la regla y logarítimica L en la reglilla, la configuración de la potencia B^p resulta como la que correspondería a la multiplicación $B \times p$, pero aquí con el significado de potencia, en vez de producto.

Figura 7

La última operación que se analizará es la potenciación, con ayuda de la figura 7. En escala doblemente logarítmica, que se simboliza LL, el punto de referencia de la regla debe ser el antilogaritmo del antilogaritmo base *b* de cero, que es *b*. Con esta disposición, tómese la potencia aritmética $B^p = x$ y apliquemos la logaritmación en cualquier base *b*, resultando la igualdad indicada a continuación:

$$log_b B^p = p \times log_b B = log_b x$$

En la identidad anterior calculemos nuevamente el logaritmo base *b*, se tendrá:

$$log_b(p \times log_b B) = log_b p + log_b(log_b B) = log_b(log_b x)$$

184

Sean los segmentos $S_1 = \{[log_b(log_b B)]\ U\}$ y $S_2 = [(log_b p)\ U]$. La adición geométrica de estos dos segmentos queda representada por la forma analítica y el razonamiento expuestos en la siguiente secuencia lógica:

$$S_1 \oplus S_2 = \{[log_b(log_b B)]\ U\} \oplus [(log_b p)\ U] = \{[log_b(log_b B) + log_b p]\ U\} =$$

$$= \{[log_b(log_b x)]\ U\} = \{[log_b(log_b B^p)]\ U\}$$

En conclusión, la configuración resultante indica que, tomando sobre la regla en escala doblemente logarítmica LL la base B de la potencia, moviendo la reglilla en escala logarítmica L hasta situar su origen sobre B y leyendo en ella la cantidad p, la potencia $x = B^p$ se encontrará sobre la regla LL en la vertical de p.

La regla de cálculo fue posible tras la invención de los logaritmos en 1614 por el matemático escocés John Napier. En 1622 el inglés William Aughtred, el primero en utilizar la letra griega π para simbolizar el cociente constante entre la longitud de toda circunferencia y su diámetro, fue quien supo aplicar las propiedades de los logaritmos para operar con números y es considerado el inventor de la regla de cálculo. Hasta 1972, cuando aparecen las primeras calculadoras de bolsillo electrónicas, la regla de cálculo fue un instrumento emblemático e imprescindible para los técnicos de todos los ámbitos, fue ampliamente utilizada en Europa y EEUU. Su fabricación se detuvo a partir de 1975 por el avance de los equipos informáticos modernos al alcance de todos.

Para operar con la regla de cálculo se requiere entender los fundamentos matemáticos en que se basa, así como se precisa habilidad en cálculo mental, a fin de poder establecer las partes entera y decimal de los resultados para las diferentes operaciones y configuraciones. De ahí que sea un instrumento de gran potencial pedagógico. Aunque sus prestaciones prácticas han quedado atrás por los avances de la electrónica, no es menos cierto que su estrecha relación con la Matemática y, en especial, con el álgebra diádica aditiva, la convierten en una herramienta docente de primer orden. Y ello sin olvidar que se trata de un instrumento icónico y asistente muy potente de los técnicos y científicos hasta los años setenta del siglo XX, con cuya aportación fue posible el vigoroso desarrollo social de la reciente modernidad.

Section XIX

DEFINITION OF SCALAR PRODUCTS
AND VECTOR OF VECTOR DYADS [11]

In Physics, vector magnitudes use mathematical vectors and products between vectors called scalar product and vector product. It is usual to indicate the scalar with a mathematical point «·» and the vector with the same cross «×» used as a symbol of the various multiplications, by the principle of symbolic economy; although here with the meaning of this kind of product between vectors. We know that this should not be misleading, because, depending on which elements are multiplied, the meaning of the multiplication sign will be established, even if it is the same for different operations. Thus, if the operators are real numbers, the cross will indicate the multiplication in R; if scalar dyads are multiplied, it will refer to the product of these entities; if a real number and a vector are composed, the multiplication indicated by «×» will be the external law of the vector space R^3 over R; if the factors are vectors, this same cross will indicate the vector product of vectors; or, if vector concretes are multiplied, the product will be the concrete vector, which we are going to define here. And something similar can be indicated about the sign of the scalar product with a point «·», which can adopt different meanings depending on the context of the factors.

---

[11] The scalar and vector products of vectors of R^3 or V^3 or E^3, which are three more or less synonymous and isomorphic notations, as well as their properties, can be found in «Lesson 5» of Mathematize 2. Roughly speaking we have both Spaces of the points and geometric vectors, when they are associated with each other, the affine space is shaped. If the scalar product or interior connection of vectors is also defined, the Euclidean space results, all in three dimensions; and, by simple abstract generalization, they are defined for any dimension n.

However, in what follows we will distinguish each specific symbol, for greater pedagogical clarity. The mathematical point «·» will indicate the operation scalar product of vectors, the angle «∧» will indicate the vector product of vectors, and for the dyadic homonyms of quantities of vector magnitudes we will reserve the circle with a point «⊙» for the scalar dyadic product and the circle with an asterisk «⊛» for the vector dyadic product.

There is no better way to define both the scalar product and the vector product of vector magnitudes than in terms of their counterparts of mathematical vectors. Let's start with the dot product of two vector dyads $\overline{a}\,U_1$ and $\overline{b}\,U_2$ of the dyadic or concrete vector spaces $\{R^3,U_1\}$ and $\{R^3,U_2\}$. We distinguish here the first elements $\overline{a}$ and $\overline{b}$ the pairs as mathematical vectors of R^3, because for scalars neither the scalar product nor the vector make sense. We have to define the dot product of said vector concretes with the following definition equation:

$$\overline{a}\,U_1 \odot \overline{b}\,U_2 = (\overline{a}\cdot\overline{b})\,(U_1 * U_2) \qquad [19.1]$$

That is, by definition, the scalar product of two vector dyads measures a scalar magnitude with the scalar concrete of the set $\{R,U_1 * U_2\}$ such that its primary is the scalar product of the vector primaries of the factors and the unit is the product dyadic of the factor units.

Regarding the vector dyadic product, we will have the following definition equation in an analogous way:

$$\overline{a}\,U_1 \circledast \overline{b}\,U_2 = (\overline{a}\wedge\overline{b})\,(U_1 * U_2) \qquad [19.2]$$

In this case, by definition, the vector product of two vector dyads measures a vector magnitude with the dyad of the set $\{R^3,U_1 * U_2\}$ such that its primary is another vector equal to the vector product of the vectors that integrate the primaries of the given concrete or dyads, and whose unit is the dyadic product of the factor units.

The definition formulas [19.1] and [19.2] will facilitate the correct interpretation of the physical equations involving these composition laws, such as the work magnitude of the dyad force, in the case of the scalar product, or the moment magnitude related to the dyad force, for the vector product.

The scalar product of vectors of R^3 is commutative, associative with respect to the scalars of R and distributive with respect to vector addition, properties of the vector space R^3 that are reproduced in the vector dyadic algebra, as can be easily verified, as is done in what follows.

The commutative property of the scalar product is derived from the definition equation [19.1]. The commutativity of the scalar product of vectors is such that we have $\overline{a} \cdot \overline{b} = \overline{b} \cdot \overline{a}$. The axiomatic commutative property [13.1] of the multiplication of units $U_1 * U_2 = U_2 * U_1$, exposed at the beginning of section XIII, produces as a consequence the following chain of equalities:

$$\overline{a}\, U_1 \odot \overline{b}\, U_2 = (\overline{a} \cdot \overline{b})(U_1 * U_2) = (\overline{b} \cdot \overline{a})(U_2 * U_1) = \overline{b}\, U_2 \odot \overline{a}\, U_1$$

The associative property in relation to the product by a scalar p of R on the left, analogously on the right, is established by the reasoning that follows, based on the associative behavior of the corresponding operations with scalars and vectors, thus as in the definition [19.1] of the dot product of vector dyads:

$$p \circ (\overline{a}\, U_1 \odot \overline{b}\, U_2) = p \circ [(\overline{a} \cdot \overline{b})(U_1 * U_2)] \qquad [19.3]$$

The definitions [9.1] and [9.2] of multiplication of a scalar by a scalar dyad, or their explicit [9.3] and [9.4], justify that:

$$p \circ [(\overline{a} \cdot \overline{b})(U_1 * U_2)] = [p \times (\overline{a} \cdot \overline{b})](U_1 * U_2)$$

For vectors, we have that $p \times (\overline{a} \cdot \overline{b}) = (p \bullet \overline{a}) \cdot \overline{b}$, which allows us to write:

$$[p \times (\overline{a} \cdot \overline{b})](U_1 * U_2) = [(p \bullet \overline{a}) \cdot \overline{b})](U_1 * U_2) \qquad [19.4]$$

The definition [19.1] of the dot product of dyads or vector leads us to:

$$[(p \bullet \overline{a}) \cdot \overline{b})](U_1 * U_2) = (p \bullet \overline{a})\, U_1 \odot \overline{b} U_2$$

And again the definitions [9.3] and [9.4] authorize us to write the following equality:

$$(p \bullet \overline{a})\, U_1 \odot \overline{b} U_2 = [p \circ (\overline{a}\, U_1)] \odot \overline{b} U_2 \qquad [19.5]$$

Therefore, the first member of [19.3] and the second of [19.5] are equal, we have the first associative property of the scalar dyadic product, which is described by the following expression:

$$p \circ (\overline{a}\, U_1 \odot \overline{b}\, U_2) = [p \circ (\overline{a}\, U_1)] \odot \overline{b} U_2$$

In turn, for vectors, we have that $p \times (\overline{a} \cdot \overline{b}) = \overline{a} \cdot (p \bullet \overline{b})$, which allows us to write the first member of [19.4] like this:

$$[p \times (\overline{a} \cdot \overline{b})]\, (U_1 * U_2) = [\overline{a} \cdot (p \bullet \overline{b})]\, (U_1 * U_2)$$

The definition [19.1] of the scalar product of vector dyads leads us to:

$$[\overline{a} \cdot (p \bullet \overline{b})]\, (U_1 * U_2) = \overline{a}\, U_1 \odot (p \bullet \overline{b})\, U_2$$

Again the definition [9.3] and [9.4] leads us to write the following equality:

$$\overline{a}\, U_1 \odot (p \bullet \overline{b})\, U_2 = \overline{a}\, U_1 \odot [p \circ (\overline{b}\, U_2)] \qquad [19.6]$$

And, the first member of [19.3] and the second of [19.6] being equal, we have the second associative property of the scalar product:

$$p \circ (\overline{a}\, U_1 \odot \overline{b}\, U_2) = \overline{a}\, U_1 \odot [p \circ (\overline{b}\, U_2)]$$

The distributive property of the dyadic scalar product with respect to the addition of vector dyads is also derived from the corresponding property of the vectors of R^3. Reasoning is described from the left, similarly it would be spun from the right. The definition [5.4] of addition of vector dyads serves to give the first step of the reasoning:

$$\overline{a}\, U_1 \odot (\overline{b}\, U_2 \oplus \overline{c}\, U_2) = \overline{a}\, U_1 \odot [(\overline{b} + \overline{c})\, U_2] \qquad [19.7]$$

Note that the principle of uniformity of addition requires that the addends refer to the same unit U_2.

The definition [19.1] of the dot product of vector dyads allows us to write the second member of [19.7] with the form:

$$\overline{a}\ U_1\odot[(\overline{b}+\overline{c})\ U_2]=[\overline{a}\cdot(\overline{b}+\overline{c})]\ (U_1*U_2)$$

The distributive property of the scalar product of vectors of R^3 is $\overline{a}\cdot(\overline{b}+\overline{c})=(\overline{a}\cdot\overline{b})+(\overline{a}\cdot\overline{c})$, which leads us to:

$$[\overline{a}\cdot(\overline{b}+\overline{c})]\ (U_1*U_2)=[(\overline{a}\cdot\overline{b})+(\overline{a}\cdot\overline{c})]\ (U_1*U_2)$$

The definition [5.3] of addition of scalar dyads justifies the equality:

$$[(\overline{a}\cdot\overline{b})+(\overline{a}\cdot\overline{c})]\ (U_1*U_2)=[(\overline{a}\cdot\overline{b})\ (U_1*U_2)]\oplus[(\overline{a}\cdot\overline{c})\ (U_1*U_2)]$$

And by virtue of the definition [19.1] of the dot product of vector dyads we arrive at:

$$[(\overline{a}\cdot\overline{b})\ (U_1*U_2)]\oplus[(\overline{a}\cdot\overline{c})\ (U_1*U_2)]=$$

$$=[(\overline{a}\ U_1)\odot(\overline{b}\ U_2)]\oplus[(\overline{a}\ U_1)\odot(\overline{c}\ U_2)] \qquad [19.8]$$

As the first member of [19.7] and the second of [19.8] are equal, the distributive property of the scalar product of vector dyads with respect to their dyadic addition remains explicit, with the analytic form:

$$\overline{a}\ U_1\odot(\overline{b}\ U_2\oplus\overline{c}\ U_2)=[(\overline{a}\ U_1)\odot(\overline{b}\ U_2)]\oplus[(\overline{a}\ U_1)\odot(\overline{c}\ U_2)]$$

In turn, for the vector product of dyads we must use the properties of the vector multiplication of vectors of R^3, which we know is not commutative, but anticommutative or antisymmetric, which is not associative, and which does verify the distributive property. The anticommutativity of the vector product of vectors is written $\overline{a}\wedge\overline{b}=-\overline{b}\wedge\overline{a}$, together with the definition [19.2] and by the commutative axiom [13.1] for the multiplication of units $U_1*U_2=U_2*U_1$, we have the following logical reasoning of dyadic algebra or concrete in relation to this lack of symmetry:

$$\overline{a}\ U_1\circledast\overline{b}\ U_2=(\overline{a}\wedge\overline{b})\ (U_1*U_2)=-(\overline{b}\wedge\overline{a})\ (U_2*U_1)=-\overline{b}\ U_2\circledast\overline{a}\ U_1$$

We take the opportunity to develop another associative form, which is also verified with the vector product and which is verified here in a similar way, which is the associative property in relation to the product by two scalars p and q of R on the left, it is analogous on the right. It is established by the sequence set out below, based on the associative behavior of the corresponding operations with scalars and vectors, as well as on definitions [9.3] and [9.4] of the product by scalars of vector concretes:

$$(p \circ \overline{a} \; U_1) \circledast (q \circ \overline{b} \; U_2) = [(p \bullet \overline{a}) \; U_1] * [(q \bullet \overline{b}) \; U_2] \qquad [19.9]$$

The definition [19.2] of the vector dyadic product of vector dyads justifies that:

$$[(p \bullet \overline{a}) \; U_1] * [(q \bullet \overline{b}) \; U_2] = [(p \bullet \overline{a}) \wedge (q \bullet \overline{b})] \; (U_1 * U_2)$$

In vector algebra we have $(p \bullet \overline{a}) \wedge (q \bullet \overline{b}) = (p \times q) \bullet (\overline{a} \wedge \overline{b})$, with the meaning to be given to the multiplication signs, depending on the elements that make up: in $(p \bullet \overline{a})$ and $(q \bullet \overline{b})$ it is about the multiplication of scalars by vectors, in $(p \times q)$ it indicates the multiplication of scalars of R and in $\overline{a} \wedge \overline{b}$ it describes the vector product of vectors. In these conditions, we have:

$$[(p \bullet \overline{a}) \wedge (q \bullet \overline{b})] \; (U_1 * U_2) = [(p \times q) \bullet (\overline{a} \wedge \overline{b})] \; (U_1 * U_2)$$

The composition law defined by [9.3] and [9.4] or multiplication of a scalar by a vector dyad leads us to:

$$[(p \times q) \bullet (\overline{a} \wedge \overline{b})] \; (U_1 * U_2) = (p \times q) \circ [(\overline{a} \wedge \overline{b}) \; (U_1 * U_2)]$$

The definition [19.2] of a dyadic vector product of vector dyads allows us to write:

$$(p \times q) \circ [(\overline{a} \wedge \overline{b}) \; (U_1 * U_2)] =$$

$$= (p \times q) \circ [(\overline{a} \; U_1) \circledast (\overline{b} \; U_2)] \qquad [19.20]$$

And, the first member of [19.19] and the second of [19.20] being equal, we have the analytical form of the investigated associative property:

$$(p \circ \overline{a} \; U_1) \circledast (q \circ \overline{b} \; U_2) = (p \times q) \circ [(\overline{a} \; U_1) \circledast (\overline{b} \; U_2)] \qquad [19.21]$$

As regards the distributive property of the vector product of vector dyads, it can be deduced with similar ease by the following reasoning, that we start with the definition [5.4] of the addition of uniform vector dyads applied to the first member of [19.22]:

$$\overline{a}\, U_1 \circledast (\overline{b}\, U_2 \oplus \overline{c}\, U_2) = \overline{a}\, U_1 \circledast [(\overline{b} + \overline{c})\, U_2] \qquad [19.22]$$

The definition [19.2] of the vector product itself legitimizes us to write:

$$\overline{a}\, U_1 \circledast [(\overline{b} + \overline{c})\, U_2] = [\overline{a} \wedge (\overline{b} + \overline{c})]\, (U_1 * U_2)$$

The distributive property of the vector product of vectors of R^3 is $\overline{a} \wedge (\overline{b} + \overline{c}) = (\overline{a} \wedge \overline{b}) + (\overline{a} \wedge \overline{c})$, with which:

$$[\overline{a} \wedge (\overline{b} + \overline{c})]\, (U_1 * U_2) = [(\overline{a} \wedge \overline{b}) + (\overline{a} \wedge \overline{c})]\, (U_1 * U_2)$$

Definition [5.4] of addition of vector dyads leads us to the following line of reasoning:

$$[(\overline{a} \wedge \overline{b}) + (\overline{a} \wedge \overline{c})]\, (U_1 * U_2) = [(\overline{a} \wedge \overline{b})\, (U_1 * U_2)] \oplus [(\overline{a} \wedge \overline{c})\, (U_1 * U_2)]$$

With the definition [19.2] of the vector product of vector dyads we arrive at:

$$[(\overline{a} \wedge \overline{b})\, (U_1 * U_2)] \oplus [(\overline{a} \wedge \overline{c})\, (U_1 * U_2)] =$$
$$= [(\overline{a}\, U_1) \circledast (\overline{b}\, U_2)] \oplus [(\overline{a}\, U_1) \circledast (\overline{c}\, U_2)] \qquad [19.23]$$

And, resulting that the first member of [19.22] is equal to the second of [19.23], we arrive at the analytic form of the distributive property:

$$\overline{a}\, U_1 \circledast (\overline{b}\, U_2 \oplus \overline{c}\, U_2) =$$
$$= [(\overline{a}\, U_1) \circledast (\overline{b}\, U_2)] \oplus [(\overline{a}\, U_1) \circledast (\overline{c}\, U_2)] \qquad [19.24]$$

Note, as already pointed out before, that in the distributive properties dyads have been added referring to the same unit U_2, because we have established that the dyadic addition requires that the addends be uniform.

In this section, the principle of symbolic economy has not been deliberately applied in an absolute way, in order to highlight the various composition laws that intervene in the properties that relate them. However, in scientific or educational practice it is usual to indicate all additive operations with the same signs of addition, just as all multiplicative laws are identified with the same multiplication graph, and the same can be said about subtractions or divisions. However, the spell that this causes leads to reason under the effect of the symbolic illusion that this simplification produces, since the elements with which it operates appear as if they were entities of R, and the custom associated with the properties of the real numbers induces unconscious reasoning with the error of subliminally admitting this fantasy, causing that, although correct conclusions are reached, in reality the logical argument is not valid, because the true algebraic nature of the related composition laws is ignored by the equations.

When the principle of symbolic economy is applied, it is necessary to make an effort to observe that the same signs used for different operations represent different composition laws depending on the nature of the elements between which they are located. Thus, when written between scalars, they refer to the addition, multiplication, subtraction, or quotient of R; when they are found between vectors, they will indicate the vector addition or subtraction, the scalar product or the vector product in R^3; and, situated between dyadic entities, they will refer to dyadic operations of $\{R,U\}$ or $\{R^3,U\}$. Otherwise, the algebraic reasoning would not have been truly understood, although the conclusion may appear to be symbolically correct.

Apartado XIX

DEFINICIÓN DE LOS PRODUCTOS ESCALAR Y VECTORIAL DE DÍADAS VECTORIALES [12]

En Física las magnitudes vectoriales se sirven de los vectores matemáticos y de productos entre vectores denominados producto escalar y producto vectorial. Es usual indicar el escalar con un punto matemático «·» y el vectorial con la misma aspa «×» utilizada como símbolo de las diversas multiplicaciones, por el principio de economía simbólica; aunque aquí con el significado de esta especie de producto entre vectores. Sabemos que ello no debe inducir a error, porque, en virtud de cuáles sean los elementos multiplicados, el significado del signo de multiplicación quedará establecido, aunque sea el mismo para operaciones diferentes. Así, si los operadores son números reales, el aspa indicará la multiplicación en R; si se multiplican concretos escalares, se referirá al producto de estos entes; si se compone un número real y un vector, la multiplicación señalada con «×» será la ley externa del espacio vectorial R^3 sobre R; si los factores son vectores, esta misma aspa indicará el producto vectorial de vectores; o, si se multiplican concretos vectoriales, el producto será el vectorial concreto, que nos disponemos a definir aquí. Y algo parecido cabe indicar sobre el signo del producto escalar con un punto «·», que puede adoptar diversos significados en función del contexto de los factores. No obstante, en lo que sigue distinguiremos cada símbolo

[12] Los productos escalar y vectorial de vectores de R^3 o V^3 o E^3, que son tres notaciones más o menos sinónimas e isomorfas, así como sus propiedades, se pueden encontrar en la «Lección 5» de *Matematizar 2*. A *grosso modo* tenemos los dos espacios de los puntos y vectores geométricos, cuando se los asocia entre sí se da forma al espacio afín, si además se define el producto escalar o conexión interior de vectores resulta el espacio euclidiano, todo ello en tres dimensiones; y, por simple generalización abstracta, se definen para cualquier dimensión n.

195

específico, para mayor claridad pedagógica. El punto matemático «·» indicará la operación producto escalar de vectores, el ángulo «∧» señalará el producto vectorial de vectores, y para los homónimos diádicos de cantidades de magnitudes vectoriales reservaremos el círculo con un punto «⊙» para el producto diádico escalar y el círculo con asterisco «⊛» para el producto diádico vectorial.

Tanto el producto escalar como el vectorial de magnitudes vectoriales no hay mejor manera de definirlos que en función de sus homónimos de los vectores matemáticos. Comencemos por el producto escalar de dos díadas vectoriales $\overline{a}\ U_1$ y $\overline{b}\ U_2$ de los espacios vectoriales diádicos o concretos $\{R^3, U_1\}$ y $\{R^3, U_2\}$. Distinguimos aquí los primeros elementos $\overline{a}$ y $\overline{b}$ de los pares como vectores matemáticos de R^3, porque para los escalares ni el producto escalar ni el vectorial tienen sentido. Tenemos que definir el producto escalar de dichos concretos vectoriales con la siguiente ecuación de definición:

$$\overline{a}\ U_1 \odot \overline{b}\ U_2 = (\overline{a} \cdot \overline{b})\,(U_1 * U_2) \qquad [19.1]$$

Es decir, por definición, el producto escalar de dos concretos vectoriales mide una magnitud escalar con el concreto escalar del conjunto $\{R, U_1 * U_2\}$ tal que su primario sea el producto escalar de los primarios vectoriales de los factores y la unidad el producto diádico o concreto de las unidades de los factores.

En cuanto al producto vectorial tendremos por su parte y de manera análoga la siguiente ecuación de definición:

$$\overline{a}\ U_1 \circledast \overline{b}\ U_2 = (\overline{a} \wedge \overline{b})\,(U_1 * U_2) \qquad [19.2]$$

En este caso, por definición, el producto vectorial de dos concretos vectoriales mide una magnitud vectorial con el concreto del conjunto $\{R^3, U_1 * U_2\}$ tal que su primario sea otro vector igual al producto vectorial de los vectores que integran los primarios de los concretos o díadas dados, y cuya unidad sea el producto diádico de las unidades de los factores.

Las fórmulas de definición [19.1] y [19.2] facilitarán la interpretación correcta de las ecuaciones físicas en que intervengan estas leyes de composición, como la magnitud trabajo del concreto fuerza, en el caso del producto escalar, o la magnitud momento atinente a la díada fuerza, para el producto vectorial.

El producto escalar de vectores de R^3 es conmutativo, asociativo respecto de los escalares de R y distributivo respecto de la adición vectorial, propiedades del espacio vectorial R^3 que se reproducen en el álgebra concreta vectorial, como se puede comprobar con facilidad, tal como se hace en lo que sigue.

La propiedad conmutativa del producto escalar se deduce a partir de la ecuación de definición [19.1]. La conmutatividad del producto escalar de vectores es tal que se tiene $\overline{a} \cdot \overline{b} = \overline{b} \cdot \overline{a}$. La propiedad conmutativa axiomática [13.1] de la multiplicación de unidades $U_1 * U_2 = U_2 * U_1$, expuesta al principio del apartado XIII, produce como consecuencia la siguiente cadena de igualdades:

$$\overline{a}\ U_1 \odot \overline{b}\ U_2 = (\overline{a} \cdot \overline{b})\ (U_1 * U_2) = (\overline{b} \cdot \overline{a})\ (U_2 * U_1) = \overline{b}\ U_2 \odot \overline{a}\ U_1$$

La propiedad asociativa en relación con el producto por un escalar p de R por la izquierda, análogamente resulta por la derecha, queda establecida por los razonamientos que se exponen a continuación, basados en el comportamiento asociativo de las correspondientes operaciones con escalares y vectores, así como en la definición [19.1] del producto escalar de concretos vectoriales:

$$p \circ (\overline{a}\ U_1 \odot \overline{b}\ U_2) = p \circ [(\overline{a} \cdot \overline{b})\ (U_1 * U_2)] \qquad [19.3]$$

Las definiciones [9.1] y [9.2] de multiplicación de un escalar por un concreto escalar, o sus explícitas [9.3] y [9.4], justifican que:

$$p \circ [(\overline{a} \cdot \overline{b})\ (U_1 * U_2)] = [p \times (\overline{a} \cdot \overline{b})]\ (U_1 * U_2)$$

Para vectores, se tiene que $p \times (\overline{a} \cdot \overline{b}) = (p \bullet \overline{a}) \cdot \overline{b}$, lo que permite escribir:

$$[p \times (\overline{a} \cdot \overline{b})]\ (U_1 * U_2) = [(p \bullet \overline{a}) \cdot \overline{b})]\ (U_1 * U_2) \qquad [19.4]$$

La definición [19.1] del producto escalar de díadas o concretos vectoriales nos lleva a:

$$[(p\bullet\overline{a})\cdot\overline{b})]\,(U_1*U_2)=(p\bullet\overline{a})\,U_1\odot\overline{b}\,U_2$$

Y nuevamente las definiciones [9.3] y [9.4] nos autorizan a escribir la siguiente igualdad:

$$(p\bullet\overline{a})\,U_1\odot\overline{b}\,U_2=[p\circ(\overline{a}\,\,U_1)]\odot\overline{b}\,U_2 \qquad [19.5]$$

Con lo que, resultando iguales el primer miembro de [19.3] y el segundo de [19.5], se tiene la primera propiedad asociativa del producto escalar, que queda descrita mediante la siguiente expresión:

$$p\circ(\overline{a}\,\,U_1\odot\overline{b}\,\,U_2)=[p\circ(\overline{a}\,\,U_1)]\odot\overline{b}\,U_2$$

A su vez, para vectores, se tiene que $p\times(\overline{a}\cdot\overline{b})=\overline{a}\cdot(p\bullet\overline{b})$, lo que permite escribir el primer miembro de [19.4] así:

$$[p\times(\overline{a}\cdot\overline{b})]\,(U_1*U_2)=[\overline{a}\cdot(p\bullet\overline{b})]\,(U_1*U_2)$$

La definición [19.1] del producto escalar de concretos vectoriales nos conduce a:

$$[\overline{a}\cdot(p\bullet\overline{b})]\,(U_1*U_2)=\overline{a}\,\,U_1\odot(p\bullet\overline{b})\,U_2$$

Otra vez la definición [9.3] y [9.4] nos lleva a escribir la siguiente igualdad:

$$\overline{a}\,\,U_1\odot(p\bullet\overline{b})\,U_2=\overline{a}\,\,U_1\odot[p\circ(\overline{b}\,\,U_2)] \qquad [19.6]$$

Y, resultando iguales el primer miembro de [19.3] y el segundo de [19.6], tenemos la segunda propiedad asociativa del producto escalar:

$$p\circ(\overline{a}\,\,U_1\odot\overline{b}\,\,U_2)=\overline{a}\,\,U_1\odot[p\circ(\overline{b}\,\,U_2)]$$

La propiedad distributiva del producto escalar diádico respecto de la adición de concretos vectoriales se deriva igualmente de la correspondiente de los vectores de $\mathbb{R}^3$. Se describe el razonamiento por la izquierda, análogamente se hilaría por la derecha. La definición [5.4] de adición de concretos vectoriales sirve para dar el primer paso del razonamiento:

$$\overline{a}\,\,U_1\odot(\overline{b}\,\,U_2\oplus\overline{c}\,\,U_2)=\overline{a}\,\,U_1\odot[(\overline{b}+\overline{c})\,U_2] \qquad [19.7]$$

Nótese que el principio de uniformidad de la adición requiere que los sumandos se refieran a la misma unidad U_2.

La definición [19.1] del producto escalar de concretos vectoriales nos autoriza a escribir el segundo miembro de [19.7] con la forma:

$$\overline{a}\ U_1 \odot [(\overline{b}+\overline{c})\ U_2] = [\overline{a}\cdot(\overline{b}+\overline{c})]\ (U_1 * U_2)$$

La propiedad distributiva del producto escalar de vectores de R^3 es $\overline{a}\cdot(\overline{b}+\overline{c}) = (\overline{a}\cdot\overline{b})+(\overline{a}\cdot\overline{c})$, lo que nos lleva a:

$$[\overline{a}\cdot(\overline{b}+\overline{c})]\ (U_1 * U_2) = [(\overline{a}\cdot\overline{b})+(\overline{a}\cdot\overline{c})]\ (U_1 * U_2)$$

La definición [5.3] de adición de concretos escalares justifica la igualdad:

$$[(\overline{a}\cdot\overline{b})+(\overline{a}\cdot\overline{c})]\ (U_1 * U_2) = [(\overline{a}\cdot\overline{b})\ (U_1 * U_2)] \oplus [(\overline{a}\cdot\overline{c})\ (U_1 * U_2)]$$

Y en virtud de la definición [19.1] del producto escalar de díadas vectoriales llegamos a:

$$[(\overline{a}\cdot\overline{b})\ (U_1 * U_2)] \oplus [(\overline{a}\cdot\overline{c})\ (U_1 * U_2)] =$$
$$= [(\overline{a}\ U_1)\odot(\overline{b}\ U_2)] \oplus [(\overline{a}\ U_1)\odot(\overline{c}\ U_2)] \qquad [19.8]$$

Resultando iguales el primer miembro de [19.7] y el segundo de [19.8], queda explícita la propiedad distributiva del producto escalar de concretos vectoriales respecto de su adición diádica, con la forma analítica:

$$\overline{a}\ U_1 \odot (\overline{b}\ U_2 \oplus \overline{c}\ U_2) = [(\overline{a}\ U_1)\odot(\overline{b}\ U_2)] \oplus [(\overline{a}\ U_1)\odot(\overline{c}\ U_2)]$$

A su vez, para el producto vectorial de concretos debemos utilizar las propiedades de la multiplicación vectorial de vectores de R^3, que sabemos no es conmutativa, sino anticonmutativa o antisimétrica, que no es asociativa, y que sí verifica la propiedad distributiva. La anticonmutatividad del producto vectorial de vectores se escribe $\overline{a}\wedge\overline{b} = -\overline{b}\wedge\overline{a}$, junto con la definición [19.2] y por el axioma conmutativo [13.1] para la multiplicación de unidades $U_1 * U_2 = U_2 * U_1$, tenemos el siguiente razonamiento lógico de álgebra diádica o concreta en relación con esta falta de simetría:

$$\overline{a}\ U_1 \circledast \overline{b}\ U_2 = (\overline{a} \wedge \overline{b})\ (U_1 * U_2) = -(\overline{b} \wedge \overline{a})\ (U_2 * U_1) = -\overline{b}\ U_2 \circledast \overline{a}\ U_1$$

Aprovechamos para desarrollar otra forma asociativa, que también se verifica con el producto vectorial y que se comprueba aquí de manera similar, que es la propiedad asociativa en relación con el producto por dos escalares p y q de R por la izquierda, resulta análoga por la derecha. Queda establecida por la ilación que se expone seguidamente, basada en el comportamiento asociativo de las correspondientes operaciones con escalares y vectores, así como en las definiciones [9.3] y [9.4] del producto por escalares de concretos vectoriales:

$$(p \circ \overline{a}\ U_1) \circledast (q \circ \overline{b}\ U_2) = [(p \bullet \overline{a})\ U_1] * [(q \bullet \overline{b})\ U_2] \qquad [19.9]$$

La definición [19.2] del producto diádico vectorial de concretos vectoriales justifica que:

$$[(p \bullet \overline{a})\ U_1] * [(q \bullet \overline{b})\ U_2] = [(p \bullet \overline{a}) \wedge (q \bullet \overline{b})]\ (U_1 * U_2)$$

En álgebra de vectores se tiene $(p \bullet \overline{a}) \wedge (q \bullet \overline{b}) = (p \times q) \bullet (\overline{a} \wedge \overline{b})$, con el significado que debe darse a los signos de multiplicación, según cuáles sean los elementos que compongan: en $(p \bullet \overline{a})$ y $(q \bullet \overline{b})$ se trata de la multiplicación de escalares por vectores, en $(p \times q)$ señala la multiplicación de escalares de R y en $\overline{a} \wedge \overline{b}$ describe el producto vectorial de vectores. En estas condiciones, tenemos:

$$[(p \bullet \overline{a}) \wedge (q \bullet \overline{b})]\ (U_1 * U_2) = [(p \times q) \bullet (\overline{a} \wedge \overline{b})]\ (U_1 * U_2)$$

La ley de composición definida por [9.3] y [9.4] o multiplicación de un escalar por un concreto vectorial nos lleva a:

$$[(p \times q) \bullet (\overline{a} \wedge \overline{b})]\ (U_1 * U_2) = (p \times q) \circ [(\overline{a} \wedge \overline{b})\ (U_1 * U_2)]$$

La definición [19.2] de producto vectorial concreto de díadas vectoriales nos permite escribir:

$$(p \times q) \circ [(\overline{a} \wedge \overline{b})\ (U_1 * U_2)] =$$
$$= (p \times q) \circ [(\overline{a}\ U_1) \circledast (\overline{b}\ U_2)] \qquad [19.20]$$

Y, resultando iguales el primer miembro de [19.19] y el segundo de [19.20], tenemos la forma analítica de la propiedad asociativa investigada:

$$(p \circ \overline{a} \ U_1) \circledast (q \circ \overline{b} \ U_2) = (p \times q) \circ [(\overline{a} \ U_1) \circledast (\overline{b} \ U_2)] \quad [19.21]$$

Por lo que se refiere a la propiedad distributiva del producto vectorial de díadas vectoriales, se deduce con facilidad análoga mediante el siguiente razonamiento, que comenzamos con la definición [5.4] de la adición de díadas vectoriales uniformes aplicada al primer miembro de [19.22]:

$$\overline{a} \ U_1 \circledast (\overline{b} \ U_2 \oplus \overline{c} \ U_2) = \overline{a} \ U_1 \circledast [(\overline{b} + \overline{c}) \ U_2] \quad [19.22]$$

La definición [19.2] del propio producto vectorial nos legitima para escribir:

$$\overline{a} \ U_1 \circledast [(\overline{b} + \overline{c}) \ U_2] = [\overline{a} \wedge (\overline{b} + \overline{c})] \ (U_1 * U_2)$$

La propiedad distributiva del producto vectorial de vectores de R^3 es $\overline{a} \wedge (\overline{b} + \overline{c}) = (\overline{a} \wedge \overline{b}) + (\overline{a} \wedge \overline{c})$, con lo cual:

$$[\overline{a} \wedge (\overline{b} + \overline{c})] \ (U_1 * U_2) = [(\overline{a} \wedge \overline{b}) + (\overline{a} \wedge \overline{c})] \ (U_1 * U_2)$$

La definición [5.4] de adición de concretos vectoriales nos lleva a la siguiente línea del razonamiento:

$$[(\overline{a} \wedge \overline{b}) + (\overline{a} \wedge \overline{c})] \ (U_1 * U_2) = [(\overline{a} \wedge \overline{b}) \ (U_1 * U_2)] \oplus [(\overline{a} \wedge \overline{c}) \ (U_1 * U_2)]$$

Con la definición [19.2] del producto vectorial de concretos vectoriales llegamos a:

$$[(\overline{a} \wedge \overline{b}) \ (U_1 * U_2)] \oplus [(\overline{a} \wedge \overline{c}) \ (U_1 * U_2)] =$$
$$= [(\overline{a} \ U_1) \circledast (\overline{b} \ U_2)] \oplus [(\overline{a} \ U_1) \circledast (\overline{c} \ U_2)] \quad [19.23]$$

Y, resultando que el primer miembro de [19.22] es igual al segundo de [19.23], llegamos a la forma analítica de la propiedad distributiva:

$$\overline{a} \ U_1 \circledast (\overline{b} \ U_2 \oplus \overline{c} \ U_2) =$$
$$= [(\overline{a} \ U_1) \circledast (\overline{b} \ U_2)] \oplus [(\overline{a} \ U_1) \circledast (\overline{c} \ U_2)] \quad [19.24]$$

Nótese, como ya se ha apuntado antes, que en las propiedades distributivas se han sumado concretos referidos a la misma unidad U_2, porque hemos establecido que la adición diádica exija que los sumandos sean uniformes.

En este apartado no se ha aplicado deliberadamente de manera absoluta el principio de economía simbólica, para poner de manifiesto las diversas leyes de composición que intervienen en las propiedades que las relacionan. Sin embargo, en la práctica científica o educacional es usual señalar con los mismos signos de adición todas las operaciones aditivas, así como se identifican con el mismo grafo de multiplicación todas la leyes multiplicativas, y lo mismo cabe decir sobre las restas o las divisiones. No obstante, el hechizo que ello provoca lleva a razonar bajo el efecto de la ilusión simbólica que esta simplificación produce, puesto que los elementos con los que se opera se aparecen como si fuesen entes de R, y la costumbre asociada a las propiedades de los números reales induce a razonar inconscientemente con el error que supone admitir subliminalmente esta fantasía, provocando que, aunque se llegue a conclusiones correctas, en realidad el argumento lógico no es válido, porque se pasa por alto la verdadera naturaleza algebraica de las leyes de composición relacionadas por las ecuaciones.

Cuando se aplica el principio de economía simbólica es obligado esforzarse en observar que los mismos signos utilizados para operaciones diversas representan leyes de composición diferentes según la naturaleza de los elementos entre los que se sitúen. Así, cuando se escriben entre escalares, se refieren a la adición, a la multiplicación, a la resta o al cociente de R; cuando se encuentren entre vectores, indicarán la adición o sustracción vectorial, el producto escalar o el producto vectorial en R^3; y, situados entre entes diádicos, se referirán a las operaciones diádicas de $\{R, U\}$ o $\{R^3, U\}$. De otro modo, no se habría entendido de verdad el razonamiento algebraico, aunque en apariencia la conclusión pueda resultar acertada simbólicamente.

Section XX

PRODUCT DEFINITION BETWEEN SCALAR AND VECTOR DYADS

Another composition law that must be defined, because it appears constantly in physical equations, is the multiplication of scalar dyads by other vectors. To do this, let be the scalar dyad $a\,U_1$ of $\{R,U_1\}$ and the vector $\overline{b}U_2$ of $\{R^3,U_2\}$. Let us use the sign «$\odot$» for this composition law. We have to establish the definition of this product with the external law of the vector space R^3 over R and dyadic multiplication, so there is no better formulation than with the following two definition equations:

$$a\,U_1 \odot \overline{b}U_2 = (a \bullet \overline{b})\,(U_1 * U_2) \qquad [20.1]$$

$$\overline{b}U_2 \odot a\,U_1 = (\overline{b} \bullet a)\,(U_2 * U_1) \qquad [20.2]$$

In [20.1] and [20.2] the sign «$\odot$» of the first member symbolizes the composition law that we are defining, the product of a scalar dyad by another vector, the sign «$\bullet$» placed in the factor $(a \bullet \overline{b})$ of the second member indicates the external law of R^3 over R, application of $R \times R^3$ in R^3, or product of a scalar by a vector, and the multiplications of the terms $(U_1 * U_2)$ and $(U_2 * U_1)$ indicate the dyadic product of two scalar dyadic, defined in section XII. Although it is usual to use the same sign «×» for all these multiplicative laws.

Due to the algebraic structure of the vector space R^3, as well as the commutative axiom [13.1] of the multiplication of concrete units, it can be written:

$$a\,U_1 \odot \overline{b}U_2 = (a \bullet \overline{b})\,(U_1 * U_2) = (\overline{b} \bullet a)\,(U_2 * U_1) \qquad [20.3]$$

The definition [20.2] of this new composition law leads to identity:

$$(\overline{b} \bullet a)\,(U_2 * U_1) = \overline{b}U_2 \circledcirc a\,U_1 \qquad [20.4]$$

And, the first member of [20.3] and the second of [20.4] are equal, the commutative property of the operation defined by [20.1] and [20.2] is concluded:

$$a\,U_1 \circledcirc \overline{b}U_2 = \overline{b}U_2 \circledcirc a\,U_1$$

This new operation contains many possibilities. For example, the real number a could indicate the scalar product of two vectors $a = \overline{a}_1 \cdot \overline{a}_2$, and this would be the case of the scalar product of the vector dyadic entities $\overline{a}_1 A_1$ and $\overline{a}_2 A_2$, where the unit U_1 would be given by the compound $U_1 = A_1 * A_2$. Under these conditions, equation [20.1] would become:

$$[(\overline{a}_1 A_1) \circledcirc (\overline{a}_2 A_2)] \circledcirc \overline{b}U_2 = [(\overline{a}_1 \cdot \overline{a}_2) \bullet \overline{b}]\,(A_1 * A_2 * U_2) \qquad [20.5]$$

All its own vector properties can be applied to the product $(\overline{a}_1 \cdot \overline{a}_2) \bullet \overline{b}$ in R³.

For its part, the vector dyad $\overline{b}U_2$ can be such that it is given by a vector dyadic product $\overline{b}U_2 = \overline{b}_1 B_1 \circledast \overline{b}_2 B_2$, which would have the form for [20.1]:

$$a\,U_1 \circledcirc (\overline{b}_1 B_1 \circledast \overline{b}_2 B_2) = [a \bullet (\overline{b}_1 \wedge \overline{b}_2)]\,(U_1 * B_1 * B_2) \qquad [20.6]$$

All the proper properties of this space can be applied to the cross product $a \bullet (\overline{b}_1 \wedge \overline{b}_2)$ in R³.

In both hypotheses of [20.5] and [20.6] the compound units have been written taking into account their associative property [13.4].

The simple verification is left to the reader that the dyadic composition law defined in this section verifies all associative and distributive forms. In turn, definitions [20.1] and [20.2] allow to deduce the corresponding divisions of this product without more than taking the second members as a dividend and one of the factors of the first members as a divisor, resulting in the scalar quotient between a vector dividend and a vector divisor, and the vector quotient between a vector dividend and a scalar divisor.

204

Apartado XX

DEFINICIÓN DE PRODUCTO DIÁDICO ENTRE
UN CONCRETO ESCALAR Y OTRO VECTORIAL

Otra ley de composición que es preciso definir, porque aparece constantemente en las ecuaciones físicas, es la multiplicación de concretos escalares por otros vectoriales. Para ello, sea la díada escalar $a\ U_1$ de $\{R, U_1\}$ y la vectorial $\overline{b}\ U_2$ de $\{R^3, U_2\}$. Utilicemos para esta ley de composición el signo «$\circledcirc$». Contamos para establecer la definición de este producto con la ley externa del espacio vectorial R^3 sobre R y la multiplicación diádica, por lo que no cabe mejor formulación que con las dos ecuaciones de definición siguientes:

$$a\ U_1 \circledcirc \overline{b}\ U_2 = (a \bullet \overline{b})\ (U_1 * U_2) \qquad [20.1]$$

$$\overline{b}\ U_2 \circledcirc a\ U_1 = (\overline{b} \bullet a)\ (U_2 * U_1) \qquad [20.2]$$

En [20.1] y [20.2] el signo «$\circledcirc$» del primer miembro simboliza la ley de composición que estamos definiendo, el producto de un concreto escalar por otro vectorial, el signo «$\bullet$» puesto en el factor $(a \bullet \overline{b})$ del segundo miembro señala la ley externa de R^3 sobre R, aplicación de $R \times R^3$ en R^3, o producto de un escalar por un vector, y las multiplicaciones de los términos $(U_1 * U_2)$ y $(U_2 * U_1)$ señalan el producto diádico de dos concretos escalares, definido en el apartado XII. Aunque sea usual usar el mismo signo «$\times$» para todas estas leyes multiplicativas.

En razón de la estructura algebraica del espacio vectorial R^3, así como por el axioma conmutativo [13.1] de la multiplicación de unidades concretas, se puede escribir:

$$a\ U_1 \circledcirc \overline{b}\ U_2 = (a \bullet \overline{b})\ (U_1 * U_2) = (\overline{b} \bullet a)\ (U_2 * U_1) \qquad [20.3]$$

La definición [20.2] de esta nueva ley de composición conduce a la identidad:

$$(\overline{b} \bullet a)\,(U_2 * U_1) = \overline{b}\,U_2 \odot a\,U_1 \qquad [20.4]$$

Y, resultando iguales el primer miembro de [20.3] y el segundo de [20.4], se concluye la propiedad conmutativa de la operación definida por [20.1] y [20.2]:

$$a\,U_1 \odot \overline{b}\,U_2 = \overline{b}\,U_2 \odot a\,U_1$$

Esta nueva operación encierra muchas posibilidades. Por ejemplo, el número real a podría indicar el producto escalar de dos vectores $a = \overline{a}_1 \cdot \overline{a}_2$, y este sería el caso del producto escalar de los entes diádicos vectoriales $\overline{a}_1\,A_1$ y $\overline{a}_2\,A_2$, donde la unidad U_1 vendría dada por la compuesta $U_1 = A_1 * A_2$. En estas condiciones, la ecuación [20.1] se convertiría en:

$$[(\overline{a}_1\,A_1) \odot (\overline{a}_2\,A_2)] \odot \overline{b}\,U_2 = [(\overline{a}_1 \cdot \overline{a}_2) \bullet \overline{b}\,]\,(A_1 * A_2 * U_2) \quad [20.5]$$

Al producto $(\overline{a}_1 \cdot \overline{a}_2) \bullet \overline{b}$ en R^3 se le pueden aplicar todas las propiedades vectoriales que le son propias.

Por su parte, la díada vectorial $\overline{b}\,U_2$ puede ser tal que venga dada por un producto diádico vectorial $\overline{b}\,U_2 = \overline{b}_1\,B_1 \circledast \overline{b}_2\,B_2$, con lo que se tendría para [20.1] la forma:

$$a\,U_1 \odot (\overline{b}_1\,B_1 \circledast \overline{b}_2\,B_2) = [a \bullet (\overline{b}_1 \wedge \overline{b}_2)]\,(U_1 * B_1 * B_2) \quad [20.6]$$

Al producto vectorial $a \bullet (\overline{b}_1 \wedge \overline{b}_2)$ en R^3 se le pueden aplicar todas las propiedades propias de este espacio.

En ambas hipótesis de [20.5] y [20.6] se han escrito las unidades compuestas teniendo en cuenta su propiedad asociativa [13.4].

Se deja al lector la sencilla comprobación de que la ley de composición diádica definida en este apartado verifica todas las formas asociativas y distributivas.

A su vez, las definiciones [20.1] y [20.2] permiten deducir las correspondientes divisiones de este producto sin más que tomar los segundos miembros como dividendo y uno de los factores de los primeros miembros como divisor, resultando el cociente escalar entre un dividendo vectorial y un divisor vectorial, y el cociente vectorial entre un dividendo vectorial y un divisor escalar.

Section XXI

DEFINITION OF DYADIC ENTITIES
IMAGINARIES AND THEIR LAWS OF COMPOSITION

We have used the algebraic structures of R and R^3 to define the entities and composition laws of scalar and vector dyads. However, we still have a very interesting algebraic structure, the C field of complex or imaginary numbers[13]. Recall that imaginary numbers are defined as pairs of real numbers x and y related to the number $i = \sqrt{-1}$ and symbolized with the form $z = x + i \times y$. So we agree to define on the basis of them the imaginary concrete entities as those in which the primary is an element $z \in C$ and in whose secondary a unit is arranged, which must necessarily be scalar, given the axiom established for this purpose in the section III. We will symbolize this new entity with the notation $z\,U$, and the set of homogeneous imaginary concretes will be written $\{C,U\}$.

Once the imaginary dyad and its sets $\{C,U\}$ have been created, the definitions of its internal and external laws of composition must be undertaken. In the first place, following the usual order, one must begin with addition: let the imaginary dyads $z_1\,U$ and $z_2\,U$ be, which must be homogeneous, given the axiom of uniformity, which is also required in this case, because we know that imaginary numbers operate as vectors in the R^2 plane. So we define the additive internal law «⊕» or application of the Cartesian product $\{C,U\} \times \{C,U\}$ in $\{C,U\}$, by means of this definition equation:

---

[13] In «Lesson 42» of Mathematize 1 the algebraic structure of the field C of imaginary numbers with their composition laws is analyzed.

$$z_1 U \oplus z_2 U = (z_1 + z_2) U$$

The addition of the first member corresponds to the sum of imaginary dyads defined here and differs from the additives of scalar or vector dyadics, although they are all marked with the same sign «⊕»; while the addition of the term $z_1 + z_2$ is the sum of C, which also differs from the addition of R and is commutative and associative, given the abelian additive group structure of C, which entails the commutative and associative properties of the addition of imaginary dyads, which we have already succinctly exposed, so as not to result excessively repetitive in the reasoning:

$$z_1 U \oplus z_2 U = (z_1 + z_2) U = (z_2 + z_1) U = z_2 U \oplus z_1 U$$

$$z_1 U \oplus (z_2 U \oplus z_3 U) = z_1 U \oplus (z_2 + z_3) U = [z_1 + (z_2 + z_3)] U =$$

$$= [(z_1 + z_2) + z_3] U = (z_1 + z_2) U \oplus z_3 U = (z_1 U \oplus z_2 U) \oplus z_3 U$$

In an analogous way we can and do define the multiplicative external generating law «*», or application of the Cartesian product set $\{C, U_1\} \times \{C, U_2\}$ in $\{C, U_1 * U_2\}$, with the defining equation:

$$z_1 U_1 * z_2 U_2 = (z_1 \times z_2)(U_1 * U_2)$$

The multiplication of the first member is the dyadic of imaginary dyads that is defined in the equation itself and differs from the dyadic with real entities, although it is indicated with the same sign «*», while the product of the term $z_1 \times z_2$ is the multiplication of C, not R, and is commutative and associative, given the commutative field structure of C. All this, together with the commutative axiom [13.1] and the associative [13.4] of the product of units, leads us to the properties commutative and associative of the dyadic product «⊛» of imaginary measurements:

$$z_1 U_1 * z_2 U_2 = (z_1 \times z_2)(U_1 * U_2) = (z_2 \times z_1)(U_2 * U_1) = z_2 U_2 * z_1 U_1$$

$$z_1 U_1 * (z_2 U_2 * z_3 U_3) = z_1 U_1 * (z_2 \times z_3)(U_2 * U_3) =$$

$$= [z_1 \times (z_2 \times z_3)](U_1 * U_2 * U_3) = [(z_1 \times z_2) \times z_3](U_1 * U_2 * U_3) =$$

$$=(z_1 \times z_2)\,(U_1 * U_2) * z_3\,U_3 = (z_1\,U_1 * z_2\,U_2) * z_3\,U_3$$

In a totally isomorphic way with the reasoning followed for {R,U}, since C has the same algebraic structure as R, each of these two numerical sets with its own additive and multiplicative internal laws, and even in analogy with {R³,U}, since C also behaves as a vector space over R, the rest of the properties are deduced and the other composition laws are defined for {C,U}. With this and remembering that the additive part has to refer to the same unit U, given the axiom of uniformity, we will have the distributive property of the product with respect to the addition of imaginary dyads:

$$z_1\,U_1 * (z_2\,U \oplus z_3\,U) = z_1\,U_1 * (z_2 + z_3)\,U =$$

$$=[z_1 \times (z_2 + z_3)]\,(U_1 * U) = [(z_1 \times z_2) + (z_1 \times z_3)]\,(U_1 * U) =$$

$$=(z_1 \times z_2)\,(U_1 * U) + (z_1 \times z_3)\,(U_1 * U) = (z_1\,U_1 * z_2\,U) \oplus (z_1\,U_1 * z_3\,U)$$

The definition of multiplication by a scalar p of C, which in turn could be singularly an element of R, because R⊂C, must be associated with a map of the Cartesian product C×{C,U} in {C,U} on the left, and {C,U}×C on {C,U} on the right, with the definition equations:

$$p \circ (z\,U) = (p \times z)\,U$$

$$(z\,U) \circ p = (z \times p)\,U$$

Since in C we have the commutative property $p \times z = z \times p$, it is immediate that the imaginary concrete multiplication by an imaginary or real scalar is commutative:

$$p \circ (z\,U) = (z\,U) \circ p$$

The properties of C guarantee the associative behavior of this dyadic multiplication by two scalars p and q of C, or singularly of R, since R⊂C:

$$(p \times q) \circ (z\,U) = [(p \times q) \times z]\,U = [p \times (q \times z)]\,U = p \circ [(q \times z)\,U]$$

Various distributive properties such as the following are also verified for this law:

$$p\circ(z_1 U\oplus z_2 U)=p\circ[(z_1+z_2)U]=[p\times(z_1+z_2)]U=$$

$$[(p\times z_1)+(p\times z_2)]U=(p\times z_1)U\oplus(p\times z_2)U=$$

$$=[p\circ(z_1 U)]\oplus[p\circ(z_2 U)]$$

$$(p+q)\circ(z U)=[(p+q)\times z]U=[(p\times z)+(q\times z)]U=$$

$$=(p\times z)U\oplus(q\times z)U=[p\circ(z U)]\oplus[q\circ(z U)]$$

In sum, we also observe here how the daily algebraic structures of the primaries are accommodated to those already established for the mathematical entities of R, R^3 or C, so that the secondary or dimensional part respond independently to the generalization of the algebra of the geometric segments, with the **affinity postulate**, derived from the possibility of establishing one-to-one correspondence between the quantities of length of the geometric segments and the quantities of any other magnitude, thereby justifying operations with magnitudes affine to length based on the geometric algebra.

In turn, the symbolic forms of the usual rules are maintained with all the terms, including the symbols of the units, as a consequence of the definitions and properties of the multiple laws of composition that configure the various daily algebraic structures.

Apartado XXI

DEFINICIÓN DE ENTES DIÁDICOS O CONCRETOS IMAGINARIOS Y DE SUS LEYES DE COMPOSICIÓN

Nos hemos servido de las estructuras algebraicas de R y de R^3 para definir los entes y leyes de composición de las díadas escalares y vectoriales. Sin embargo, todavía disponemos de una estructura algebraica de sumo interés, el cuerpo C de los números complejos o imaginarios[14]. Recordemos que los números imaginarios se definen como pares de números reales x e y relacionados con el número $i = \sqrt{-1}$ y simbolizados con la forma $z = x + i \times y$. Así que convenimos en definir en base a ellos los entes concretos imaginarios como aquellos en los que el primario sea un elemento $z \in C$ y en cuyo secundario se disponga una unidad, que habrá de ser necesariamente escalar, dado el axioma establecido al efecto en el apartado III. Este nuevo ente lo simbolizaremos con la notación $z\ U$, y el conjunto de los concretos imaginarios homogéneos se escribirá $\{C, U\}$.

Una vez creada la díada imaginaria y sus conjuntos $\{C, U\}$, deben acometerse las definiciones de sus leyes de composición internas y externas. En primer lugar, siguiendo el orden habitual, debe empezarse por la adición: sean los concretos imaginarios $z_1\ U$ y $z_2\ U$, que habrán de ser homogéneos, dado el axioma de uniformidad, exigible también en este caso, porque sabemos que los números imaginarios operan como vectores en el plano R^2. Así que definimos la ley interna aditiva «$\oplus$» o aplicación del producto cartesiano $\{C, U\} \times \{C, U\}$ en $\{C, U\}$, mediante esta ecuación de definición:

---

[14] En la «Lección 42» de *Matematizar 1* se analiza la estructura algebraica del cuerpo C de los números imaginarios con sus leyes de composición.

$$z_1\ U \oplus z_2\ U = (z_1 + z_2)\ U$$

La adición del primer miembro corresponde a la suma de concretos imaginarios definida aquí y difiere de las aditivas de diádicas escalares o vectoriales, aunque se señalen todas con el mismo signo «$\oplus$»;mientras que la adición del término $z_1 + z_2$ es la suma de C, que también difiere de la adición de R y que es conmutativa y asociativa, dada la estructura de grupo aditivo abeliano de C, lo que conlleva las propiedades conmutativa y asociativa de la adición de concretos imaginarios, que exponemos ya sucintamente, para no resultar en exceso reiterativos en los razonamientos:

$$z_1\ U \oplus z_2\ U = (z_1 + z_2)\ U = (z_2 + z_1)\ U = z_2\ U \oplus z_1\ U$$

$$z_1\ U \oplus (z_2\ U \oplus z_3\ U) = z_1\ U \oplus (z_2 + z_3)\ U = [z_1 + (z_2 + z_3)]\ U =$$

$$= [(z_1 + z_2) + z_3]\ U = (z_1 + z_2)\ U \oplus z_3\ U = (z_1\ U \oplus z_2\ U) \oplus z_3\ U$$

De manera análoga podemos definir y definimos la ley externa generatriz multiplicativa «$*$», o aplicación del conjunto producto cartesiano $\{C, U_1\} \times \{C, U_2\}$ en $\{C, U_1 * U_2\}$ que responde a la ecuación de definición:

$$z_1\ U_1 * z_2\ U_2 = (z_1 \times z_2)\ (U_1 * U_2)$$

La multiplicación del primer miembro es la diádica de concretos imaginarios que se define en la propia ecuación y difiere de la diádica con entes reales, aunque se indique con el mismo signo «$*$», mientras que el producto del término $z_1 \times z_2$ es la multiplicación de C, no la de R, y es conmutativa y asociativa, dada la estructura de cuerpo conmutativo de C. Todo ello, junto con el axioma conmutativo [13.1] y el asociativo [13.4] del producto de unidades, nos lleva a las propiedades conmutativa y asociativa del producto diádico «$\circledast$» de mediciones imaginarias:

$$z_1\ U_1 * z_2\ U_2 = (z_1 \times z_2)\ (U_1 * U_2) = (z_2 \times z_1)\ (U_2 * U_1) = z_2\ U_2 * z_1\ U_1$$

$$z_1\ U_1 * (z_2\ U_2 * z_3\ U_3) = z_1\ U_1 * (z_2 \times z_3)\ (U_2 * U_3) =$$

$$= [z_1 \times (z_2 \times z_3)]\ (U_1 * U_2 * U_3) = [(z_1 \times z_2) \times z_3]\ (U_1 * U_2 * U_3) =$$

$$=(z_1 \times z_2)\ (U_1 * U_2) *z_3\ U_3 =(z_1\ U_1 *z_2\ U_2)*z_3\ U_3$$

De manera totalmente isomorfa con los razonamientos seguidos para $\{R,U\}$, puesto que C tiene la misma estructura algebraica que R, cada uno de estos dos conjuntos numéricos con sus propias leyes internas aditiva y multiplicativa, e incluso en analogía con $\{R^3, U\}$, puesto que C también se comporta como espacio vectorial sobre R, se deducen el resto de propiedades y se definen las demás leyes de composición para $\{C, U\}$. Con ello y recordando que la parte aditiva se tiene que referir a la misma unidad U, dado el axioma de uniformidad, tendremos la propiedad distributiva del producto respecto de la adición de concretos imaginarios:

$$z_1\ U_1 *(z_2\ U \oplus z_3\ U) = z_1\ U_1 *(z_2 +z_3)\ U =$$

$$=[z_1 \times (z_2 +z_3)]\ (U_1 * U) = [(z_1 \times z_2)+(z_1 \times z_3)]\ (U_1 * U) =$$

$$=(z_1 \times z_2)\ (U_1 * U)+(z_1 \times z_3)\ (U_1 * U) = (z_1\ U_1 *z_2\ U)\oplus(z_1\ U_1 *z_3\ U)$$

La definición de la multiplicación por un escalar p de C, que a su vez podría ser singularmente un elemento de R, porque $R \subset C$, debe asociarse con una aplicación del producto cartesiano $C \times \{C,U\}$ en $\{C,U\}$ por la izquierda, y $\{C,U\} \times C$ en $\{C,U\}$ por la derecha, con las ecuaciones de definición:

$$p \circ (z\ U) = (p \times z)\ U$$

$$(z\ U) \circ p = (z \times p)\ U$$

Dado que en C se tiene la propiedad conmutativa $p \times z = z \times p$, es inmediato que la multiplicación concreta imaginaria por un escalar imaginario o real resulta conmutativa:

$$p \circ (z\ U) = (z\ U) \circ p$$

Las propiedades de C garantizan el comportamiento asociativo de esta multiplicación diádica por dos escalares p y q de C, o singularmente de R, ya que $R \subset C$:

$$(p \times q) \circ (z\ U) = [(p \times q) \times z]\ U = [p \times (q \times z)]\ U = p \circ [(q \times z)\ U]$$

También se verifican para esta ley diversas propiedades distributivas como las siguientes:

$$p \circ (z_1 \ U \oplus z_2 \ U) = p \circ [(z_1 + z_2) \ U] = [p \times (z_1 + z_2)] \ U =$$

$$[(p \times z_1) + (p \times z_2)] \ U = (p \times z_1) \ U \oplus (p \times z_2) \ U =$$

$$= [p \circ (z_1 \ U)] \oplus [p \circ (z_2 \ U)]$$

$$(p + q) \circ (z \ U) = [(p + q) \times z] \ U = [(p \times z) + (q \times z)] \ U =$$

$$= (p \times z) \ U \oplus (q \times z) \ U = [p \circ (z \ U)] \oplus [q \circ (z \ U)]$$

En suma, observamos también aquí cómo se acomodan las estructuras algebraicas díadicas de los primarios a las ya establecidas para los entes matemáticos de R, R^3 o C, de manera que los secundarios o parte dimensional respondan de manera independiente a la generalización del álgebra de los segmentos geométricos, con el **postulado de afinidad**, derivado de la posibilidad de establecer correspondencias biunívocas entre la cantidades de longitud de los segmentos geométricos y las cantidades de cualquier otra magnitud, con lo que se justifican las operaciones con magnitudes afines a la longitud en base al álgebra geométrica.

A su vez, las formas simbólicas de las reglas usuales se mantienen con todos los términos, incluidos los símbolos de las unidades, como consecuencia de las definiciones y propiedades de las múltiples leyes de composición que configuran las diversas estructuras algebraicas díadicas.

Section XXII

EFFECTS OF THE PRINCIPLE
OF SYMBOLIC ECONOMY

Throughout the development of this first physical algebra we have observed how the geometric experiment of the multiplication of segments or lengths is generalized in the abstract, giving rise to the generic algebra of dyads, as mathematical representatives of the quantities of physical magnitudes. We have repeatedly warned about the hypnotic effect that can be produced by availing itself of the symbolic economy, understood as the simplification of signs for the different operations of the same species, such as the additive ones, all denoted with the typical cross « + », the multiplicative ones indicated generically , for example, with the cross «×», subtraction with the hyphen «–» or divisions with the slash «/». To break this spell and warn for pedagogical purposes about how easy it is to be fascinated by it and to believe that what really remains in the dark is understood, we have made an effort of symbolic detail, to make explicit the maximum number of distinguishable composition laws, as well as the relationships that arise between them; although, given the large number of these, as the symbolism is limited and it would not be useful to take such differentiation to the absolute extreme, it is inevitable and even convenient that some share common signs, which is not an obstacle for the phenomenon to be explained with sufficient clarity didactic.

For a better overview, the symbols of the operations involved in dyadic algebra can be detailed, represented by the sign $\mathscr{D}$, unlike the structures of R, C and R^3 or any other. Such operations are those of the usual structures and those specifically defined

215

for specific entities. In this way the following synoptic diagram results:

Type of dyadic composition law Section of the *First Algebra of Magnitudes*	Ordinary number algebra (see note)		Dyadic algebra or physical	With the principle of symbolic economy
	In R y C	In R^3	In $\mathcal{D}$	
Addition (V)	+	+	⊕	+
Subtraction (VIII)	−	−	⊖	−
Multiplication by a number (IX)	×	•	∘	×
Homogeneous division (XI)	/ ÷		// ≑	/ ÷
Heterogeneous multiplication (XII)	×		✳	×
Heterogeneous division (XVI)			// ≑	/ ÷
Scalar product (XIX)		•	⊙	•
Vector product (XIX)		∧	⊛	×
Product of mixed magnitudes (XX)			◎	×

Left grouping labels: *Scalar magnitudes and vector* (Addition, Subtraction, Multiplication by a number); *Scalar magnitudes* (Homogeneous division, Heterogeneous multiplication, Heterogeneous division); *Vector magnitudes* (Scalar product, Vector product, Product of mixed magnitudes).

(Note) The symbols of the operations in R. C and R^3 obviously refer to the addition, subtraction, multiplication and division of these algebraic structures, not to the dyadic or concrete ones that are defined in the sections of the first column.

Take, for example, the expression [19.21], which is the result of previous reasoning with dyadic algebra:

$$(p \circ \overline{a}\ U_1) \circledast (q \circ \overline{b}\ U_2) = (p \times q) \circ [(\overline{a}\ U_1) \circledast (\overline{b}\ U_2)]$$

The sign «∘» indicates the multiplicative operation of a scalar of R by a vector dyad; on the other hand, «⊛» represents the composition law called the vector product of vector dyads; and, finally, with the symbol «×» the multiplication of real numbers is named. The principle of symbolic economy consists in denoting all these laws of composition of the same multiplicative species with the same character, for example, the cross «×». And with that the traditional notation results:

$$(p \times \overline{a}\ U_1) \times (q \times \overline{b}\ U_2) = (p \times q) \times [(\overline{a}\ U_1) \times (\overline{b}\ U_2)]$$

Observing this last expression, unless one has algebraic expertise, it is difficult to escape the illusion caused by the constant sign of the cross «×» and it is easy to believe that the property that describes equality is evident by the laws of R^3. However, this is not the case, because what it relates are physical dyads, and the full meaning of equality is given by the algebraic reasoning that has led to the deduction of [19.21], by virtue of the different composition laws that relate the own equation and specifically defined between the spaces $\{R^3, U_1\}$, $\{R^3, U_2\}$ and $\{R^3, U_1 * U_2\}$.

The same can be said about any other expression in dyadic algebra, such as the distributive property described in [19.24]:

$$\overline{a}\ U_1 \circledast (\overline{b}\ U_2 \oplus \overline{c}\ U_2) = [(\overline{a}\ U_1) \circledast (\overline{b}\ U_2)] \oplus [(\overline{a}\ U_1) \circledast (\overline{c}\ U_2)]$$

If the additive signs «⊕», which refers to the sum of vector dyads, and the multiplicative «⊛», which symbolizes the dyadic vector product of vector concretes, are replaced by the usual «+» and «×», we will have simplified or implicit equality:

$$\overline{a}\ U_1 \times (\overline{b}\ U_2 + \overline{c}\ U_2) = [(\overline{a}\ U_1) \times (\overline{b}\ U_2)] + [(\overline{a}\ U_1) \times (\overline{c}\ U_2)]$$

The observation of the equation, unless one knows how to distinguish each operation according to the elements that it relates, seduces the intellect almost without remedy, making it believe that the distributive property is immediately fulfilled; however, this is not the case, because we are dealing with an equality of physical algebra, for which an exhaustive reasoning has had to be followed, such as that presented instead for equation [19.24].

So the isomorphic appearance of physical algebra with the structures of R or R^3, which is appreciated when the principle of symbolic economy is applied, is not much less presumed, as in practice it has been assumed, not without a good deal of uncertainty and negligence, suggested in their writings by the best authors; although other less rigorous minds, enraptured by the symbology in the manner described above, might believe in this imposture and take the dyadic properties that require specific proof for granted. Which, not because it is common, ceases to be erroneous and clearly arbitrary, violating the most elementary logic of knowledge and forgetting or disregarding the algebraic obligation to define the laws of composition between entities destined to represent quantities of related scientific magnitudes through physical equations.

Therefore, the principle of symbolic economy would lead directly to the conclusion that to operate with quantities of magnitudes, the isomorphic appearance with the operations of R and R^3 could be relied upon, admitting as evident the following formal rule:

To compose concrete entities or physical dyads, it is enough to specify in the equations the abbreviations or symbols of the units that intervene and operate with them according to the algebraic laws of real numbers and geometric vectors, and considering that they multiply the accompanying measurements, that they can be manipulated in the same way; which is simply pretending that the units are common algebraic elements.

However, we have already sufficiently justified that, although this rule may have practical utility, it should not be admitted on the sole basis of mere symbolic logic and its appearance isomorphic with the structures of R and R^3, but is a logical consequence of physical algebra defined and developed through the various laws of composition duly configured.

Fortunately, the stated rule is correct, as physical algebra shows, but tradition has arrived at it by invalid reasoning, which is inadmissible and further justifies the need for algebra of magnitudes, if only because Physics must settle on good logical foundations, so it is not understood that to date it has remained hidden in the imposture described by an operational rule without prior proof, which we can now understand fully justified and was previously a mere whim of the most vulgar intuition[15]. But there is another reason, if possible, more transcendent to appreciate the unappealable need for dyadic or physical algebra, and that is that it reveals something of the utmost importance, which is nothing but how composite units are built, which are all the result of the abstract generalization of the geometric multiplication of segments or lengths; so, as in this abstraction a good dose of arbitrariness inseparable from the equations of definition is observed, this should move us to the greatest prudence when trying to assess the essence of any of these derived magnitudes, because in principle they have no other character than that of mathematical entities originated by composition laws defined by means of an illusion that generalizes the geometric multiplication of segments, artificially assimilating any quantity of any magnitude to a quantity of length or abstract segment[16].

On the other hand, it should be noted that the rule described has exceptions, like any intellectual simplification, because we have observed that the multiplication of dyads, defined in section XII, given its status as a law of external composition, does not allow the existence of unitary elements and inverse, which distances the structure of scalar dyads from that of the body of reals. This alerts us to the meaning to be given to the negative exponents that often appear in physical equations, because they cannot have any other meaning than to symbolize the divisors or denominators in which powers of units appear in this way; but they cannot at all indicate the inverse of any unit, because such

---

[15] In «Lesson 1» of Mathematize 3, a mathematical method of logic is developed and it is described how an invalid reasoning could lead to a correct conclusion, without assuming that such logical schemes are admissible.

[16] A somewhat more elementary exposition, but equally valid for understanding the algebra of magnitudes, can be found on the same author's syllabus, «Lesson 3» of Mathematize 3.

an entity does not exist. Therefore, the rule established here cannot in any way replace the algebra of magnitudes and should only be understood as an aid that speeds up operational practice, but without giving it greater significance, because something else could only induce serious errors in the analysis of magnitudes and physical equations.

A more vulgar reason about the need for dyadic algebra, although no less conclusive, is that to repudiate it would be something like legitimizing that the arithmetic multiplication table also remains subject to the discretion of each one, and on this it does not seem that any normal brain can have any doubt.

Furthermore, this First Algebra of Magnitudes clearly points out the possibility of constructing other more abstract and complex ones, linked to algebraic structures of dimensions greater than three, or even to other non-Euclidean metrics, which will undoubtedly contribute to the development of innovative models. of Theoretical Physics.

Meanwhile, with the composition laws defined in this monograph, all the operations that can be found in the physical equations can be based, so that, although not all possible cases have been analyzed, a task that would be difficult and even unfeasible, with analysis used could solve any imaginable assumption.

Apartado XXII

EFECTOS DEL PRINCIPIO
DE ECONOMÍA SIMBÓLICA

A lo largo del desarrollo de esta primera álgebra física hemos observado cómo se generaliza en abstracto el experimento geométrico de la multiplicación de segmentos o longitudes, dando lugar al álgebra genérica de los concretos o díadas, como representantes matemáticos de las cantidades de las magnitudes físicas. Hemos advertido reiteradamente sobre el efecto hipnótico que puede producir acogerse a la economía simbólica, entendida como la simplificación de signos para las distintas operaciones de la misma especie, tales como las aditivas, denotadas todas con la típica cruz «+», las multiplicativas indicadas genéricamente, por ejemplo, con el aspa «×», las restas con el guión «–» o las divisiones con la barra «/». Para romper ese hechizo y advertir a efectos pedagógicos sobre lo fácil que resulta fascinarse por él y creer que se comprenda lo que realmente permanezca en la oscuridad, hemos hecho un esfuerzo de detalle simbólico, para explicitar el máximo número de leyes de composición distinguibles entre sí, así como las relaciones que surgen entre ellas; aunque, dado el gran número de estas, como la simbología es limitada y tampoco tendría utilidad llevar al extremo absoluto tal diferenciación, es inevitable y hasta conveniente que algunas compartan signos comunes, lo cual no es obstáculo para que el fenómeno pueda explicarse con suficiente claridad didáctica.

Para una mejor visión de conjunto se pueden detallar los símbolos de las operaciones que intervienen en el álgebra diádica, representada con el signo $\mathscr{D}$, a diferencia de las estructuras de R, C y R^3 o cualquier otra. Tales operaciones son las propias de las estructuras usuales y las definidas específicamente para los entes concretos. De este modo resulta el siguiente esquema sinóptico:

Tipo de ley de composición diádica / Apartado de la *Primera álgebra de magnitudes*	Álgebra numérica ordinaria (ver nota)		Álgebra diádica o física	Con el principio de economía simbólica
	En R y C	En R³	En $\mathscr{D}$	
Magnitudes escalares y vectoriales — Adición (V)	+	+	$\oplus$	+
Sustracción (VIII)	−	−	$\ominus$	−
Multiplicación por un número (IX)	×	•	○	×
División homogénea (XI)	/ ÷		// ÷	/ ÷
Magnitudes escalares — Multiplicación heterogénea (XII)	×		$*$	×
División heterogénea (XVI)			// ÷	/ ÷
Magnitudes vectoriales — Producto escalar (XIX)		•	$\odot$	•
Producto vectorial (XIX)		$\wedge$	$\circledast$	×
Producto de magnitudes mixtas (XX)			$\circledcirc$	×

(Nota) Los símbolos de las operaciones en R, C y R³ obviamente se refieren a la adición, sustracción, multiplicación y división propias de estas estructuras algebraicas, no a las diádicas o concretas que se definen en los apartados de la primera columna.

Tomemos, por ejemplo, la expresión [19.21], que es el resultado de un razonamiento previo con el álgebra diádica:

$$(p \circ \overline{a}\ U_1) \circledast (q \circ \overline{b}\ U_2) = (p \times q) \circ [(\overline{a}\ U_1) \circledast (\overline{b}\ U_2)]$$

El signo «○» indica la operación multiplicativa de un escalar de R por una díada vectorial; por su parte, con «$\circledast$» se representa la ley de composición denominada producto vectorial de díadas vectoriales; y, finalmente, con el símbolo «×» se nombra la multiplicación de números reales. El principio de economía simbólica consiste en denotar todas estas leyes de composición de la misma especie multiplicativa con el mismo carácter, por ejemplo, el aspa «×». Y con ello resulta la notación tradicional:

$$(p \times \overline{a}\ U_1) \times (q \times \overline{b}\ U_2) = (p \times q) \times [(\overline{a}\ U_1) \times (\overline{b}\ U_2)]$$

Observando esta última expresión, salvo que se tenga pericia algebraica, resulta difícil sustraerse de la ilusión que provoca el signo constante del aspa «×» y se tiende a creer con facilidad que la propiedad que describe la igualdad sea evidente por las leyes propias de R^3. Sin embargo, no es así, porque lo que relaciona son díadas físicas, y el significado completo de la igualdad viene dado por el razonamiento algebraico que ha conducido a la deducción de [19.21], en virtud de las diferentes leyes de composición que relaciona la propia ecuación y específicamente definidas entre los espacios $\{R^3, U_1\}$, $\{R^3, U_2\}$ y $\{R^3, U_1 * U_2\}$.

Lo mismo cabe decir sobre cualquier otra expresión del álgebra diádica, como por ejemplo, la propiedad distributiva descrita en [19.24]:

$$\overline{a}\ U_1 \circledast (\overline{b}\ U_2 \oplus \overline{c}\ U_2) = [(\overline{a}\ U_1) \circledast (\overline{b}\ U_2)] \oplus [(\overline{a}\ U_1) \circledast (\overline{c}\ U_2)]$$

Si se sustituyen los signos aditivo «$\oplus$», que se refiere a la suma de díadas vectoriales, y el multiplicativo «$\circledast$», que simboliza el producto vectorial diádico de concretos vectoriales, por los usuales «+» y «×», se tendrá la igualdad simplificada o implícita:

$$\overline{a}\ U_1 \times (\overline{b}\ U_2 + \overline{c}\ U_2) = [(\overline{a}\ U_1) \times (\overline{b}\ U_2)] + [(\overline{a}\ U_1) \times (\overline{c}\ U_2)]$$

La observación de la ecuación, salvo que se sepa distinguir cada operación en función de los elementos que relaciona, seduce al intelecto casi sin remedio, haciéndole creer que la propiedad distributiva se cumpla de modo inmediato; sin embargo, no es así, porque estamos ante una igualdad del álgebra física, para cuya deducción ha debido seguirse todo un razonamiento exhaustivo, como el presentado en su lugar para la ecuación [19.24].

Así que la apariencia isomorfa del álgebra física con las estructuras de R o R^3, que se aprecia cuando se aplica el principio de economía simbólica, no es ni muchos menos presumible, como en la práctica se ha venido presuponiendo no sin buena dosis de incertidumbre y negligencia, sugeridas en sus escritos por los mejores autores; aunque otras mentes menos rigurosas, embelesadas por la simbología de la manera descrita antes, pudieran creer en esa impostura y dar por obvias las propiedades diádicas que requieren prueba específica. Lo cual, no por ser común, deja de resultar erróneo y claramente arbitrario, violando la lógica más elemental del conocimiento y olvidando o despreciando la obligación algebraica de definir las leyes de composición entre entes destinados a representar cantidades de magnitudes científicas relacionadas mediante ecuaciones físicas.

Por lo tanto, el principio de economía simbólica llevaría de manera directa a la conclusión de que para operar con cantidades de magnitudes se pudiera confiar en la apariencia isomorfa con las operaciones de R y de R^3, admitiendo como evidente la siguiente regla formal:

Para componer entes concretos o díadas físicas basta con especificar en las ecuaciones las abreviaturas o símbolos de las unidades que intervengan y operar con estos según las leyes algebraicas de los números reales y de los vectores geométricos, y considerando que multipliquen a las medidas que acompañan, que se podrán manipular de igual modo; lo que supone simplemente fingir que las unidades sean elementos algebraicos comunes.

Sin embargo, ya hemos justificado suficientemente que, aunque esta regla pueda tener utilidad práctica, no debe ser admitida con el único fundamento de la mera lógica simbólica y su apariencia isomorfa con las estructuras de R y de R^3, sino que es consecuencia lógica del álgebra física definida y desarrollada a través de las diversas leyes de composición debidamente configuradas.

Afortunadamente, la regla enunciada es correcta, como lo demuestra el álgebra física, pero la tradición ha llegado a ella por un razonamiento no válido[17], lo que resulta inadmisible y justifica aún más la necesidad del álgebra de magnitudes, aunque solo fuera porque la Física debe asentarse en buenos cimientos lógicos,

---

[17] En la «Lección 1» de *Matematizar 3* se desarrolla un método matemático de la lógica y se describe cómo un razonamiento no válido podría conducir a una conclusión correcta, sin que ello suponga que tales esquemas lógicos sean admisibles.

por lo que no se comprende que hasta la fecha haya permanecido oculta en la impostura descrita por una regla operativa sin prueba previa, que ahora podemos entender justificada plenamente y antes era mero capricho de la intuición más vulgar. Pero hay otra razón si cabe más trascendente para apreciar la necesidad inapelable del álgebra diádica o física, y es que nos revela algo de suma importancia, que no es sino cómo se construyen las unidades compuestas, que son todas ellas fruto de la generalización abstracta de la multiplicación geométrica de segmentos o longitudes; así que, como en esta abstracción se observa una buena dosis de arbitrariedad inseparable de las ecuaciones de definición, ello ha de movernos a la mayor prudencia cuando se pretenda valorar la esencia de cualesquiera de esas magnitudes derivadas, porque en principio no tienen otro carácter que el de entes matemáticos originados por unas leyes de composición definidas por medio de una ilusión que generaliza la multiplicación geométrica de segmentos, asimilando artificialmente cualquier cantidad de toda magnitud a una cantidad de longitud o segmento abstracto[18].

Por otra pate, debe advertirse que la regla descrita tiene excepciones, como cualquier simplificación intelectual, porque hemos observado que la multiplicación de díadas, definida en el apartado XII, dada su condición de ley de composición externa, no permite la existencia de los elementos unitario ni inverso, lo que distancia a la estructura de las díadas escalares de la del cuerpo de los reales. Ello alerta sobre el significado que ha de darse a los exponentes negativos que aparecen a menudo en las ecuaciones físicas, porque no pueden tener otro sentido que simbolizar de esa forma los divisores o denominadores en que aparezcan potencias de unidades; pero en absoluto pueden indicar el inverso de ninguna unidad, porque tal ente no existe. Por tanto, la regla aquí establecida no puede sustituir en modo alguno el

---

[18]   Una exposición algo más elemental, pero igualmente válida para el entendimiento del álgebra de magnitudes, se encuentra en el temario del mismo autor. «Lección 3» de *Matematizar 3*.

álgebra de magnitudes y solo debe entenderse como ayuda que agiliza la práctica operativa, pero sin darle mayor trascendencia, porque otra cosa solo podría inducir errores graves en el análisis de las magnitudes y ecuaciones físicas.

Una razón más vulgar sobre la necesidad del álgebra diádica, aunque no menos concluyente, es que repudiarla sería algo así como legitimar que la tabla de multiplicar aritmética también quedase sujeta al arbitrio de cada cual, y sobre esto no parece que ningún seso normal pueda tener duda alguna.

Además, esta *Primera álgebra de magnitudes* señala con claridad la posibilidad de construir otras más abstractas y complejas, ligadas a estructuras algebraicas de dimensiones superiores a tres, o incluso a otras métricas no euclidianas, lo que sin duda habrá de contribuir al desarrollo de modelos innovadores de la Física teórica.

Entretanto, con las leyes de composición definidas en esta monografía se pueden fundamentar todas las operaciones que puedan encontrarse en las ecuaciones físicas, por lo que, aunque no hayan quedado analizados todos los casos posibles, tarea que resultaría penosa y hasta inviable, con los métodos de análisis empleados se podría resolver cualquier supuesto imaginable.

Section XXIII

CLASSES OF MAGNITUDES

We have dedicated sections I and II to the concepts of quantity, magnitude and measurement, which refer to three fundamental entities of Physics, and we have agreed that the measurements are represented by dyadic entities and their specific algebra, which allows defining some units starting from others, so that, by composing some of them, the others will be deduced according to the theory to which they belong. The magnitudess from which others are established by means of the algebra of magnitudes and **generating external laws** of sections XII to XVII are called **primary** or **simple**. Those that are expressed in terms of the former will be said to be **secondary** or **compound**. Various authors tend to name the secondary **derived** magnitudes. In turn, we will also call **fundamental** the magnitudes that are taken as the basis of a certain dimensional system, to compose them algebraically as appropriate. Finally, we will end with the introduction of a new concept of magnitude, derived from the «dysmetric» variant, which we will define at the end of this section. In relation to this criterion, we will establish the innovative concepts of **rigid** and **flexible** magnitudes.

The truth is that there is no unanimity in ordinary nomenclature, although this seems irrelevant to us, since the essential thing in each specific case will be to establish what the independent magnitudes are for abstract algebra, because these non-composite magnitudes will be the ones that can shed the most light on the physical meaning of the phenomena analyzed.

Physics is an experimental science, but its experiments, described using common language, would only serve to accumulate historical knowledge with little prospective value: it would be useless to know the trajectory of a particle if there were no way to understand the laws of its motion. However, if scientific facts are written in mathematical language, taking advantage of the entities of this abstract science to couple natural observations into them, something wonderful happens: numbers, vectors, functions and other mathematical instruments allow observations to be organized in such a way that invariant relationships emerge between the different established variables, which allow us to determine some as a function of others. These variables from the mathematical point of view are for Physics or other sciences measurements of magnitudes, in the style of linear, superficial or volumetric geometry.

It is observed in nature that there are phenomena such as time or distance that lack direction, although they can be taken in one direction or the opposite; These can be represented by scalar measurements of the type $\{R,U\}$, whose real part can be positive or negative, for which a convention on the meaning adopted as positive will suffice, and it will be possible to operate according to dyadic algebra. Other magnitudes such as mass are always positive, which constitutes a particular case of the previous ones. In turn, other phenomena such as a speed or a force, in addition to their size or quantity, are not indifferent to the direction or the sense within it, so these physical facts fit very well as vectors and will have to be referred to a adequate reference system in which its components can be determined to operate with them according to vector algebra; in the dyadic field they can be represented by the set $\{R^3,U\}$, if their scope were the three-dimensional Euclidean space, or in general $\{R^n,U\}$ for a space of dimension n. In this way, we verify that the great classification of the physical quantities that reflect certain natural properties has to be established between these two: scalars and vector; thus, once this conceptualization is admitted, it turns out that Physics is absorbed by the properties and compositions of numbers and vectors, members of their corresponding dyadic entities, with which the magic of subsuming natural facts in the appropriate laws will have been achieved pre-existing abstract mathematics, taking advantage of the general truth content of these entities to inflate them with physical meanings. Therefore, the first physical operation relevant to the measurement is to identify a magnitude and base its scalar or vector nature, to establish the type of algebra that is going to be implemented when operating with the measurements, and also in accordance with the essential dyadic algebra.

It should not be overlooked that vector algebra can refer to that which is proper to the field of complex numbers, with its specific laws of composition, because it is already known that some natural phenomena are subsumed by the algebraic structure conferred on such imaginary numbers; It seems incredible, but it is a fact that a mathematical abstraction such as complex algebra, born long before alternating current, serves to explain electrical phenomena of this nature; although sometimes, it must also be recognized that Physics stimulates the development of mathematical structures, which does not change the fact that Physics is included in Mathematics in any case. So we could metaphorically indicate that doing Physics is nothing more than reducing mathematical abstraction to a concrete form that is justified by the relevant essays, which implies that Physics can be considered as the Mathematics of experiment.

Examples of scalar magnitudes are length, area, volume, time, mass, density, temperature, work, energy, power, intensity of electric current in linear conductors, voltage, among many others. On the other hand, the displacement of a mobile, the speed, the acceleration, the force and all those that are characterized by a quantity or module, a direction and a sense of action are vector magnitudes.

The magnitudes that show their indivisible elements are called **discrete** and the element that determines them can be taken as a fundamental unit; for example, registered vehicles, inhabitants of a country, packaging of an industry, and the like are examples of discrete magnitudes. On the contrary, physical quantities such as length, weight, temperature or energy do not have this discrete characteristic and any quantity, no matter how small, is divisible into other smaller ones, such magnitudes are called **continuous** and there are no better ones referring to them than the real numbers, given their symbolized continuity on the real line. Continuous magnitudes are such that they leave no choice but to establish the pattern by means of experimental physical references that implicitly indicate a certain unit quantity that is not determinable or numerically expressible of the magnitude measured and indicated by a specific and arbitrary abstract symbol, being mathematically represented by a set dyadic of the type $\{R,U\}$ or $\{R^n,U\}$, depending on whether they are scalars or vector, respectively. For their part, discrete magnitudes, which do show natural unit elements, can be mathematically described with dyads formed with the set N of natural numbers, denoted $\{N,U\}$, if the measurements could only be positive, or that of the integers Z, referred to the concrete set $\{Z,U\}$, if the measurements could be positive and negative.

Entering the dyadic universe, we cannot escape an important and unobjectionable observation that substantially expands the mathematical field to represent physical phenomena. Every dyad is composed of two elements, a mathematical primary, number, vector or tensor, and a physical secondary indicative of a material reality that is taken as a standard unit of some magnitude, the meter, the kilogram or the second, for example. Then, all measurement is established by that pair of elements closely related to each other, which we have come to call a concrete number, a physical dyad or simply measurement. Since the beginning of time, the tendency towards the «arithmetization» of Physics has paid attention to the measurements of the quantities of magnitudes, tacitly assuming that the standard units should always have the same invariable quantity of the magnitude implicit in them. It is precisely on this invariance that the International System of Units is based. Thus, it has always been assumed that the standard meter must contain the same amount of length in any position in space and under any circumstance. And the same assumption is attributed to the standard kilogram and the second standard. We could distinguish this hypothesis with the qualifier of **isometry**. However, nothing prevents us from formulating the opposite and more general variant of imagining that the quantity of each implicit magnitude in physical patterns can vary from one point to another in space, for various reasons that are not of interest at the moment in this phase of abstract formulation and logic of the mathematical tool of dyadic algebra. We can name this new variant with the term **«dysmetry»** and it consists of the following:

Let us take any standard unit U of some magnitude, such as the meter or the kilogram. For this, it would be enough to prepare a straight rod of a certain extension indicative of a length or a material body formed by a certain grouping of matter, which we will call a weight. These material elements can present the same quantity of their associated magnitude at all points in space, isometric hypothesis, or on the contrary, the same physical bodies can vary in their quantity of magnitude depending on the spatial position, which is a «dismetric» axiom, with which the two existing logical possibilities are completed and which is more generic than the first hypothesis, which is contained in it as a particular case.

Let O be a point in space and let U_O symbolize the quantity of length or mass that the rod or weight in question contains implicit in that point. Let P be any point in space and let U_P be the amount of length or mass of the same material elements, which are the selected rod and weight, so that the dyadic ratio $U_P /\!/ U_O$ can be established. As we are dealing with homogeneous magnitudes, this ratio corresponds to the division in section XI, which we know always results in a real number and, therefore, dimensionless. Let us designate this number with the symbol δ_P. We will call this ratio the «dysmetryc» density of the quantity in question at the point P in relation to the point O. Obviously, we have the relation $U_P = \delta_P \circ U_O$, where the operation «∘» is the multiplication of a measurement by a scalar of section IX, an expression that we could also have established from the beginning without more than considering that U_p will have to be a quantity equal to $\bar{\delta}_p$ times the quantity of the same implicit magnitude in U_0, defining the real number $\bar{\delta}_p$ as the «dysmetric» density of the magnitude considered in P with respect to O.

When the «dysmetric» density $\bar{\delta}_p$ is constant, we are in an isometric space. On the other hand, when it varies in any way, we will find a «dysmetric» space. In this way, «dysmetry» was born as a more comprehensive and powerful mathematical tool than current isometry to represent natural phenomena in diverse and variable settings.

In the second volume of this work, the mathematization of «dysmetry» is developed in greater length, as well as some consequences and applications to generalize physical laws with the «dysmetric» formulation, showing that we are faced with a powerful mathematical tool that allows develop a new Physics. Here we limit ourselves to introducing the innovative concepts of **flexible magnitudes** and the **«dysmetric» spaces** resulting from the choice of variation of physical patterns.

Apartado XXIII

CLASES DE MAGNITUDES

Hemos dedicado los apartados I y II a los conceptos de cantidad, magnitud y medición, que se refieren a tres entes fundamentales de la Física, y hemos convenido que las mediciones sean representadas por los entes diádicos y su específica álgebra, que permite definir unas unidades a partir de otras, de modo que, componiendo algunas de ellas, se deducirán las demás en función de la teoría a que pertenezcan. Se llaman **primarias** o **simples** las magnitudes a partir de las cuales se establecen otras por medio del álgebra de magnitudes y las **leyes externas generatrices** de los apartados XII a XVII. Las que se expresen en función de las primeras se dirá que son **secundarias** o **compuestas**. Diversos autores suelen nombrar a las secundarias magnitudes **derivadas**. A su vez, también llamaremos **fundamentales** a las magnitudes que se tomen como base de un determinado sistema dimensional, para componerlas algebraicamente según proceda. Por último, terminaremos con la introducción de un nuevo concepto de magnitud, derivado de la variante «dismétrica», que definiremos al final de este apartado. Con relación a este criterio estableceremos los conceptos innovadores de magnitudes **rígidas** y **flexibles**.

La verdad es que no existe unanimidad en la nomenclatura ordinaria, aunque ello nos parece irrelevante, pues lo esencial en cada caso concreto será establecer cuáles sean las magnitudes independientes para el álgebra abstracta, porque estas magnitudes no compuestas serán las que más luz puedan arrojar sobre el sentido físico de los fenómenos analizados.

La Física es una ciencia experimental, pero sus experimentos, descritos mediante el lenguaje común, solo servirían para acumular conocimiento histórico con escaso valor prospectivo: de nada serviría conocer la trayectoria de una partícula si no hubiera forma de comprender las leyes de su movimiento. Sin embargo, si los hechos científicos se escriben en lenguaje matemático, aprovechando los entes de esta ciencia abstracta para acoplar en ellos las observaciones naturales, sucede algo maravilloso: los números, los vectores, las funciones y demás instrumentos matemáticos permiten organizar las observaciones de modo que brotan relaciones invariantes entre las diferentes variables establecidas, que permiten determinar unas en función de otras. Dichas variables desde el punto de vista matemático son para la Física u otras ciencias mediciones de magnitudes, al estilo de las lineales, superficiales o volumétricas de la geometría.

Se observa en la naturaleza que hay fenómenos como el tiempo o la distancia que carecen de dirección, aunque puedan tomarse en un sentido o su opuesto; estos cabe representarlos mediante mediciones escalares del tipo $\{R, U\}$, cuya parte real puede ser positiva o negativa, para lo que bastará una convención sobre el sentido adoptado como positivo, y se podrá operar de acuerdo con el álgebra diádica. Otras magnitudes como la masa se muestran siempre positivas, lo que constituye un caso particular de las anteriores. A su vez, otros fenómenos como una velocidad o una fuerza, además de su tamaño o cantidad, no son indiferentes a la dirección ni al sentido dentro de ella, por lo que estos hechos físicos encajan muy bien como vectores y habrá que referirlos a un sistema de referencia adecuado en el que se podrán determinar sus componentes para operar con ellas de acuerdo con el álgebra vectorial; en el campo diádico se podrán representar por el conjunto $\{R^3, U\}$, si su ámbito fuese el espacio euclídeo de tres dimensiones, o en general $\{R^n, U\}$ para un espacio de dimensión n. De este modo comprobamos que la gran clasificación de las magnitudes físicas que reflejan determinadas propiedades naturales se tiene que establecer entre estas dos: **escalares** y **vectoriales**; así, una vez admitida esta conceptuación, resulta que la Física queda absorbida por las propiedades y composiciones de los números y los vectores, integrantes de sus correspondientes entes diádicos, con lo cual se habrá conseguido la magia de subsumir los hechos naturales en las oportunas leyes matemáticas abstractas preexistentes, aprovechando el contenido de verdad general de estos entes para insuflarlos con significados físicos. Por consiguiente, la primera operación física atinente a la medida es identificar una magnitud y fundamentar su naturaleza escalar o vectorial, para establecer el tipo de álgebra que se va a implementar al operar con las mediciones, y de acuerdo también con la imprescindible álgebra diádica.

No se debe dejar de observar que el álgebra vectorial puede referirse a la que es propia del cuerpo de los números complejos, con sus específicas leyes de composición, porque ya se sabe que algunos fenómenos naturales quedan subsumidos por la estructura algebraica conferida a tales números imaginarios; parece increíble, pero es un hecho, que una abstracción matemática como el álgebra compleja, nacida mucho antes que la corriente alterna, sirva para explicar los fenómenos eléctricos de esa naturaleza; aunque a veces, también haya que reconocer que la Física estimula el desarrollo de estructuras matemáticas, lo cual no cambia el hecho de que la Física quede incluida en todo caso en la Matemática. Así que podríamos indicar metafóricamente que hacer Física no es sino reducir la abstracción matemática a una forma concreta que se justifica por los ensayos pertinentes, lo que supone que la Física pueda considerarse como la Matemática del experimento.

Son ejemplos de magnitudes escalares la longitud, el área, el volumen, el tiempo, la masa, la densidad, la temperatura, el trabajo, la energía, la potencia, la intensidad de corriente eléctrica en conductores lineales, el voltaje, entre otras muchas. Por su parte, son magnitudes vectoriales el desplazamiento de un móvil, la velocidad, la aceleración, la fuerza y todas las que queden caracterizadas por una cantidad o modulo, una dirección y un sentido de acción.

Las magnitudes que muestran a sus elementos indivisibles se llaman **discretas** y como unidad fundamental se puede tomar el propio elemento que las determina; por ejemplo, los vehículos matriculados, los habitantes de un país, los envases de una industria y similares son ejemplos de magnitudes discretas. Por el contrario, las magnitudes físicas como la longitud, el peso, la temperatura o la energía no tienen esta característica discreta y cualquier cantidad por pequeña que sea resulta divisible en otras más pequeñas, tales magnitudes se denominan **continuas** y no hay mejores referentes a ellas que los números reales, dada su continuidad simbolizada en la recta real. Las magnitudes continuas son tales que no dejan otra opción que establecer el patrón mediante referencias físicas experimentales que indican implícitamente una cierta cantidad unitaria no determinable ni expresable numéricamente de la magnitud medida y señalada mediante un símbolo abstracto específico y arbitrario, quedando representadas matemáticamente por un conjunto diádico del tipo $\{R, U\}$ o $\{R^n, U\}$, según que sean escalares o vectoriales, respectivamente. Por su parte, las magnitudes **discretas**, que sí muestran elementos unitarios naturales, se pueden describir matemáticamente con díadas formados con el conjunto N de los números naturales, denotado $\{N, U\}$, si las medidas solo pudieran ser positivas, o el de los enteros Z, referido al conjunto concreto $\{Z, U\}$, si las medidas pudieran ser positivas y negativas.

Adentrándonos en el universo diádico, no se nos puede escapar una observación importante e inobjetable que amplía de modo sustancial el campo matemático para representar fenómenos físicos. Toda díada está compuesta de dos elementos, un primario matemático, número, vector o tensor, y un secundario físico indicativo de una realidad material que se toma como unidad patrón de alguna magnitud, el metro, el kilogramo o el segundo, por ejemplo. Entonces, toda medición queda establecida por ese par de elementos estrechamente vinculados entre sí, que hemos dado en llamar número concreto, díada física o simplemente medición. Desde el principio de los tiempos la tendencia a la «aritmetización» de la Física a puesto la atención en las medidas de las cantidades de magnitudes, asumiendo tácitamente que las unidades patrón debieran tener siempre la misma cantidad invariable de la magnitud implícita en ellas. En esta invariancia se basa precisamente el Sistema Internacional de Unidades. Así, siempre se ha presupuesto que el metro patrón haya de contener la misma cantidad de longitud en cualquier posición del espacio y bajo cualquier circunstancia. Y la misma suposición se atribuye al kilogramo patrón y al segundo patrón. Podríamos distinguir esta hipótesis con el calificativo de **isometría**. Sin embargo, nada nos impide formular la previsión opuesta y más general de imaginar que la cantidad de cada magnitud implícita en los patrones físicos puedan variar de un punto a otro del espacio, por diversas causas que no interesan de momento en esta fase de formulación abstracta y lógica de la herramienta matemática del álgebra diádica. Esta variante nueva podemos nombrarla con el término **«dismetría»** y consiste en lo siguiente:

Tomemos una unidad patrón cualquiera U de alguna magnitud, como podrían ser el metro o el kilogramo. Para ello bastaría preparar un varilla recta de cierta extensión indicativa de una longitud o un cuerpo material formado por una cierta agrupación de materia, que llamaremos pesa. Estos elementos materiales pueden presentar la misma cantidad de su magnitud asociada en todos los puntos del espacio, hipótesis isométrica, o por el contrario, los mismos cuerpos físicos pueden variar en su cantidad de magnitud dependiendo de la posición espacial, que es axioma «dismetrico» con el que se completan las dos posibilidades lógicas existentes y que es mas genérico que la primera hipótesis, que queda contenida en él como caso particular.

Sea un punto O del espacio y simbolicemos U_O la cantidad de longitud o de masa que la varilla o la pesa consideradas contengan implícitas en dicho punto. Sea un punto P cualquiera del espacio y designemos U_P la cantidad de longitud o de masa de los mismos elementos materiales, que son la varilla y la pesa seleccionadas, de modo que se podrá establecer la razón diádica $U_P /\!/ U_O$. Como estamos ante magnitudes homogéneas, esta razón corresponde a la división del apartado XI, que sabemos da como resultado siempre un número real y, por tanto, adimensional. Designemos a este número con el símbolo δ_P. Llamaremos a este cociente densidad «dismétrica» de la magnitud en cuestión en el punto P

en relación con el punto O. Obviamente, se tiene la relación $U_P = \delta_P \circ U_O$, donde la operación «$\circ$» es la multiplicación de una medición por un escalar del apartado IX, expresión esta que también podríamos haber establecido desde el principio sin más que considerar que U_P habrá de ser una cantidad igual a δ_P veces la cantidad de la misma magnitud implícita en U_O, definiendo el número real δ_P como la densidad «dismétrica» de la magnitud considerada en P respecto de O.

Cuando la densidad «dismétrica» δ_P sea constante, nos hallaremos en un espacio isométrico. En cambio, cuando varíe de alguna forma nos encontraremos con un espacio «dismétrico». De este modo nace la «dismetría» como herramienta matemática más amplia y potente que la isometría actual para representar fenómenos naurales en ámbitos diversos y variables.

En el segundo volumen de este trabajo se desarrolla con más extensión la matematización de la «dismetría» así como algunas consecuencias y aplicaciones para generalizar las leyes físicas con la formulación «dismétrica», poniendo de manifiesto que nos hallamos ante una poderosa herramienta matemática que permite desarrollar una nueva Física. Aquí nos limitamos a introducir los innovadores conceptos de **magnitudes flexibles** y los **espacios «dismétricos»** resultantes de la opción de variación de los patrones físicos.

EQUALITY, IDENTITY, EQUATION AND PHYSICAL LAW

In Physics, the experiments aim to determine **invariant relations** expressed with dyadic algebra between quantities of different magnitudes and possible **constants**[19] that relate them and it is usual to apply the principle of symbolic economy with operations of the same kind. For example, Newton's second law states that the ratio between the amount of force applied to a body and the amount of acceleration it gives it is constant, and this constant is precisely what is called the mass of inertia of the body. This law or invariant relationship is symbolized by the abbreviated dyadic equation $\overline{F} = m \odot \overline{a}$, the dyadic product between a scalar and a vector dyadic, or economically $\overline{F} = m \times \overline{a}$, where $\overline{F}$ and $\overline{a}$ are vectors, which we have been distinguishing with the upper dash, although any other could serve symbology, and m is a positive real number, and all the factors accompanied by their inseparable units.

Equations that equate vector dyads like $\overline{F} = m \odot \overline{a}$ should actually be written explicitly $\overline{F} \, U_F = m \, U_m \odot \overline{a} \, U_a$, where U_F is the unit of force, U_m the unit of mass, and U_a the unit of acceleration. Due to homogeneity, these units cannot be independent and the relation $U_F = U_m * U_a$ must exist between them, economically $U_F = U_m \times U_a$, since otherwise the dyads of the first and second member could not be equal, according to the equality criterion in section IV. In turn, any vector dyadic equation can always be reduced to scalar forms, using figure 8: let $\overline{e}_F$ be the versor or

---

[19] We will not question here the existence of physical constants, in the ordinary sense of the term, although the second volume of this work explains the reasons why this notion should be revised in «dysmetric» spaces.

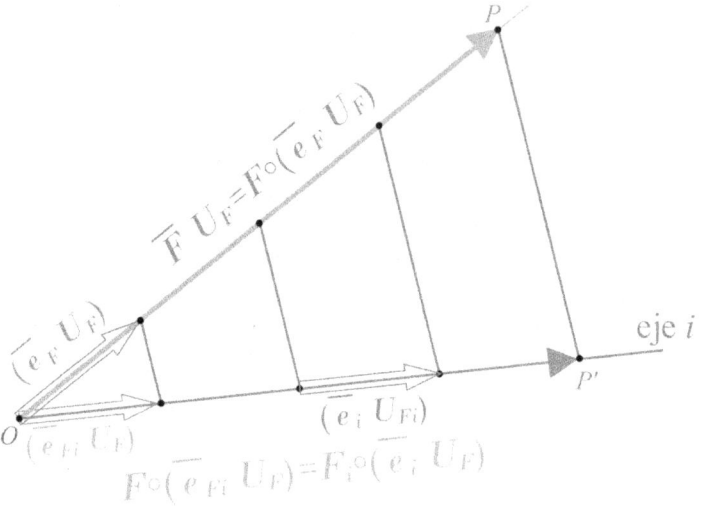

Geometrically, if OP', a parallel and generally oblique projection of OP on any axis i, is measured by the unit U_{Fi}, the dyad $F\ U_{Fi}$ results. If the same segment OP' is measured with the unit U_F, the dyad $F_i\ U_F$ results. Then, so that OP' is referred to the same unit U_F of OP, it is enough to express the measure through the projection F_i of OP.

Figure 8

unit vector in the direction of $\overline{F}$, it will be linked to the vector dyad $(\overline{e}_F\ U_F)$ and thus we will have the equality $\overline{F}U_F = F\circ(\overline{e}_F\ U_F)$, where F is the component of on $\overline{e}_F$ and the module of $\overline{F}$ represents an abstract length equal to the unit of force U_F. Let R^3 be any basis, whose versores are denoted $\overline{e}_1$, $\overline{e}_2$ and $\overline{e}_3$, abbreviated $\{\overline{e}_i\}$, with i = 1, 2, 3. We admit the geometric fiction that $\overline{e}_i$ has as a module the abstract unit of length representing U_F, that is to say, that the modules of $\overline{e}_F$ and $\overline{e}_i$ are equal and we pretend that they represent the unit U_F. We define $\overline{e}_{Fi}$ as the projection or component on the i axis of the versor $\overline{e}_F$, its module will represent the quantity U_{Fi}, or the fake projection of U_F on the same axis i, because U_F is not a length. Let us designate by F_i the

240

components of $\overline{F}$ on each axis i. Thales' Theorem guarantees the following: given a segment divided into lengths equal to a certain unit of length, projecting the divisions in parallel onto another line, the same number of projected units result as in the origin segment, so that the measure of the projected segment in projected units is the same as the original segment measurement in the first few units[20]. Applying this geometric property to the vector $\overline{F}$ and its projection F_i on an axis i, we will have that F_i will be represented by the scalar concrete $F\,U_{Fi}$; we want to find out how much F_i has to be worth for the projection to be represented by the dyad $F_i\,U_F$. The vectors $\overline{e}_{Fi}$ and $\overline{e}_i$ are collinear, so there will be a scalar k_i such that $\overline{e}_{Fi}=k_i \bullet \overline{e}_i$, with which, the relationship between U_{Fi} and U_F must be admitted to be the same, with $U_{Fi}=k_i \circ U_F$, a result which also we would arrive by means of the axiom of continuity of section IV. Under these conditions, you will have:

$$F_i\,U_F=F\,U_{Fi}=F\circ(k_i\circ U_F)=(k_i\times F)\,U_F$$

In accordance with the equality criterion in section IV, it is concluded that for all i, it is:

$$F_i=k_i\times F$$

Therefore, the measure F_i of the segment OP' in the unit U_F in relation to the measure F of OP with the same unit is in the same ratio k_i as the collinear vectors $\overline{e}_i$ and $\overline{e}_{Fi}$, projection of $\overline{e}_{Fi}$. And so it is justified that to determine the projection on any axis of every vector dyad $\overline{F}U_F$ it is sufficient to find the component $F_i=k_i\times F$ of $\overline{F}_i$, resulting in the projected dyad $\overline{F}_i U_F$.

And so is the important forgotten property that we could state like this: **given a vector measurement, its projection on any axis is in the same dyadic ratio as the versor of the dyad projected on said axis between the versor of this same axis.**

---

[20] The statement, the meaning and the geometric deduction of Thales' Theorem can be found in «Lesson 26» of Mathematize 1.

Applying this property to the acceleration vector, the dyadic vector equation of Newton's second law $\overline{F}\, U_F = m\, U_m \odot \overline{a}\, U_a$ is divided into its three scalar components, referred to the same units U_F, U_m and U_a:

$$F_i\, U_F = m\, U_m * a_i\, U_a \text{ con } i = 1, 2, 3$$

We insist, the preceding confusing reasoning of dyadic algebra shows that the criterion for assigning units to a vector formula is far from obvious, as it is negligently shown in educational settings. This is manifested by the fact that the segment OP measures F times U_F and the segment OP' measures F times U_{Fi}, as a consequence of Thales' Theorem. The same measure F, but different units U_F and U_{Fi}. Hence, to express OP' in the same U_F unit as OP, the coefficient k_i, given by trigonometry, must be taken into account.

Leaving aside such uncertainties, we axiomatize the previous criterion and thus every vector physical equation can be transformed into its corresponding scalars, one for each reference axis, and with the same units, so from here on we will limit ourselves to analyzing only equations of scalar dyadic nature. And thus, the form of equality of the equations implicitly assumes that the dyads of the first and second member must be homogeneous, in accordance with the criterion established by [4.1], with which it will be generally assumed that $a_1\, U_1 = a_2\, U_2$. Multiplying with the operation of section IX by the real number a_1^{-1} inverse of a_1, the following dyadic expression immediately results:

$$a_1^{-1} \circ (a_1\, U_1) = a_1^{-1} \circ (a_2\, U_2)$$

According to the algebra of R, with definition [3.1] and with the multiplication of a dyad by a scalar, an operation defined in section IX, we will have:

$$U_1 = (a_1^{-1} \times a_2)\, U_2$$

Which can also be written with the fractional notation of real numbers:

$$U_1 = \frac{a_2}{a_1} \, U_2$$

Finally, by virtue of the definition of the division of homogeneous units, analyzed in section XI, we will have as a conclusion:

$$\frac{a_2}{a_1} = \frac{U_1}{U_2} \qquad [24.1]$$

Equation [24.1] would seem obvious, if it were admitted without more than with the symbols of the units, it would operate as if they were elements of R; but they are not, so the second member does not indicate an arithmetic but a dyadic quotient between units of homogeneous magnitudes, the one in section XI, so formula [24.1] requires, as we have justified here, to apply a dyadic algebra as the one developed in this monograph, so that only after having justified it with the foundation of the laws of composition of scalar dyads, applied to the equality of homogeneous dyads, does it acquire the important meaning that has been attributed to it since Fourier, although taking into account here the dyadic quotient of the second member counts: **the arithmetic quotient of the measurements of the same quantity of a certain magnitude expressed with homogeneous units is equal to the inverse dyadic ratio between the units.**

Precisely such a ratio between the units is the quotient that dimensional analysis schemes forget to justify, because of the absolute absence of a physical algebra, since they limit themselves to operating with the unit symbols as if they were elements of R without dealing with why they compose them in that way and admitting without more what subjective intuition may dictate in this regard; hence, based on this prejudice, the various theories of dimensional analysis start with equation [24.1] «arithmetized», interpreting the quotient between units as a numerical quotient; while we have previously traveled a long and rigorous algebraic path to base that same formula and give testimony of its unequivocal meaning, rejecting any arbitrary operational form, which we have tried to save in this work with the generic postulation of the algebra of magnitudes and specifically [24.1] in its section XI, precisely defining the quotient between homogeneous units.

The equations of Physics such as the «arithmetized» classical vector $\overline{F} = m \times \overline{a}$ or in their complete dyadic formulation $\overline{F} \, U_F = m \, U_m \odot \overline{a} \, U_a$ are called **universal laws**, because they have the characteristic that they are mathematical symbolic forms that subsume empirical observation, therefore, In order to have an unequivocal meaning, its elements, in this case $\overline{F}$, m and $\overline{a}$, must be established through appropriate epistemic definitions, in order to build an exact knowledge of the observations, well founded, and methodically and rationally worked out.

On the other hand, other equations, such as the one that establishes the speed as a function of the distance traveled in a certain time, or in mathematical terms the derivative of the position vector with respect to time, are the only consequence of the arbitrariness of physical thought to compose magnitudes. In this case, the velocity would be a magnitude derived from the magnitudes of length and time, so an epistemic definition of the velocity is not required, but only a definition equation as a function of these primary magnitudes. Hence, these types of physics equations must be called **definition equations**.

Both the universal laws and the equations of definition are equalities between dyadic entities, so when operating on them it will be necessary to adhere to the guidelines of the algebra of these elements. Thus, for example, for Newton's second law in abbreviated notation $\overline{F} = m \odot \overline{a}$, $\overline{F}$ it is indicated by a vector dyad of $\{R^3, N\}$, where N is the symbol of the unit of force called newton, assuming that it operates in the System International; for its part, the mass m will indicate another scalar dyad of the type $\{R, kg\}$, with the unit of mass called the standard kilogram; and the acceleration will belong to the dyadic set $\{R^3, m /\!/ s^2\}$, referred to the compound unit called standard meter divided by the second pattern raised to the dyadic square, all of this dyadically. In turn, the universal law $\overline{F} = m \odot \overline{a}$ will be divided into three equations, one per coordinate, which will have the form $F_i = m * a_i$, with $i = 1, 2, 3$, and so that F_i, m and a_i will be elements respectively of $\{R,N\}$, $\{R,kg\}$ and $\{R,m/\!/s^2\}$, so it must always be borne in mind that such equations relate concrete entities and that they must be operated with through the algebra of magnitudes. Note that the operation «$\odot$» is the multiplication of section XX, the «$*$» is the one defined in section XII and the one indicated by «$/\!/$» is the division of section XVI, all of them **generating laws**.

However, despite the clarity of what has been said, it is notorious and striking that all texts from any scientific field, even the most reputable ones, absolutely forget this evidence and develop their expositions and theories omitting any reference to the laws of composition of the physical units, with which scientific equations are presented as if they related real numbers or vectors; something totally erroneous and inappropriate, because we have already sufficiently justified up to now that the basic mathematical elements of the physical sciences are the dyadic entities, requiring a specific algebra, which cannot be left to the subjective discretion of each one.

So it is necessary and healthy to save that unanimous vice with which texts undertake operations with units without any preamble or algebraic motivation, trusting that readers or students are enlightened by a kind of epiphany that guides them along the correct path of operations with magnitudes.

We hope with this monographic work to shed light on this matter and contribute to unearthing the forgotten pillar of science, laying the foundations for an algebra of magnitudes that provides objective criteria to judge and manipulate these entities according to their true physical nature. In particular, in relation to the physical equations object of this section, their generic form will be represented by an equality like [4.1] of homogeneous scalar concrete entities $a_1 U_1 = a_2 U_2$ or by an equality like [4.2] for vector entities $\overline{a}_1 U_1 = \overline{a}_2 U_2$. The uniformity axiom [4.3] makes it possible to ensure in both cases that there exists $k \in R$ such that $U_2 = k \circ U_1$. Substituting, we will have:

$$a_1 U_1 = a_2 U_2 = a_2 (k \circ U_1)$$

$$\overline{a}_1 U_1 = \overline{a}_2 U_2 = \overline{a}_2 (k \circ U_1)$$

Where «$\circ$» is the multiplication of section IX. The properties of this operation allow us to write these scalar and vector equations in the following way:

$$a_1 U_1 = a_2 U_2 = (a_2 \times k) U_1$$

$$\overline{a}_1 U_1 = \overline{a}_2 U_2 = (\overline{a}_2 \bullet k) U_1$$

The operation «$\times$» is the multiplication of R and «$\bullet$» is the multiplication of a real number by a vector. Once both members are uniform, the equality criterion in section IV establishes the identity of the primaries:

$$a_1 = a_2 \times k = k \times a_2$$

$$\overline{a}_1 - \overline{a}_2 \bullet k = k \bullet \overline{a}_2$$

In summary, in terms of the algebra of magnitudes, every physical equation is represented by the equality of two dyadic entities, scalar or vector, which must be homogeneous without the need for them to be uniform, and the equality of dyads is split into relations between its primaries and secondaries, according to the following scalar and vector reasoning schemes:

$$a_1 U_1 = a_2 U_2 \Rightarrow \text{Si } U_2 = k \circ U_1 \Rightarrow a_1 = k \times a_2 \qquad [24.2]$$

$$\overline{a}_1 U_1 = \overline{a}_2 U_2 \Rightarrow \text{Si } U_2 = k \circ U_1 \Rightarrow \overline{a}_1 = k \bullet \overline{a}_2 \qquad [24.3]$$

In the particular case that the physical equations identify uniform dyads, the relationships between their primaries and secondaries will correspond to the particular case $k = 1$.

The decomposition of dyadic equalities, described by [24.2] and [24.3], will be called the **doubling theorem**. We will verify that it is of utmost importance for the analysis of any scientific equation, as will be seen in section XXVI on physical constants, where some highly relevant universal laws are valued and unfolded.

In order to practice with the algebra of magnitudes and observe its functionality, let's proceed to deduce some physical equations of importance and some complexity. Let's start with the **continuity equation for hydrodynamics**. Let's imagine a fluid in stationary motion. Its state will be represented by a dyadic vector field with primary defined by an application of R^3 in R^3, independent of time and symbolized $\overline{v} = v(x,y,z)$, so that each point $P(x,y,z)$ of the fluid present a velocity $\overline{v}$ as a function of the coordinates (x,y,z) of the point, with its unit in the secondary.

Suppose that the density ρ of the fluid is variable, it will be represented by a dyadic scalar field with the primary described by the function $\rho = \rho(x,y,z,t)$, which represents an application of R^4 in R, with its unit of measured in the secondary. Let us admit that both functions $\overline{v}$ and ρ are differentiable.

Figure 9 represents in ordinary geometric space an elementary parallelepiped from a generic point, taking the increments of the variables Δx, Δy and Δz. Let the versores $\overline{i}$, $\overline{j}$ and $\overline{k}$ be an orthonormal basis of R^3 and the components of $\overline{v}$ in this basis $v_1(x,y,z)$, $v_2(x,y,z)$ and $v_3(x,y,z)$. Let us observe how the mass of the fluid inside said parallelepiped varies at any given instant t.

To fix ideas we will use the units of the International System. The quantity of fluid per unit of time that crosses the face of the parallelepiped that passes through P and is normal to the versor $\overline{i}$ will be described by the following product of four dyadic entities:

$$[\rho(x,y,z,t)\ kg/\!/m^3]*[v_1(x,y,z)\ m/\!/s]*[\Delta y\ m]*[\Delta z\ m]\ [24.4]$$

According to the definition of multiplication in section XII, its properties and the division in section XVI, the quantity [24.4] becomes the dyad:

$$[\rho(x,y,z,t)\times v_1(x,y,z)\times \Delta y\times \Delta z]\ kg/\!/s \qquad [24.5]$$

Analysis of the variation of a vector
field through an elementary
parallelepiped at any point P

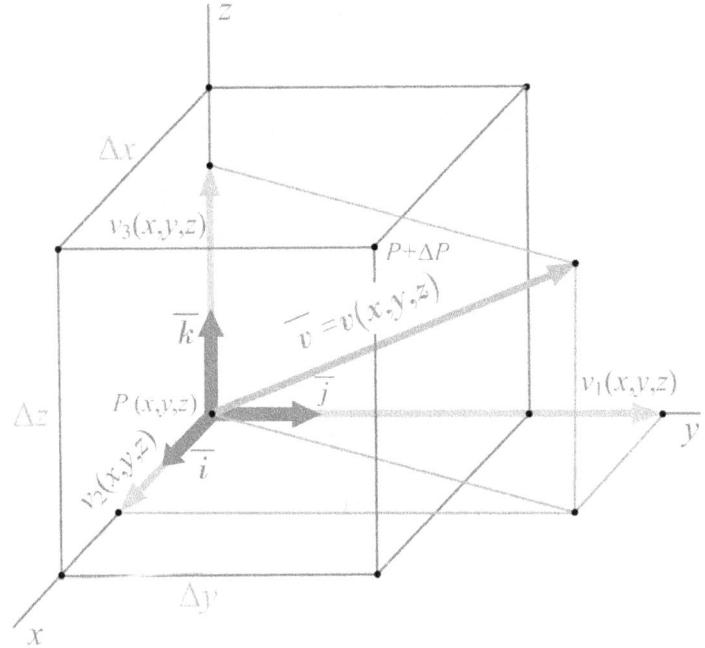

Figure 9

Similarly, the amount of fluid that passes through the face of
the parallelepiped parallel to the previous one per unit of time
and that passes through the point $P+\Delta P$ of coordinates
$(x+\Delta x, y+\Delta y, z+\Delta z)$ will be given by:

$$[\rho(x+\Delta x, y, z, t) \times v_1(x+\Delta x, y, z) \times \Delta y \times \Delta z] \; kg/\!/s \qquad [24.6]$$

Finding the difference between [24.6] and [24.5], dividing and
multiplying it by Δx, so that it will not vary, and by definition of a
partial derivative, in the limit when Δx tends to zero, we arrive at
the variation of the amount of fluid that crosses both parallel
faces of the parallelepiped at the point $P(x,y,z)$, which will be
represented by the expression:

$$\left[\frac{\partial\left[\rho(x,y,z,t)\times v_1(x,y,z)\right]}{\partial x}\times \Delta x\times \Delta y\times \Delta z\right]\frac{kg}{s} \quad [24.7]$$

Similarly, the amount of fluid that passes through the two faces parallel to each other and normal to the versor $\overline{j}$ will be given by:

$$\left[\frac{\partial\left[\rho(x,y,z,t)\times v_2(x,y,z)\right]}{\partial y}\times \Delta x\times \Delta y\times \Delta z\right]\frac{kg}{s} \quad [24.8]$$

And, finally, the amount of fluid that passes through the two normal faces to the versor $\overline{k}$ will be:

$$\left[\frac{\partial\left[\rho(x,y,z,t)\times v_3(x,y,z)\right]}{\partial z}\times \Delta x\times \Delta y\times \Delta z\right]\frac{kg}{s} \quad [24.9]$$

The dyadic sum of the uniform quantities [24.7], [24.8] and [24.9] will indicate the quantity of fluid entering or leaving the elementary parallelepiped per unit time at the point $P(x,y,z)$. Dividing this total quantity by the dyad [$(\Delta x\times\Delta y\times\Delta z)$ m^3], according to the definition of division in section XVI, we will have said quantity per unit of geometric volume in P indicated by the following concrete:

$$\left[\frac{\partial\left[\rho(x,y,z,t)\times v_1(x,y,z)\right]}{\partial x}+\frac{\partial\left[\rho(x,y,z,t)\times v_2(x,y,z)\right]}{\partial y}+ \right.$$
$$\left. +\frac{\partial\left[\rho(x,y,z,t)\times v_3(x,y,z)\right]}{\partial z}\right]\frac{kg}{s*m^3} \quad [24.10]$$

In mathematical field theory, the div divergence of a vector field $\overline{A}=A_1(x,y,z)\times\overline{i}+A_2(x,y,z)\times\overline{j}+A_3(x,y,z)\times\overline{k}$ is defined with the expression:

$$div\ \overline{A}=\frac{\partial A_1(x,y,z)}{\partial x}+\frac{\partial A_2(x,y,z)}{\partial y}+\frac{\partial A_3(x,y,z)}{\partial z}$$

With this notation, applied to the vector field product $\rho(x,y,z,t) \times \overline{v}(x,y,z)$, the dyad [24.10], assuming the coordinates x, y, z and t, for simplicity, and recalling since the analysis is being carried out for a certain instant t, the variation of the amount of fluid in the parallelepiped with the following dyad can be written in a synthetic way:

$$[div\ (\rho \times \overline{v})]\ kg /\!/ (s * m^3) \qquad [24.11]$$

So far we have only taken into account the variation in mass due to the instantaneous movement of the fluid, but the density being generally variable with time, this effect must also be introduced. To do this, we must observe that the change in mass due to the change in density in the elementary parallelepiped at point P between the instants t and t y t+Δt will have the dyadic form:

$$[[\rho(x,y,z,t+\Delta t) - \rho(x,y,z,t)] \times (\Delta x \times \Delta y \times \Delta z)]\ kg$$

Dividing by the measurement [Δt s] and taking the limit when ?t tends to zero, we arrive at the dyad expressed with the partial derivative with respect to t for the change in mass in the parallelepiped per unit time:

$$\left[\frac{\partial \left[\rho(x, y, z, t) \right]}{\partial t} \times \Delta x \times \Delta y \times \Delta z \right] \frac{kg}{s}$$

Dividing this quantity by the dyad [(Δx×Δy×Δz) m^3] with the operation in section XVI, we arrive at the variation in mass per unit of time and volume:

$$\left[\frac{\partial \left[\rho(x, y, z, t) \right]}{\partial t} \right] \frac{kg}{s * m^3} \qquad [24.12]$$

Adding [24.11] and [24.12] we arrive at the variation in total unit mass, which we write in abbreviated form, regardless of the coordinates, which are understood, by the following dyadic quantity:

$$\left[div\left(\rho \times \bar{v} \right) + \frac{\partial \rho}{\partial t} \right] \frac{kg}{s*m^3} \qquad [24.13]$$

Imagining two scalar fields that represent the contribution and loss of fluid at each point $P(x,y,z)$ per unit of time and volume, which we can represent by $\varphi(x,y,z)$ for the sources and $\sigma(x,y,z)$ for drains or sinks, grouping them into a single scalar field function represented by the form $\psi(x,y,z) = \varphi(x,y,z) - \sigma(x,y,z)$, where the minus sign must have the meaning of decrease in mass, equaling [24.13], we arrive at the physical equation known as **hydrodynamic continuity**:

$$\left[div\left(\rho \times \bar{v} \right) \right] \frac{kg}{s*m^3} = \left[\psi - \frac{\partial \rho}{\partial t} \right] \frac{kg}{s*m^3} \qquad [24.14]$$

Note that, according to the equality criterion in section IV, it has been assumed that the first and second members of [24.14] are uniform, although it would suffice if they were homogeneous, as we will analyze next. Under these conditions, the doubling theorem [24.2] with $k=1$ will allow us to write the algebraic equality without the secondaries:

$$div\left(\rho \times \bar{v} \right) = \psi - \frac{\partial \rho}{\partial t}$$

In general, if the units of length, mass and time were U_{L1}, U_{m1} and U_{t1} for the first member, and U_{L2}, U_{m2} and U_{t2} for the second, equation [24.4] will be formulated as follows:

$$\left[div\left(\rho \times \bar{v} \right) \right] \frac{U_{m1}}{U_{t1}*U_{L1}^3} = \left[\psi - \frac{\partial \rho}{\partial t} \right] \frac{U_{m2}}{U_{t2}*U_{L2}^3} \qquad [24.15]$$

To simplify, designating by U_1 and U_2 the compound units of the first and second members, respectively, and assuming $U_2 = k \circ U_1$, the doubling theorem [24.2] leads us to the generic algebraic equation:

$$div \left(\rho \times \overline{v} \right) = k \times \left[\psi - \frac{\partial \rho}{\partial t} \right] \qquad [24.16]$$

Formula [24.15] is the universal dyadic expression of the hydrodynamic continuity equation and [24.16] is its form of unfolded ordinary algebra.

In the particular case that the fluid was incompressible, the density would be constant in space and in time, the corresponding partial derivative would cancel out and we would have:

$$\rho \times div \ \overline{v} = k \times \psi$$

If there were also no sources or drains or sinks, we would have at all points ψ=0, with which the continuity equation would have the form:

$$div \ \overline{v} = 0$$

Needless to remember again that, if the units of the first and second members of the concrete equation were uniform, it is enough to take k=1 in the doubling.

Let's look at a second example of a physical equation that refers to the phenomenon of **heat conduction**. Let us consider a dyadic vector field defined by an application of R³ in R³, independent of time and symbolized $\overline{Q}$=Q(x,y,z), so that at each point P(x,y,z) there is a given heat transfer by the vector, which indicates the **heat current** in units of heat per unit of time and unit of area in the normal direction, which will be a function of the coordinates (x,y,z) of the point. Suppose that the temperature T of the fluid is variable, it will be represented by a dyadic scalar field described by a function of R⁴ in R denoted T=T(x,y,z,t) and its unit of measurement. Let us admit that both functions $\overline{Q}$ and T are differentiable.

Let's look at the same figure 6, in which we will now have to substitute $\overline{v}$ for $\overline{Q}$, which are silent or indifferent symbols for the purposes of analysis. The figure represents an elementary parallelepiped in ordinary geometric space from a generic point, taking the increments of the variables Δx, Δy and Δz. Fixing ideas, we will use the units of the International System for the different magnitudes that intervene in the phenomenon.

Under these conditions, the change in heat due to the field $\overline{Q}$ in the elementary parallelepiped and per unit volume will be described by the same equation [24.11], without more than substituting the field $\rho \times \overline{v}$ for it $\overline{Q}$, associating the corresponding units with it, placing in the numerator the calorie as a unit of heat, or any other, which leads us to the following concrete:

$$[div\ \overline{Q}]\ cal\,/\!/(s*m^3) \tag{24.17}$$

Next, it is necessary to assess the contribution to heat by the variation of the thermal field at a fixed point and with respect to time t. Thus, being c with its units the specific heat of the body, the variation of heat in the elemental parallelepiped due to the change in temperature with time will be determined by the product of the following dyadic entities:

$$\left[\rho \frac{kg}{m^3}\right]*\left[\Delta x\ m\right]*\left[\Delta y\ m\right]*\left[\Delta z\ m\right]*\left[c\frac{cal}{kg*K}\right] \times \left[\frac{\partial T}{\partial t}\frac{K}{s}\right]$$

Dividing the previous quantity between the dyad that defines the volume of the parallelepiped $[(\Delta x \times \Delta y \times \Delta z)\,m^3]$ and operating with the dyadic algebra, we arrive at the expression of the heat variation at point P due to the thermal variation:

$$\left[\rho \times c \times \frac{\partial T}{\partial t}\right]\frac{cal}{s*m^3} \tag{24.18}$$

Adding the uniform quantities [24.17] and [24.18], we arrive at the total change in heat at point P per unit of time and volume with the dyad:

$$\left[div\ \overline{Q} + \rho \times c \times \frac{\partial T}{\partial t}\right]\frac{cal}{s*m^3}$$

Assuming that there are no heat sources or sinks, it is enough to set the previous quantity to zero to obtain the conduction equation:

$$\left[div\ \overline{Q} + \rho \times c \times \frac{\partial\ T}{\partial\ t} \right] \frac{cal}{s*m^3} = 0 \qquad [24.19]$$

The law of heat conduction axiomatizes the experience that the relationship between heat current and the change in temperature is given by:

$$\left[\overline{Q}\ \frac{cal}{s*m^2} \right] = -\left[\lambda\ \frac{cal}{s*m*K} \right]*\left[grad\ T\ \frac{K}{m} \right] \quad [24.20]$$

Where λ is called the **coefficient of thermal conductivity** and is specific to each substance. Its secondary is expressed by the fundamental units indicated in [24.20]. In turn, the operator *grad* or «nabla» ∇ of a scalar field *T* is the gradient vector field, which in the theory of mathematical field analysis is defined on an orthonormal basis $(\overline{i}, \overline{j}, \overline{k})$ by the equation:

$$grad\ T = \nabla T = \frac{\partial\ T}{\partial\ x} \times \overline{i} + \frac{\partial\ T}{\partial\ y} \times \overline{j} + \frac{\partial\ T}{\partial\ z} \times \overline{k}$$

Substituting [24.20] in [24.19], we have the following dyadic equation:

$$\left[div\left(-\lambda \times grad\ T \right) + \rho \times c \times \frac{\partial\ T}{\partial\ t} \right] \frac{cal}{s*m^3} = 0$$

In mathematical field theory, the **Laplacian** operator is defined, which is represented by the same letter delta Δ of the increments for variables and means the divergence of the gradient vector field. With which, the previous equation is symbolically transformed into the well-known fundamental equation of heat conduction:

$$\left[-\lambda \times \Delta T + \rho \times c \times \frac{\partial\ T}{\partial\ t} \right] \frac{cal}{s*m^3} = 0$$

254

In the foregoing we have developed two significant physical equations, which illustrate the deduction procedure to be followed to analyze physical phenomena based on the algebra of magnitudes, through the operations defined in this monograph for the dyadic entities of Physics.

The two phenomena analyzed have no relation to each other, because they refer to very different material realities; however, the mathematical components of the primaries of the magnitudes described are very similar and in some respects appear identical or at least isomorphic. On the contrary, the different nature between both cases is evident in the units of the secondary ones, which are different because they refer to different magnitudes. This shows that the mathematical apparatus of the primary ones can acquire different meanings depending on the context of the fact under investigation, and hence the importance of not losing sight of the secondary ones and their units, which determine the intervening magnitudes in each phenomenon, otherwise it would be impossible to fully understand the physical meaning of the derived equations.

Thus, the importance of operating with dyads instead of ordinary algebraic entities has become evident, since otherwise, apart from the objective incorrectness that this implies, the equations lose much of their significance, becoming mere very elegant symbologies mathematical in nature, though physically abstruse.

On the contrary, operations with dyads allow logical steps to be established without the slightest confusion, as can be seen in the dyadic division that gives rise to the quantity [24.10], which unequivocally deduces a quotient between measurements, both with their two primary elements and secondary, being established unambiguously by means of that composition law. With all this, the convenient **generator procedure** is revealed to systematically determine scientific formulations, operating in the first place not only with simple common algebraic entities, but with quantities of magnitudes, to later equate those that are homogeneous and that must be identified with each other, by virtue of its own observed nature, to **create a physical equation**.

We observe in this process the decisive intervention of the **generating external laws of composition**, without which Physics would be emptied of content. And this is how the algebra of magnitudes decisively helps the best understanding and exact formulation of the analyzed natural phenomena, while at the same time establishing the essential composition laws so that physical language is complete and univocal, safe from subjective, intuitive interpretations or arbitrary.

Let us finally clarify that we have established the **criterion of dyadic equality** with the general condition that equal dyads represent the same quantity of magnitude. This definition opens up three possibilities: first, that the dyads of both members are homogeneous, in which case we will speak of **identity**; second, that the dyad of one of the members is generated by those of the other through generative operations, in which case equality will represent a **physical equation**; and third, that a physical equation has been established experimentally, which gives rise to an equality that we will call a **physical law**.

For example, all equality of homogeneous quantities such as $150\ cm = 1.50\ m$ are identities. Equalities that relate dyads through generating operations, such as the definition of surface $3\ m * 2\ m = 6\ m^2$ or speed $6\ m /\!/ 2\ s = 3\ m /\!/ s$, are physical equations. And, finally, generating empirical relations such as Newton's second law, $2\ kg * 6\ m /\!/ s^2 = 12\ N$, are physical laws.

Therefore, the definitions of additive operations are identities (sections V to XI); the definitions of the generating multiplicative operations are equations (sections XII to XX); and formulations such as *Newton's second law* are physical laws. However, for convenience in practice we use the terms equality, identity and equation as synonyms, a simplification that lacks physical-mathematical relevance.

Apartado XXIV

IGUALDAD, IDENTIDAD, ECUACIÓN Y LEY FÍSICA

En Física los experimentos tienen por objeto determinar **relaciones invariantes** expresadas con el álgebra diádica entre cantidades de diversas magnitudes y posibles **constantes**[21] que las relacionen y es usual aplicar el principio de economía simbólica con operaciones de la misma especie. Por ejemplo, la *segunda ley de Newton* establece que la razón entre la cantidad de fuerza aplicada a un cuerpo y la cantidad de aceleración que le confiere es constante, y esta constante es precisamente lo que se denomina masa de inercia del cuerpo. Esta ley o relación invariante se simboliza mediante la ecuación diádica abreviada $\overline{F} = m \odot \overline{a}$, producto diádico entre un concreto escalar y otro vectorial, o económicamente $\overline{F} = m \times \overline{a}$, donde $\overline{F}$ y $\overline{a}$ son vectores, que venimos distinguiendo con el guión superior, aunque pudiera servir cualquier otra simbología, y m es un número real positivo, y todos los factores acompañados de sus unidades inseparables.

Las ecuaciones que igualan díadas vectoriales semejantes a $\overline{F} = m \times \overline{a}$ en realidad deberían escribirse en forma explícita, $\overline{F} \ U_F = m \ U_m \odot \overline{a} \ U_a$, donde U_F sea la unidad de fuerza, U_m la unidad de masa y U_a la unidad de aceleración. Por homogeneidad, estas unidades no pueden ser independientes y habrá de existir entre ellas la relación $U_F = U_m * U_a$, económicamente $U_F = U_m \times U_a$, pues de otro modo las díadas del primer y del segundo miembro no se podrían igualar, de acuerdo con el criterio de igualdad del apartado IV. A su vez, toda ecuación diádica vectorial se podrá reducir siempre a formas escalares, sirviéndonos de la figura 8: sea

---

[21] No cuestionaremos aquí la existencia de las constantes físicas, en el sentido ordinario del término, aunque en el segundo volumen de este trabajo se motivan las razones por las que en los espacios «dismétricos» se debe revisar esta noción.

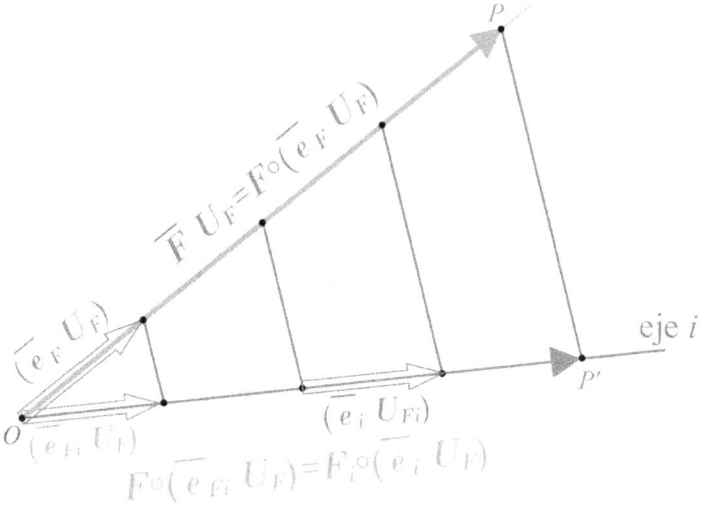

Geométricamente se tiene que, si *OP'*, proyección paralela y en general oblicua de *OP* sobre un eje *i* cualquiera, se mide mediante la unidad U_{Fi}. resulta la díada $F\ U_{Fi}$. Si el mismo segmento *OP'* se mide con la unidad U_F, resulta la díada $F_i\ U_F$. Luego, para que *OP'* quede referido a la misma unidad U_F de *OP*, basta expresar la medida mediante la proyección F_i de *OP*.

Figura 8

$\overline{e}_F$ el versor o vector unitario en la dirección de $\overline{F}$, estará ligado a la díada vectorial $(\overline{e}_F\ U_F)$ y se tendrá así que la igualdad $\overline{F}\,U_F = F\circ(\overline{e}_F\ U_F)$, donde F sea la componente de $\overline{F}$ sobre $\overline{e}_F$ y el módulo de $\overline{e}_F$ represente una longitud abstracta igual a la unidad de fuerza U_F. Sea una base cualquiera de R^3, cuyos versores se indican $\overline{e}_1$, $\overline{e}_2$ y $\overline{e}_3$, abreviadamente $\{\overline{e}_i\}$, con $i=1, 2, 3$. Admitimos la ficción geométrica de que $\overline{e}_i$ tenga por módulo la unidad de longitud abstracta en representación de U_F, es decir, que los módulos de $\overline{e}_F$ y $\overline{e}_i$ son iguales y fingimos que representen la unidad U_F. Definimos $\overline{e}_{Fi}$ como la proyección o componente sobre el eje *i* del versor $\overline{e}_F$, su módulo representará la cantidad U_{Fi}, o fingida proyección de U_F sobre el mismo eje *i*, porque U_F no es una longitud. Designemos por F_i las componentes de $\overline{F}$ sobre cada eje *i*. El *Teorema de Tales* garantiza lo siguiente: dado un segmento dividido en longitudes iguales a cierta unidad de longitud, proyectando paralelamente las divisiones sobre otra recta, resultan igual cantidad de unidades proyectadas que en el segmento origen, de modo que la medida del segmento proyectado en unidades proyectadas es la misma que la medida del segmento original en las primeras unidades[22]. Aplicando esta propiedad geométrica al vector $\overline{F}$ y su proyección F_i sobre un eje *i*, se tendrá que F_i quedará representada por el concreto escalar $F\ U_{Fi}$; queremos averiguar cuánto ha de valer F_i para que la proyección quede representada por el concreto $F_i\ U_F$. Los vectores $\overline{e}_{Fi}$ y $\overline{e}_i$ son colineales, por lo que existirá un escalar k_i tal que $\overline{e}_{Fi}=k_i\bullet\overline{e}_i$, con lo cual, la relación entre U_{Fi} y U_F debe admitirse que sea la misma, con $U_{Fi}=k_i\circ U_F$, resultado al que también llegaríamos mediante el axioma de continuidad del apartado IV. En estas condiciones, se tendrá:

$$F_i\ U_F=F\ U_{Fi}=F\circ(k_i\circ U_F)=(k_i\times F)\ U_F$$

De acuerdo con el criterio de igualdad del apartado IV, se llega a la conclusión de que para todo *i*, es:

$$F_i=k_i\times F$$

Por tanto, la medida F_i del segmento OP' en la unidad U_F en relación con la medida F de OP con igual unidad está en la misma razón k_i que los vectores colineales $\overline{e}_i$ y $\overline{e}_{Fi}$, proyección de $\overline{e}_{Fi}$. Y así se justifica que para determinar la proyección sobre un eje cualquiera de toda díada vectorial $\overline{F}U_F$ es suficiente hallar la componente $F_i=k_i\times F$ de $\overline{F}_i$, resultando la díada proyectada $\overline{F}_iU_F$.

Y así resulta la importante propiedad olvidada que podríamos enunciar así: **dada una medición vectorial, su proyección sobre un eje cualquiera está en la misma razón diádica que el versor de la díada proyectado sobre dicho eje entre el versor de este mismo eje.**

---

[22] El enunciado, el significado y la deducción geométrica del *Teorema de Tales* se puede encontrar en la «Lección 26» de *Matematizar 1*.

Aplicando esta propiedad al vector aceleración, la ecuación concreta vectorial de la *segunda ley de Newton* $\overline{F}\ U_F = m\ U_m \textcircled{\odot} \overline{a}\ U_a$ se desdobla en sus tres componentes escalares, referidas a las mismas unidades U_F, U_m y U_a:

$$F_i\ U_F = m\ U_m * a_i\ U_a \text{ con } i = 1, 2, 3$$

Insistimos, el lioso razonamiento precedente de álgebra diádica pone de manifiesto que el criterio de asignación de unidades a una fórmula vectorial no es ni mucho menos evidente, como se muestra negligentemente en los ámbitos docentes. Así se manifiesta por el hecho de que el segmento OP mide F veces U_F y el segmento OP' mide F veces U_{Fi}, debido al *Teorema de Tales*. La misma medida F, pero distintas unidades U_F y U_{Fi}. De ahí que, para expresar OP' en la misma unidad U_F que OP, se deba tener en cuenta el coeficiente k_i, dado por la trigonometría.

No obstante, está claro en todo caso que toda ecuación física vectorial se podrá transformar en sus correspondientes escalares, una por cada eje de referencia, y **con las mismas unidades**, por lo que a partir de aquí nos limitaremos a analizar únicamente ecuaciones de naturaleza diádica escalar. Y así, la forma de igualdad de las ecuaciones supone implícitamente que las díadas del primer y del segundo miembro deban ser homogéneas, de conformidad con el criterio establecido por [4.1], con lo cual se tendrá en general que $a_1\ U_1 = a_2\ U_2$. Multiplicando por el número real a_1^{-1} inverso de a_1, resulta inmediatamente la ecuación concreta:

$$a_1^{-1} \circ (a_1\ U_1) = a_1^{-1} \circ (a_2\ U_2)$$

De acuerdo con el álgebra de R, con la definición [3.1] y con la multiplicación de un concreto por un escalar, operación definida en el apartado IX, tendremos:

$$U_1 = (a_1^{-1} \times a_2)\ U_2$$

Que también se puede escribir con la notación fraccionaria de los números reales:

$$U_1 = \frac{a_2}{a_1}\ U_2$$

Finalmente, en virtud de la definición de división de unidades homogéneas, analizada en el apartado XI, tendremos como conclusión:

$$\frac{a_2}{a_1} = \frac{U_1}{U_2}$$

[24.1]

La ecuación [24.1] parecería obvia, si se admitiese sin más que con los símbolos de las unidades se opere como si fueran elementos de R; pero no lo son, por lo que el segundo miembro no indica un cociente aritmético sino diádico entre unidades de magnitudes homogéneas, el del apartado XI, por lo que la fórmula [24.1] requiere, tal como hemos justificado aquí, aplicar un álgebra diádica como la desarrollada en esta monografía, de modo que solo después de haberla justificado con el fundamento de las leyes de composición de las díadas escalares, aplicadas a la igualdad de díadas homogéneas, adquiere el importante significado que se le viene atribuyendo desde Fourier, aunque teniendo en cuenta aquí el cociente diádico del segundo miembro: **el cociente aritmético de las medidas de una misma cantidad de cierta magnitud expresada con unidades homogéneas es igual a la razón diádica inversa entre las unidades.**

Precisamente tal razón entre las unidades es el cociente que se olvidan justificar los esquemas de análisis dimensional, a causa de la ausencia absoluta de un álgebra física, pues se limitan a operar con los símbolos de unidades como si fuesen elementos de R sin ocuparse del porqué los componen de esa manera y admitiendo sin más lo que la intuición subjetiva pueda dictar al respecto; de ahí que, en base a este prejuicio, las diversas teorías de análisis dimensional empiecen por la ecuación [24.1] «aritmetizada», interpretando el cociente entre unidades como un cociente numérico; mientras que nosotros hemos recorrido previamente todo un largo camino algebraico y riguroso para fundamentar esa misma fórmula y dar testimonio de su significado inequívoco, rechazando toda forma operacional arbitraria, que hemos pretendido salvar en este trabajo con la postulación genérica del álgebra de magnitudes y específicamente para [24.1] en su apartado XI, definiendo en forma precisa el cociente entre unidades homogéneas.

Las ecuaciones de la Física como la vectorial clásica «aritmetizada» $\overline{F} = m \times \overline{a}$ o en su formulación diádica completa $\overline{F}\ U_F = m\ U_m \odot \overline{a}\ U_a$ se denominan **leyes universales**, porque tienen la característica de que son formas simbólicas matemáticas que subsumen la observación empírica, por lo que, para que tengan sentido inequívoco, sus elementos, en este caso $\overline{F}$, m y $\overline{a}$, deben ser establecidos mediante oportunas definiciones epistémicas, para construir un conocimiento exacto de las observaciones, bien fundamentado, y labrado metódica y racionalmente.

En cambio, otras ecuaciones, como por ejemplo la que establece la velocidad en función de la distancia recorrida en cierto tiempo, o en términos matemáticos la derivada del vector posición respecto del tiempo, son consecuencia única de la arbitrariedad del pensamiento físico para componer magnitudes. En este caso, la velocidad sería una magnitud derivada de las magnitudes de longitud y tiempo, por lo que no se requiere una definición epistémica de la velocidad, sino solo una ecuación de definición en función de dichas magnitudes primarias. De ahí que a este tipo de ecuaciones de la Física deban llamarse **ecuaciones de definición**.

Tanto las leyes universales como las ecuaciones de definición son igualdades entre entes diádicos, por lo que al operar sobre ellas habrá que atenerse a las directrices del álgebra de estos elementos. Así, por ejemplo, para la *segunda ley de Newton* en notación abreviada $\overline{F} = m \odot \overline{a}$, la $\overline{F}$ indica una díada vectorial de $\{R^3, N\}$, donde N es el símbolo de la unidad de fuerza llamada newton, suponiendo que se opere en el Sistema Internacional; por su parte, la masa m indicará otra díada escalar del tipo $\{R, kg\}$, con la unidad de masa llamada kilogramo patrón; y la aceleración $\overline{a}$ pertenecerá al conjunto diádico $\{R^3, m /\!/ s^2\}$, referido a la unidad compuesta denominada metro patrón dividido por el segundo patrón elevado al cuadrado diádico, todo ello diádicamente. A su vez, la ley universal $\overline{F} = m \odot \overline{a}$ se desdoblará en tres ecuaciones, una por coordenada, que tendrán la forma $F_i = m * a_i$, con $i = 1, 2, 3$, y de modo que F_i, m y a_i serán respectivamente elementos de $\{R,N\}$, $\{R,kg\}$ y $\{R,m/\!/s^2\}$, por lo que habrá que tener siempre presente que tales ecuaciones relacionan entes concretos y que deberá operarse con ellos mediante el álgebra de magnitudes. Obsérvese que la operación «⊚» es la multiplicación del apartado XX, la «∗» es la definida en el apartado XII y la indicada por «/» es la división del apartado XVI, todas ellas **leyes generatrices**.

Sin embargo, a pesar de la claridad de lo dicho, resulta notorio y llamativo que todos los textos de cualquier ámbito científico, incluso los más reputados, olviden absolutamente esta evidencia y desarrollen sus exposiciones y teorías omitiendo toda referencia a las leyes de composición de las unidades físicas, con lo que presentan las ecuaciones científicas como si relacionasen números reales o vectores; algo totalmente erróneo e inapropiado, porque ya hemos justificado suficientemente hasta aquí que los elementos matemáticos básicos de las ciencias físicas son los entes diádicos, requiriendo de un álgebra específica, que no puede dejarse al arbitrio subjetivo de cada cual.

De modo que es necesario y saludable salvar ese vicio unánime con que los textos acometen sin ningún preámbulo ni motivación algebraica las operaciones con unidades, confiando que los lectores o estudiantes sean iluminados por una especie de epifanía que les guíe por la senda correcta de las operaciones con magnitudes.

Esperamos con este trabajo monográfico aportar luz en esta materia y contribuir a desenterrar el pilar olvidado de la ciencia, sentando las bases de un álgebra de magnitudes que proporcione criterios objetivos para juzgar y manipular estos entes con arreglo a su verdadera naturaleza física. En particular, en relación con las ecuaciones físicas objeto de este apartado, su forma genérica vendrá representada por una igualdad como [4.1] de entes concretos escalares homogéneos $a_1\ U_1 = a_2\ U_2$ o por una igualdad como [4.2] para entes vectoriales $\overline{a}_1\ U_1 = \overline{a}_2\ U_2$. El axioma de uniformidad [4.3] permite asegurar en ambos casos que exista $k \in R$ tal que $U_2 = k \circ U_1$. Sustituyendo, tendremos:

$$a_1\ U_1 = a_2\ U_2 = a_2\ (k \circ U_1)$$

$$\overline{a}_1 \ U_1 = \overline{a}_2 \ U_2 = \overline{a}_2 \ (k \circ U_1)$$

Donde «○» es la multiplicación del apartado IX. Las propiedades de esta operación nos permiten escribir estas ecuaciones escalar y vectorial de la siguiente forma:

$$a_1 \ U_1 = a_2 \ U_2 = (a_2 \times k) \ U_1$$

$$\overline{a}_1 \ U_1 = \overline{a}_2 \ U_2 = (\overline{a}_2 \bullet k) \ U_1$$

La operación «×» es la multiplicación de R y «•» es la multiplicación de un número real por un vector. Una vez que ambos miembros son uniformes, el criterio de igualdad del apartado IV establece la identidad de los primarios:

$$a_1 = a_2 \times k = k \times a_2$$

$$\overline{a}_1 = \overline{a}_2 \bullet k = k \bullet \overline{a}_2$$

En resumen, en términos del álgebra de magnitudes, toda ecuación física queda representada por la igualdad de dos entes diádicos, escalares o vectoriales, que han de ser homogéneos sin necesidad de que sean uniformes, y la igualdad de concretos se desdobla en sendas relaciones entre sus primarios y secundarios, de acuerdo con los siguientes esquemas de razonamiento escalar y vectorial:

$$a_1 \ U_1 = a_2 \ U_2 \Rightarrow \text{Si } U_2 = k \circ U_1 \Rightarrow a_1 = k \times a_2 \qquad [24.2]$$

$$\overline{a}_1 \ U_1 = \overline{a}_2 \ U_2 \Rightarrow \text{Si } U_2 = k \circ U_1 \Rightarrow \overline{a}_1 = k \bullet \overline{a}_2 \qquad [24.3]$$

En el caso particular de que las ecuaciones físicas identifiquen concretos uniformes, las relaciones entre sus primarios y secundarios corresponderán al caso particular $k=1$.

La descomposición de las igualdades diádicas, descrita por [24.2] y [24.3], será denominada **teorema del desdoblamiento**. Comprobaremos que es de suma importancia para el análisis de cualquier ecuación científica, como se observará en el apartado XXVI sobre las constantes físicas, donde se valoran y desdoblan algunas leyes universales de gran relevancia.

Análisis de la variación de un campo
vectorial a través de un paralelepípedo
elemental en un puno *P* cualquiera

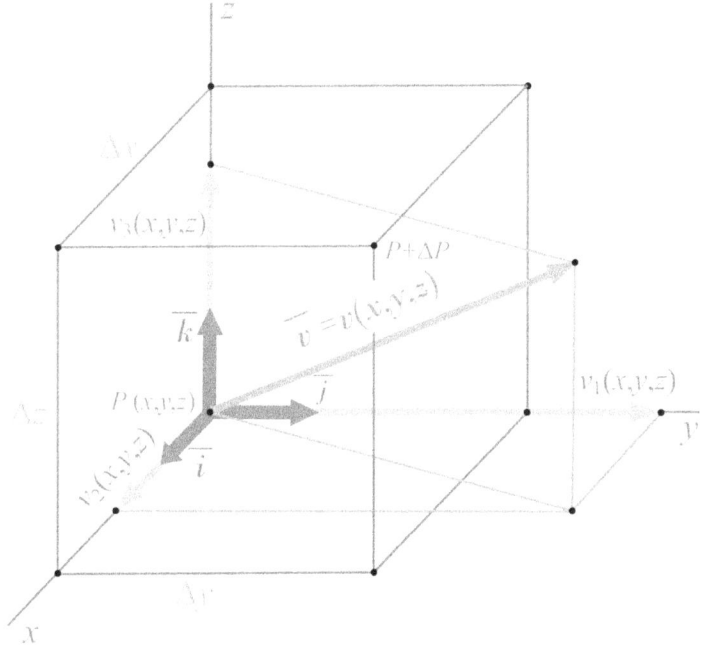

Figura 9

A fin de practicar con el álgebra de magnitudes y observar su funcionalidad, pasemos a deducir algunas ecuaciones físicas de importancia y cierta complejidad. Empecemos con la **ecuación de continuidad de la hidrodinámica**. Imaginemos un fluido en movimiento estacionario. Su estado quedará representado por un campo vectorial diádico con primario definido por una aplicación de R^3 en R^3, independiente del tiempo y simbolizada $\overline{v} = v(x,y,z)$, de modo que cada punto $P(x,y,z)$ del fluido presente una velocidad $\overline{v}$ función de las coordenadas (x,y,z) del punto, con su unidad de medida en el secundario.

Supongamos que la densidad ρ del fluido sea variable, quedará representada por un campo escalar diádico con el primario descrito por la función $\rho = \rho(x,y,z,t)$, que representa una aplicación de R^4 en R, con su unidad de medida en el secundario. Admitamos que ambas funciones $\overline{v}$ y ρ sean derivables.

La figura 9 representa en el espacio geométrico ordinario un paralelepípedo elemental a partir de un punto genérico, tomando los incrementos de las variables Δx, Δy e Δz. Sean los versores $\overline{i}$, $\overline{j}$ y $\overline{k}$ una base ortonormal de R^3 y las componentes de $\overline{v}$ en esta base $v_1(x,y,z)$, $v_2(x,y,z)$ y $v_3(x,y,z)$. Observemos cómo varía la masa del fluido en el interior del citado paralelepípedo en un cierto instante cualquiera t.

Para fijar ideas utilizaremos las unidades del Sistema Internacional. La cantidad de fluido por unidad de tiempo que atraviesa la cara del paralelepípedo que pasa por P y es normal al versor $\overline{i}$ será descrita por el siguiente producto de cuatro entes diádicos:

$$[\rho(x,y,z,t)\ kg/\!/m^3] * [v_1\ (x,y,z)\ m/\!/\text{s}] * [\Delta y\ m] * [\Delta z\ m] \quad [24.4]$$

De acuerdo con la definición de multiplicación del apartado XII, de sus propiedades y de la división del apartado XVI, la cantidad [24.4] se convierte en la díada:

$$[\rho(x,y,z,t) \times v_1\ (x,y,z) \times \Delta y \times \Delta z]\ kg/\!/s \quad\quad [24.5]$$

Análogamente, la cantidad de fluido que atraviesa por unidad de tiempo la cara del paralelepípedo paralela a la anterior y que pasa por el punto $P + \Delta P$ de coordenadas $(x + \Delta x, y + \Delta y, z + \Delta z)$ será dada por:

$$[\rho(x + \Delta x, y, z, t) \times v_1\ (x + \Delta x, y, z) \times \Delta y \times \Delta z]\ kg/\!/s \quad [24.6]$$

Hallando la diferencia entre [24.6] y [24.5], dividiéndola y multiplicandola por Δx, por lo que no variará, y por definición de derivada parcial, en el límite cuando Δx tiende a cero, se llega a la variación de la cantidad de fluido que atraviesa ambas caras paralelas del paralelepípedo en el punto $P(x,y,z)$, que estará representada por la expresión:

$$\left[\frac{\partial \left[\rho\left(x, y, z, t\right) \times v_1\left(x, y, z\right)\right]}{\partial x} \times \Delta x \times \Delta y \times \Delta z \right] \frac{kg}{s} \quad [24.7]$$

De manera análoga, la cantidad de fluido que atraviesa las dos caras paralelas entre sí y normales al versor $\overline{j}$ será dada por:

$$\left[\frac{\partial \left[\rho\left(x, y, z, t\right) \times v_2\left(x, y, z\right)\right]}{\partial y} \times \Delta x \times \Delta y \times \Delta z \right] \frac{kg}{s} \quad [24.8]$$

Y, finalmente, la cantidad de fluido que atraviesa las dos caras normales al versor $\overline{k}$ será:

$$\left[\frac{\partial \left[\rho\left(x, y, z, t\right) \times v_3\left(x, y, z\right)\right]}{\partial z} \times \Delta x \times \Delta y \times \Delta z \right] \frac{kg}{s} \quad [24.9]$$

La suma diádica de las cantidades uniformes [24.7], [24.8] y [24.9] indicará la cantidad de fluido que entra o sale del paralelepípedo elemental por unidad de tiempo en el punto $P(x,y,z)$. Dividiendo esa cantidad total entre el concreto $[(\Delta x \times \Delta y \times \Delta z)\ m^3]$, de acuerdo con la definición de división del apartado XVI, tendremos dicha cantidad por unidad de volumen geométrico en P indicada por el concreto siguiente:

$$\left[\frac{\partial \left[\rho\left(x, y, z, t\right) \times v_1\left(x, y, z\right)\right]}{\partial x} + \frac{\partial \left[\rho\left(x, y, z, t\right) \times v_2\left(x, y, z\right)\right]}{\partial y} + \frac{\partial \left[\rho\left(x, y, z, t\right) \times v_3\left(x, y, z\right)\right]}{\partial z} \right] \frac{kg}{s*m^3} \quad [24.10]$$

En la teoría matemática de campos se define la divergencia *div* de un campo vectorial $\overline{A} = A_1(x,y,z) \times \overline{i} + A_2(x,y,z) \times \overline{j} + A_3(x,y,z) \times \overline{k}$ con la expresión:

$$div\ \overline{A} = \frac{\partial A_1\left(x, y, z\right)}{\partial x} + \frac{\partial A_2\left(x, y, z\right)}{\partial y} + \frac{\partial A_3\left(x, y, z\right)}{\partial z}$$

Con esta notación, aplicada al campo vectorial producto $\rho(x,y,z,t) \times \overline{v}(x,y,z)$, el concreto [24.10], dando por puestas las coordenadas x, y, z y t, para simplificar, y recordando que el análisis se está haciendo para un cierto instante t cualquiera, se podrá escribir de forma sintética la variación de la cantidad de fluido en el paralelepípedo con el concreto siguiente:

$$[div\,(\rho \times \overline{v})]\ kg /\!/(s * m^3) \qquad [24.11]$$

Hasta aquí solo hemos tenido en cuenta la variación de masa por causa del movimiento instantáneo del fluido, pero siendo en general variable la densidad con el tiempo, hay que introducir también este efecto. Para ello, debemos observar que la variación de masa por cambio de la densidad en el paralelepípedo elemental del punto P entre los instantes t y $t+\Delta t$ tendrá la forma diádica:

$$[[\rho(x,y,z,t+\Delta t) - \rho(x,y,z,t)] \times (\Delta x \times \Delta y \times \Delta z)]\ kg$$

Dividiendo por la medición [$\Delta t\ s$] y tomando límite cuando Δt tiende a cero, llegamos la díada expresada con la derivada parcial respecto de t para la variación de masa en el paralelepípedo por unidad de tiempo:

$$\left[\frac{\partial\left[\rho(x,y,z,t)\right]}{\partial t} \times \Delta x \times \Delta y \times \Delta z \right] \frac{kg}{s}$$

Dividiendo esa cantidad entre el concreto [$(\Delta x \times \Delta y \times \Delta z)\ m^3$] con la operación del apartado XVI, llegamos a la variación de masa por unidad de tiempo y de volumen:

$$\left[\frac{\partial\left[\rho(x,y,z,t)\right]}{\partial t} \right] \frac{kg}{s * m^3} \qquad [24.12]$$

Sumando [24.11] y [24.12] llegamos a la variación de masa unitaria total, que escribimos abreviadamente prescindiendo de las coordenadas, que se sobreentienden:

$$\left[div\left(\rho \times \overline{v}\right) + \frac{\partial \rho}{\partial t} \right] \frac{kg}{s * m^3} \qquad [24.13]$$

Imaginando dos campos escalares que representen la aportación
y la pérdida de fluido en cada punto $P(x,y,z)$ por unidad de tiempo
y de volumen, que podemos representar mediante $\varphi(x,y,z)$ para
las fuentes y $\sigma(x,y,z)$ para los desagües o sumideros, agrupándolas
en una única función de campo escalar representada con la forma
$\psi(x,y,z) = \varphi(x,y,z) - \sigma(x,y,z)$, donde el signo menos debe tener el
significado de disminución de masa, igualando a [24.13], llegamos
a la ecuación física conocida como de **continuidad hidrodinámica**:

$$\left[div\left(\rho \times \bar{v} \right) \right] \frac{kg}{s*m^3} = \left[\psi - \frac{\partial \rho}{\partial t} \right] \frac{kg}{s*m^3} \qquad [24.14]$$

Observemos que, de acuerdo con el criterio de igualdad del
apartado IV, se ha supuesto que el primer y el segundo miembro
de [24.14] sean uniformes, aunque bastaría que fuesen
homogéneos, como analizaremos enseguida. En estas condiciones,
el teorema del desdoblamiento [24.2] con $k=1$ permitirá escribir la
igualdad algebraica sin los secundarios:

$$div\left(\rho \times \bar{v} \right) = \psi - \frac{\partial \rho}{\partial t}$$

En general, si las unidades de longitud, masa y tiempo fuesen
U_{L1}, U_{m1} y U_{t1} para el primer miembro, y U_{L2}, U_{m2} y U_{t2} para el
segundo, la ecuación [24.4] quedará formulada así:

$$\left[div\left(\rho \times \bar{v} \right) \right] \frac{U_{m1}}{U_{t1}*U_{L1}^3} = \left[\psi - \frac{\partial \rho}{\partial t} \right] \frac{U_{m2}}{U_{t2}*U_{L2}^3} \qquad [24.15]$$

Para simplificar, designando por U_1 y U_2 las unidades
compuestas del primer y del segundo miembro, respectivamente,
y suponiendo $U_2 = k \circ U_1$, el teorema del desdoblamiento [24.2] nos
lleva a la ecuación algebraica genérica:

$$div\left(\rho \times \bar{v} \right) = k \times \left[\psi - \frac{\partial \rho}{\partial t} \right] \qquad [24.16]$$

La fórmula [24.15] es la expresión concreta universal de la ecuación de continuidad hidrodinámica y la [24.16] es su forma de álgebra ordinaria desdoblada.

En el caso particular de que el fluido fuese incompresible, se tendría la densidad constante en el espacio y en el tiempo, la derivada parcial correspondiente se anularía y tendríamos:

$$\rho \times div \ \overline{v} = k \times \psi$$

Si además no hubiese fuentes ni desagües o sumideros, se tendría en todo punto $\psi=0$, con lo cual la ecuación de continuidad quedaría con la forma:

$$div \ \overline{v} = 0$$

Sobra recordar otra vez que, si las unidades del primer y del segundo miembro de la ecuación concreta fuesen uniformes, basta tomar $k=1$ en el desdoblamiento.

Veamos un segundo ejemplo de ecuación física que se refiere al fenómeno de la **conducción del calor**. Consideremos un campo vectorial diádico definido por una aplicación de R^3 en R^3, independiente del tiempo y simbolizada $\overline{Q}=Q(x,y,z)$, de modo que en cada punto $P(x,y,z)$ se presente una transferencia de calor dada por el vector $\overline{Q}$, que indica la **corriente calorífica** en unidades de calor por unidad de tiempo y unidad de área en dirección normal, que será función de las coordenadas (x,y,z) del punto. Supongamos que la temperatura T del fluido sea variable, quedará representada por un campo escalar diádico descrito por una función de R^4 en R denotada $T=T(x,y,z,t)$ y su unidad de medida. Admitamos que ambas funciones $\overline{Q}$ y T sean derivables.

Observemos la misma figura 6, en la que ahora tendremos que sustituir $\overline{v}$ por $\overline{Q}$, que son símbolos mudos o indiferentes a efectos del análisis. La figura representa en el espacio geométrico ordinario un paralelepípedo elemental a partir de un punto genérico, tomando los incrementos de las variables Δx, Δy e Δz. Fijando ideas, utilizaremos las unidades del Sistema Internacional para las diferentes magnitudes que intervienen en el fenómeno.

En estas condiciones, la variación de calor debida al campo $\overline{Q}$ en el paralelepípedo elemental y por unidad de volumen vendrá descrita por la misma ecuación [24.11], sin más que sustituir en ella el campo $\rho \times \overline{v}$ por el $\overline{Q}$, asociando a este las unidades que correspondan, situando en el numerador la caloría como unidad de calor, o cualquier otra, lo que nos lleva al concreto siguiente:

$$[div \ \overline{Q}] \ cal/\!/(s*m^3) \tag{24.17}$$

A continuación es preciso valorar la contribución al calor por la variación del campo térmico en un punto fijo y respecto del tiempo t. Así, siendo c con sus unidades el calor específico del cuerpo, la variación de calor en el paralelepípedo elemental debido al cambio de la temperatura con el tiempo vendrá determinado por el producto de los siguientes entes diádicos:

$$\left[\rho \ \frac{kg}{m^3} \right] * [\Delta x \ m] * [\Delta y \ m] * [\Delta z \ m] * \left[c \frac{cal}{kg*K} \right] \times \left[\frac{\partial T}{\partial t} \frac{K}{s} \right]$$

Dividiendo la anterior cantidad entre el concreto que define el volumen del paralelepípedo $[(\Delta x \times \Delta y \times \Delta z) \ m^3]$ y operando con el álgebra diádica, llegamos a la expresión de la variación de calor en el punto P debida a la variación térmica:

$$\left[\rho \times c \times \frac{\partial T}{\partial t} \right] \frac{cal}{s*m^3} \tag{24.18}$$

Sumando las cantidades uniformes [24.17] y [24.18], llegamos a la variación total de calor en el punto P por unidad de tiempo y de volumen:

$$\left[div \ \overline{Q} + \rho \times c \times \frac{\partial T}{\partial t} \right] \frac{cal}{s*m^3}$$

Suponiendo que no existan fuentes ni sumideros de calor, basta igualar a cero la cantidad anterior para tener la ecuación de conducción:

$$\left[div \ \overline{Q} + \rho \times c \times \frac{\partial T}{\partial t} \right] \frac{cal}{s*m^3} = 0 \tag{24.19}$$

La *ley de la conducción del calor* axiomatiza la experiencia de que la relación entre corriente calorífica y la variación de temperatura es dada por:

$$\left[\overline{Q} \, \frac{cal}{s*m^2} \right] = - \left[\lambda \, \frac{cal}{s*m*K} \right] * \left[grad \ T \, \frac{K}{m} \right] \quad [24.20]$$

Donde λ se denomina **coeficiente de conductividad térmica** y es propio de cada sustancia. Su secundario viene expresado por las unidades fundamentales que se indican en [24.20]. A su vez, el operador *grad* o «nabla» ∇ de un campo escalar T es el campo vectorial **gradiente**, que en la teoría del análisis matemático de campos se define en una base ortonormal $(\overline{i}, \overline{j}, \overline{k})$ por la ecuación:

$$grad \, T = \nabla T = \frac{\partial T}{\partial x} \times \overline{i} + \frac{\partial T}{\partial y} \times \overline{j} + \frac{\partial T}{\partial z} \times \overline{k}$$

Sustituyendo [24.20] en [24.19], tenemos la ecuación diádica siguiente:

$$\left[div \left(- \lambda \times grad \ T \right) + \rho \times c \times \frac{\partial T}{\partial t} \right] \frac{cal}{s*m^3} = 0$$

En la teoría matemática de campos se define el operador **laplaciano**, que se representa con la misma letra delta Δ de los incrementos para variables y significa la divergencia del campo vectorial gradiente. Con lo cual, la ecuación anterior se transforma simbólicamente en la conocida ecuación fundamental de la conducción de calor:

$$\left[- \lambda \times \Delta T + \rho \times c \times \frac{\partial T}{\partial t} \right] \frac{cal}{s*m^3} = 0$$

En lo que precede hemos desarrollado dos ecuaciones físicas significativas, que ilustran el procedimiento de deducción que ha de seguirse para analizar en función del álgebra de magnitudes los fenómenos físicos, mediante las operaciones definidas en esta monografía para los entes diádicos de la Física.

Los dos fenómenos analizados no tienen relación alguna entre sí, porque se refieren a realidades materiales muy diferentes; sin embargo, las componentes matemáticas de los primarios de las magnitudes descritas son muy parecidas y en algunos aspectos aparecen idénticas o al menos isomorfas. Por el contrario, la diferente naturaleza entre ambos casos queda patente en las unidades de los secundarios, que son distintas porque se refieren a magnitudes dispares. Ello evidencia que el aparato matemático de los primarios puede adquirir diferentes significados en función del contexto del hecho que se investigue, y de ahí la importancia de no perder de vista los secundarios y sus unidades, que determinan las magnitudes intervinientes en cada fenómeno, pues de otro modo resultaría imposible entender con plenitud el sentido físico de las ecuaciones deducidas.

Así se ha puesto de manifiesto la importancia de operar con díadas en vez de con los entes algebraicos ordinarios, pues de otro modo, aparte de la incorrección objetiva que ello supone, las ecuaciones pierden gran parte de su significación, deviniendo en meras simbologías muy elegantes de índole matemática, aunque físicamente abstrusas.

Por el contrario, las operaciones con díadas permiten establecer pasos lógicos sin la menor confusión, como se puede observar en la división diádica que da lugar a la cantidad [24.10], que deduce inequívocamente un cociente entre mediciones, ambos con sus dos elementos primario y secundario, quedando establecidos sin ambigüedad por medio de esa ley de composición. Con todo ello se pone de manifiesto el **procedimiento generador** conveniente para determinar sistemáticamente las formulaciones científicas, operando en primer lugar no solo con simples entes algebraicos vulgares, sino con cantidades de magnitudes, para después igualar las que sean homogéneas y que deban identificarse entre sí, en virtud de su propia naturaleza observada, para **crear una ecuación física**.

Observamos en ese proceso la decisiva intervención de las **leyes de composición externas generatrices**, sin las cuales la Física se vaciaría de contenido. Y así es como el álgebra de magnitudes ayuda decisivamente a la mejor comprensión y exacta formulación de los fenómenos naturales analizados, quedando al mismo tiempo establecidas las leyes de composición imprescindibles para que el lenguaje físico sea completo y unívoco, a salvo de interpretaciones subjetivas, intuitivas o arbitrarias.

Aclaremos finalmente que el **criterio de igualdad diádica** lo hemos establecido con la condición general de que las díadas iguales representen la misma cantidad de magnitud. Esta definición nos abre tres posibilidades: primera, que las díadas de ambos miembros sean homogéneas, en cuyo caso se hablará de **identidad**; segunda, que la díada de uno de los miembros sea generada por las del otro por medio de las operaciones generatrices, en cuyo supuesto la igualdad representará una **ecuación física**; y tercera, que una ecuación física haya sido establecida experimentalmente, lo que da lugar a una igualdad que llamaremos **ley física**.

Por ejemplo, toda igualdad de cantidades homogéneas como 150 $cm = 1,50$ m son identidades. Las igualdades que relacionan díadas mediante operaciones generatrices, tales como la definición de superficie 3 $m * 2$ $m = 6$ m^2 o la velocidad 6 $m /\!/ 2$ $s = 3$ $m /\!/ s$, son ecuaciones físicas. Y, finalmente, relaciones empíricas generatrices como la *segunda ley de Newton*, 2 $kg * 6$ $m /\!/ s^2 = 12$ N, son leyes físicas.

Por tanto, las definiciones de las operaciones aditivas son identidades (apartados V a XI); las definiciones de las operaciones multiplicativas generatrices son ecuaciones (apartados XII a XX); y formulaciones tales como la *segunda ley de Newton* son leyes físicas. No obstante, por comodidad en la práctica usamos como sinónimos los términos igualdad, identidad y ecuación, simplificación que carece de relevancia físico-matemática.

Section XXV

DEFINITION OF DIMENSIONS
OF THE PHYSICAL MAGNITUDES

Lord Kelvin used to say: «There is something extremely interesting in the fact that we can establish a metric system based on a unit of length and a unit of time. There is nothing new in it, since it has been known since Newton's time, but it retains all its interest and relevance».

Although Kelvin reduced the dimensional basis to two magnitudes, length and time, to which today mass has been incorporated, his reflection is nonetheless valid. A rational or coherent system of units must be such that it includes the minimum number of fundamental magnitudes from which all the others are derived. Hence, three systems of units have traditionally been formed: Cegesimal System, International System and Technical or Terrestrial System. The **Cegesimal System**, known by the acronym CGS, initials of the units centimeter, gram and second, which are adopted as primary units of the fundamental magnitudes of length, mass and time; It should be noted that the gram, symbolized g, is one thousandth of the standard kilogram. For its part, the **International System** or MKS, by the initials of meter, kilogram and second, which are the primary units adopted by this system, is recommended by the International Committee of Weights and Measures and was implemented in Spain by law in 1967 Finally, the **Terrestrial or Technical System** differs from the previous ones in that it does not use mass as a fundamental magnitude, but rather force, which is the action that interests technical applications, which is why it is often used in the field engineering; the other two fundamental magnitudes correspond to length and time, with their primary units the meter and the second, and the kilopond is established as the primary unit of force, defined as the force with which the Earth attracts a mass of one kilogram in a point of latitude equal to 45 degrees at sea level, where the acceleration of gravity is 980.665 centimeters per second squared, abbreviated 980.665 $cm//s^2$ «arithmetizing» 980.665 cm/s^2.

The above allows to approach the concept of dimension and dimensional equation of a universal law or a definition equation. To specify, let us assume as the International system of units, in which the fundamental magnitudes are length, mass and time, which will be denoted for dimensional purposes with the initials L, M and T. Suppose we want to find the form symbolic that represents the composition of the magnitudes of a surface without quantities or units, only specifying the magnitudes concerned; the dimensional shape of a surface will be symbolized [S], the brackets to refer to the dimensional equation and the S to indicate the proper magnitude of a surface; It is clear that every unit of surface, by the definition of dyadic multiplication, is the geometric product of two lengths, which can be written, making abstraction of the units and paying attention only to the magnitudes, with $[S] = L * L = L^2$, and this is, by definition, the form of the equation dimensional of the magnitude called surface, described by the square of two symbolic lengths. The square brackets indicate the meaning that the equation that follows does not address specific values of magnitudes, but is limited to composing magnitudes in the abstract of the quantities indicated in the second member. For the example of the area, this means that this magnitud derives from the length and is equivalent to a length multiplied geometrically by another length, a relationship that the respective units of area and length must respect. Remember that geometric multiplication has nothing to do with arithmetic, for the reasons widely explained in this work. Therefore, it should not be forgotten that this product refers to that of dyadic entities, although it commonly coincides in nomenclature with the multiplication of real numbers, with the current force of the false hypothesis of the International System of Units, which erroneously attributes the abelian multiplicative group structure to magnitudes, as we have already noted previously. With the volume V we also have the symbolic dimensional equation $[V] = L * L * L = L^3$, so that the dimension of the magnitude affine to the volume is the geometric cube of the length quantity, as before the dyadic cube, not that of R. Similarly, the density D, as an expression of the dyadic quotient between the unit of mass M and that of volume V of a body, will have as its dimensional form the symbolized geometric quotient $[D] = M//V = M * L^{-3}$, using negative exponents with the usual meaning of divisors or denominators of the same positive power, which means that the density magnitude will be derived from the mass and the length with the indicated dimensional expression with respect to dyadic algebra. In general, every dimensional equation will have the form of the monomial $[X] = L^{\alpha} * M^{\beta} * T^{\gamma}$, where α, β and γ are whole or fractional numbers, therefore, positive or negative, and the derived magnitude X is said to be α dimensions relative to L, β dimensions relative to M and γ dimensions relative to T.

On the other hand, it does not seem doubtful whether the **principle of physical homogeneity** should be admitted or the axiomatic assumption that **the physical formulas symbolize laws that must be admitted are independent of the units in which they are expressed,** which requires that in an equation that includes specific entities the dimensional shapes of the first and second members must coincide, because they refer to the same magnitude, although the respective units may differ, but without ceasing to be homogeneous with respect to the fundamental magnitudes that compose them; in this way the dimensional equations are revealed as a check of the homogeneity of the units that intervene in the members of a formula, so that a homogeneity defect will always reveal a calculation error in its deduction. And this is indicated by stating in synthesis that every physical equation must be dimensionally consistent, which means, in harmony with [4.1], that only two dyadic entities can be equated if they are expressed in homogeneous units, that is, in units of the same magnitude although, it is insisted, they may be different from each other, as would be, for example, the units km//h and m//s in relation to the derived magnitude called velocity.

In general, a universal law or definition equation can be imagined that is expressed in the following scalar form:

$$a\,U = (a_1\,U_1)^{\delta_1} * (a_2\,U_2)^{\delta_2} * \ldots * (a_n\,U_n)^{\delta_n} \quad [25.1]$$

Formula [25.1] is nothing more than an algebraic expression assembled with scalar dyadic entities, which in general can come from each of the components of a vector equation. If the units are changed, the same quantity $a\,U$ of the magnitude M corresponding to the unit U can be indicated with the dyad $b\,V$, where V is another unit of the same magnitude M, so by hypothesis U and V will be units homogeneous. And by the principle of physical homogeneity, the concrete $b\,V$ can be expressed with the same form of [25.1], that is, it will have:

$$b\,V = (b_1\,V_1)^{\delta_1} * (b_2\,V_2)^{\delta_2} * \ldots * (b_n\,V_n)^{\delta_n} \quad [25.2]$$

By hypothesis, the dyads $a\,U$ and $b\,V$ represent the same amount of M, so the second members of [25.1] and [25.2] can be equalized, and then operate with the dyadic algebra, and the real ones when appropriate, from in accordance with [16.2] and [17.1] or, in short, with the single rule that includes all the specific operations, described in section XXII; from which it results:

$$\left(\frac{a_1}{b_1}\right)^{\delta_1} \times \left(\frac{a_2}{b_2}\right)^{\delta_2} \times \ldots \times \left(\frac{a_n}{b_n}\right)^{\delta_n} = \left(\frac{V_1}{U_1}\right)^{\delta_1} * \left(\frac{V_2}{U_2}\right)^{\delta_2} * \ldots * \left(\frac{V_n}{U_n}\right)^{\delta_n} \quad [25.3]$$

In equation [25.3] the quotients between units $V_i\,//U_i$ are real numbers, by virtue of the axiom of continuity and the definition of division of homogeneous concretes, established in section XI; so formula [25.3] actually relates real numbers.

Let's analyze the meaning of the exponents δ_i. For this, with M being the derived magnitude associated with the dyad in the first member of [25.1], let M_i be the fundamental magnitudes that correspond to the factors a_i, U_i of the terms of the second member. The dimensional equation of M can be written symbolically as the relationship between quantities of the fundamental magnitudes, resulting in the form:

$$[M] = M_1^{\delta_1} * M_2^{\delta_2} * \ \ldots \ * M_n^{\delta_n} \qquad [25.4]$$

In conclusion, we can establish that the δ_i terms indicate the dimensions of the magnitude M in the base of fundamental magnitudes $\{M_i\}$, resulting in that the di themselves configure the equation of change of units [25.3], in which it is appreciated that the exponents of the ratios of the measures a_i / b_i are the same as the inverse ratios of the units $V_i /\!/ U_i$ when establishing a unit change from $\{U_i\}$ to $\{V_i\}$.

By dimensional base we have to understand any set of fundamental magnitudes $\{M_i\}$ such that any other derived magnitude can be composed by means of dyadic algebra with those of the base and that these are independent of each other, that is, that none of them can be composed into In no way through the others, because otherwise, according to the mathematical meaning that is attributed to the concept of the basis of a given structure, one could not speak of a basis properly.

These elementary mathematics considerations allow us to imagine two dimensional bases of magnitudes $\{M_i\}$ and $\{M'_j\}$, with i taking the values from 1 to m, and j from 1 to n. In general, there is no reason to require that m and n must be equal. We observe that any magnitude X can be composed with the elements of the two bases and we wonder what relationship will exist between the dimensions of the magnitude X in both systems of basic magnitudes. To solve this question, suppose that X in the base $\{M_i\}$ has the dimensions δ_i, taking i the values from 1 to m, with the dimensional expression:

$$[X] = M_1^{\delta_1} * M_2^{\delta_2} * \ \ldots \ * M_m^{\delta_m} \qquad [25.5]$$

In turn, the magnitude X at the base $\{M'_j\}$ will have dimensions δ'_j, with j from 1 to n, and its dimensional equation will be:

$$\left[X\right] = M'^{\delta'_1}_1 * M'^{\delta'_2}_2 * \; \ldots \; * M'^{\delta'_n}_n \qquad [25.6]$$

Since, by hypothesis, $\{M'_j\}$ is a base of the magnitude system established in the physical theory considered, the magnitudes of the first base $\{M_i\}$ can be expressed by means of their corresponding dimensional equations:

$$\left[M_1\right] = M'^{a_{11}}_1 * M'^{a_{12}}_2 * \; \ldots \; * M'^{a_{1n}}_n$$

$$\left[M_2\right] = M'^{a_{21}}_1 * M'^{a_{22}}_2 * \; \ldots \; * M'^{a_{2n}}_n$$

$$\;\cdots\cdots\cdots\cdots\cdots\cdots\cdots\cdots\cdots\cdots\cdots \qquad [25.7]$$

$$\left[M_m\right] = M'^{a_{m1}}_1 * M'^{a_{m2}}_2 * \; \ldots \; * M'^{a_{mn}}_n$$

The coefficients α_{ij} of [25.7], with i of 1 to m and j from 1 to n, indicate the dimensions of the magnitude M_i of the old base with respect to the magnitudes of the new base $\{M'_j\}$, that is, that α_{ij} is, by definition, the dimension of the magnitude M_i with respect to the magnitude M'_j. So, substituting the M_i of [25.7] in [25.5], dispensing with the brackets, given that their meaning is not transcendent for these purposes, we have:

$$\left[X\right] = \left(M'^{a_{11}}_1 * M'^{a_{12}}_2 * \; \ldots \; * M'^{a_{1n}}_n\right)^{\delta_1} *$$

$$* \left(M'^{a_{21}}_1 * M'^{a_{22}}_2 * \; \ldots \; * M'^{a_{2n}}_n\right)^{\delta_2} *$$

$$\;\cdots\cdots\cdots\cdots\cdots\cdots\cdots\cdots\cdots\cdots\cdots$$

$$* \left(M'^{a_{m1}}_1 * M'^{a_{m2}}_2 * \; \ldots \; * M'^{a_{mn}}_n\right)^{\delta_m}$$

Taking into account that the power of a power is another power whose exponent is the product of the exponents and identifying exponents with the expression [25.6] for each M'_j, the new dimensions δ_j of the magnitude X in the base result $\{M'_j\}$ as a function of the first δ_i with respect to the base $\{M_i\}$ and the dimensions α_{ij}, which are the dimensions of the magnitudes of the first base with respect to those of the second:

$$\delta'_1 = \delta_1 \times \alpha_{11} + \delta_2 \times \alpha_{21} + \ldots + \delta_m \times \alpha_{m1}$$
$$\delta'_2 = \delta_1 \times \alpha_{12} + \delta_2 \times \alpha_{22} + \ldots + \delta_m \times \alpha_{m2} \qquad [25.8]$$

$$\ldots\ldots\ldots\ldots\ldots\ldots\ldots\ldots\ldots\ldots\ldots\ldots\ldots\ldots\ldots\ldots\ldots\ldots$$

$$\delta'_n = \delta_1 \times \alpha_{1n} + \delta_2 \times \alpha_{2n} + \ldots + \delta_m \times \alpha_{mn}$$

The group of equations [25.8] are relations between real numbers, so the addition and multiplication that appear in them are the composition laws of the body R. They are similar and are deduced with an analogous illation to those that result for changes in base on vector spaces[23]. They can be written in abbreviated form with the form of summation:

$$\delta'_j = \sum_{i=1}^{m} \delta_i \times \alpha_{ij} \quad ; \quad j = 1,2,\ldots,n \qquad [25.9]$$

To symbolize these sums in algebra, it is usual to use the **concentrated or indexed notation**, which is nothing more than the convention of eliminating sums, considering that the repeated index in the factors of a monomial means or replaces the summation with respect to it, keeping the other subscripts constant. With this symbology that simplifies writing, the new dimensions with respect to the old ones will be expressed with the following synthetic formula:

$$\delta'_j = \delta_i \times \alpha_{ij}$$

$$i=1, 2, \ldots, m; j=1, 2, \ldots, n$$

It is also possible to symbolize the base change equations using **matrix notation**, resulting in the following expression:

[23] The change of base in V^n, generalizable to dimension n, is exposed analytically in «Lesson 4» of the volume Mathematize 2.

$$\begin{bmatrix} \delta'_1 & \delta'_2 & \cdots & \delta'_n \end{bmatrix} = \begin{bmatrix} \delta_1 & \delta_2 & \cdots & \delta_m \end{bmatrix} \begin{bmatrix} \alpha_{11} & \alpha_{12} & \cdots & \alpha_{1n} \\ \alpha_{21} & \alpha_{22} & \cdots & \alpha_{2n} \\ \cdots & \cdots & \cdots & \cdots \\ \alpha_{m1} & \alpha_{m2} & \cdots & \alpha_{mn} \end{bmatrix}$$

This section serves as proof of the logical connection with the usual dimensional analysis, once the gap in the algebra of magnitudes has been bridged, giving meaning and foundation to equation [24.1], which usually marks the beginning of Physics texts as an abstract principle and intuitive, and that here we have inferred based on the structures of dyadic entities with their composition laws, and thus we verify that the dimensional aspects of magnitudes and physical equations are described by the dyadic algebra model, based on the geometric algebra.

At the same time, we verify with unquestionable clarity that any dimensional equation with the form $[X] = L^{\alpha} \times M^{\beta} \times T^{\gamma}$ is nothing more than mere symbology empty of true physical-mathematical meaning, as long as the previous and precise definition of the **generating external laws of composition,** as established in sections XII et seq. of this dyadic algebra.

Apartado XXV

DEFINICIÓN DE DIMENSIONES
DE LAS MAGNITUDES FÍSICAS

Decía lord Kelvin: «Hay algo sumamente interesante en el hecho de que podamos establecer un sistema métrico basado en una unidad de longitud y en una unidad de tiempo. No hay en ello nada nuevo, pues es ya conocido desde los tiempos de Newton, pero conserva todo su interés y actualidad».

Aunque Kelvin reducía la base dimensional a dos magnitudes, la longitud y el tiempo, a las que hoy en día se les ha incorporado la masa, su reflexión es, no obstante, válida. Un sistema racional o coherente de unidades debe ser tal que incluya el mínimo número de magnitudes fundamentales de las que se deriven todas las demás. De ahí que tradicionalmente se hayan conformado tres sistemas de unidades: Sistema Cegesimal, Sistema Internacional y Sistema Técnico o Terrestre. El **Sistema Cegesimal**, conocido por las siglas CGS, iniciales de las unidades centímetro, gramo y segundo, que son las adoptadas como unidades primarias de las magnitudes fundamentales de longitud, masa y tiempo; hay que observar que el gramo, simbolizado g, es la milésima parte del kilogramo patrón. Por su parte, el **Sistema Internacional** o MKS, por las iniciales de metro, kilogramo y segundo, que son las unidades primarias adoptadas por este sistema, es el recomendado por el Comité Internacional de Pesas y Medidas y fue implantado en España por ley en 1967. Finalmente, el **Sistema Terrestre o Técnico** difiere de los anteriores en que no se sirve de la masa como magnitud fundamental, sino de la fuerza, que es la acción que interesa a las aplicaciones técnicas, por lo que suele utilizarse con frecuencia en el ámbito de la ingeniería; las otras dos magnitudes fundamentales corresponden a la longitud y el tiempo, con sus unidades primarias el metro y el segundo, y como unidad primaria de fuerza se establece el **kilopondio**, definido como la fuerza con que la Tierra atrae a una masa de un kilogramo en un punto de latitud igual a 45 grados a nivel del mar, donde la aceleración de la gravedad es de 980,665 centímetros por segundo al cuadrado, abreviadamente 980,665 $cm /\!/ s^2$ o «aritmetizando» 980,665 cm/s^2.

Lo expuesto permite abordar el concepto de **dimensión** y de **ecuación dimensional** de una ley universal o de una ecuación de definición. Para concretar, asumamos como sistema de unidades el Internacional, en el que las magnitudes fundamentales son la longitud, la masa y el tiempo, que abreviadamente se denotarán a efectos dimensionales con las iniciales L, M y T. Supóngase que se quiere hallar la forma simbólica que represente la composición de las magnitudes de una superficie sin cantidades ni unidades, solo especificando las magnitudes interesadas; la forma dimensional de una superficie se simbolizará $[S]$, los corchetes para aludir a ecuación dimensional y la S para indicar la magnitud propia de una superficie; es claro que toda unidad de superficie, por la definición de multiplicación concreta, es el producto geométrico de dos longitudes, lo que puede escribirse, haciendo abstracción de las unidades y atendiendo solo a las magnitudes, con $[S]=L*L=L^2$, y esta es, por definición, la forma de la ecuación dimensional de la magnitud denominada superficie, descrita por el cuadrado de dos longitudes simbólicas. Con los corchetes se indica el significado de que la ecuación que sigue no atiende a valores concretos de cantidades, sino que se limita a componer cantidades en abstracto de las magnitudes indicadas en el segundo miembro. Para el ejemplo de la superficie, ello significa que esta magnitud deriva de la longitud y equivale a una longitud multiplicada geométricamente por otra longitud, relación que deberán respetar las respectivas unidades de superficie y longitud. Recuérdese que la multiplicación geométrica nada tiene que ver con la aritmética, por las razones expuestas ampliamente en este trabajo. Por tanto, no debe olvidarse que este producto se refiere al de los entes diádicos, aunque vulgarmente coincida en nomenclatura con la multiplicación de los números reales, con la fuerza vigente de la hipótesis falsa del Sistema Internacional de

Unidades, que atribuye erróneamente a las magnitudes la estructura de grupo multiplicativo abeliano, como ya hemos advertido con anterioridad. Con el volumen V se tiene igualmente la ecuación dimensional simbólica $[V] = L * L * L = L^3$, de modo que la dimensión de la magnitud atinente al volumen es el cubo geométrico de la magnitud longitud, igual que antes el cubo diádico, no el de R. De manera análoga, la densidad D, como expresión del cociente diádico entre la unidad de masa M y la de volumen V de un cuerpo, tendrá como forma dimensional el cociente geométrico simbolizado $[D] = M /\!/ V = M * L^{-3}$, utilizando los exponentes negativos con el significado usual de divisores o denominadores de la misma potencia positiva, lo que significa que la magnitud densidad será derivada de la masa y de la longitud con la expresión dimensional indicada respecto del álgebra diádica. En general, toda ecuación dimensional tendrá la forma del monomio $[X] = L^{\alpha} * M^{\beta} * T^{\gamma}$, donde α, β y γ sean números enteros o fraccionarios, por tanto, positivos o negativos, y se dice que la magnitud derivada X es de α dimensiones con relación a L, de β dimensiones respecto de M y de γ dimensiones con respecto a T.

Por otra parte, no parece dudoso que deba admitirse el **principio de homogeneidad física** o supuesto axiomático de que **las fórmulas físicas simbolicen leyes que debe admitirse sean independientes de las unidades en que se expresen**, lo que exige que en una ecuación que incluya entes concretos las formas dimensionales del primer miembro y del segundo han de coincidir, porque se refieren a la misma magnitud, aunque las respectivas unidades puedan diferir, pero sin dejar de ser homogéneas respecto de las magnitudes fundamentales que las compongan; de este modo las ecuaciones dimensionales se revelan como una comprobación de la homogeneidad de las unidades que intervengan en los miembros de una fórmula, por lo que un defecto de homogeneidad revelará siempre un error de cálculo en su deducción. Y esto se indica enunciando en síntesis que **toda ecuación física debe ser dimensionalmente consistente**, lo que significa, en armonía con [4.1], que **solo se podrán igualar dos entes concretos si viniesen**

expresados en unidades homogéneas, es decir, en unidades de la misma magnitud aunque, se insiste, puedan ser diferentes entre sí, como lo serían, por ejemplo, las unidades *km⫽h* y *m⫽s* en relación con la magnitud derivada llamada velocidad.

En general, puede imaginarse una ley universal o ecuación de definición que venga expresada con la forma escalar siguiente:

$$a\,U = \left(a_1\,U_1\right)^{\delta_1} * \left(a_2\,U_2\right)^{\delta_2} * \ \ldots\ * \left(a_n\,U_n\right)^{\delta_n} \qquad [25.1]$$

La fórmula [25.1] no es otra cosa que una expresión algebraica montada con entes diádicos escalares, que en general podrá provenir de cada una de las componentes de una ecuación vectorial. Si se cambian las unidades, la misma cantidad *a U* de la magnitud *M* correspondiente a la unidad *U* se podrá indicar con el concreto *b V*, donde *V* es otra unidad de la misma magnitud *M*, por lo que por hipótesis *U* y *V* serán unidades homogéneas. Y por el principio de homogeneidad física, el concreto *b V* se podrá expresar con la misma forma de [25.1], es decir, que se tendrá:

$$b\,V = \left(b_1\,V_1\right)^{\delta_1} * \left(b_2\,V_2\right)^{\delta_2} * \ \ldots\ * \left(b_n\,V_n\right)^{\delta_n} \qquad [25.2]$$

Por hipótesis, las díadas *a U* y *b V* representan la misma cantidad de *M*, por lo que se pueden igualar los segundos miembros de [25.1] y [25.2], para luego operar con el álgebra diádica, y de los reales cuando corresponda, de acuerdo con [16.2] y [17.1] o, en suma, con la regla única que acoge todas las operaciones concretas, descrita en el apartado XXII; de lo que resulta:

$$\left(\frac{a_1}{b_1}\right)^{\delta_1} \times \left(\frac{a_2}{b_2}\right)^{\delta_2} \times \ \ldots\ \times \left(\frac{a_n}{b_n}\right)^{\delta_n} = \left(\frac{V_1}{U_1}\right)^{\delta_1} * \left(\frac{V_2}{U_2}\right)^{\delta_2} * \ \ldots\ * \left(\frac{V_n}{U_n}\right)^{\delta_n} \qquad [25.3]$$

En la ecuación [25.3] los cocientes entre unidades $V_i\,⫽U_i$ son números reales, en virtud del axioma de continuidad y de la definición de división de concretos homogéneos, establecida en el apartado XI; por lo que la fórmula [25.3], en realidad, relaciona números reales.

Analicemos el significado de los exponentes δ_i. Para ello, siendo M la magnitud derivada asociada a la díada en el primer miembro de [25.1], sean M_i las magnitudes fundamentales que correspondan a los factores $a_i \, U_i$ de los términos del segundo miembro. La ecuación dimensional de M se podrá escribir simbólicamente como la relación entre cantidades de las magnitudes fundamentales, resultando la forma:

$$\left[M \right] = M_1^{\delta_1} * M_2^{\delta_2} * \ \ldots \ * M_n^{\delta_n} \qquad [25.4]$$

En conclusión, podemos establecer que los términos δ_i indiquen las dimensiones de la magnitud M en la base de magnitudes fundamentales $\{M_i\}$, resultando que los mismos δ_i configuran la ecuación de cambio de unidades [25.3], en la que se aprecia que los exponentes de las razones de las medidas a_i / b_i son los mismos que las razones inversas de las unidades $V_i /\!/ U_i$ cuando se establece un cambio de unidades de $\{U_i\}$ a $\{V_i\}$.

Por base dimensional hemos de entender todo conjunto de magnitudes fundamentales $\{M_i\}$ tal que cualquier otra magnitud derivada se pueda componer mediante el álgebra diádica con las de la base y que estas sean independientes entre sí, es decir, que ninguna de ellas pueda componerse en modo alguno mediante las demás, porque en otro caso, de acuerdo con el significado matemático que se atribuye al concepto de base de una estructura determinada, no se podría hablar de base con propiedad.

Estas consideraciones de Matemática elemental nos permiten imaginar dos bases dimensionales de magnitudes $\{M_i\}$ y $\{M'_j\}$, con i tomando los valores de 1 a m, y j de 1 a n. En general, no hay razón para exigir que m y n deban ser iguales. Observamos que cualquier magnitud X podrá componerse con los elementos de las dos bases y nos preguntamos qué relación existirá entre las dimensiones de la magnitud X en ambos sistemas de magnitudes básicas. Para resolver esta cuestión, supongamos que X en la base $\{M_i\}$ tenga las dimensiones δ_i, tomando i los valores de 1 a m, con la expresión dimensional:

$$\left[X \right] = M_1^{\delta_1} * M_2^{\delta_2} * \ \ldots \ * M_m^{\delta_m} \qquad [25.5]$$

A su vez, la magnitud X en la base $\{M'_j\}$ presentará unas dimensiones δ'_j, con j de 1 a n, y su ecuación dimensional será:

$$\left[X\right] = M'^{\delta'_1}_1 * M'^{\delta'_2}_2 * \ldots * M'^{\delta'_n}_n \qquad [25.6]$$

Puesto que, por hipótesis, $\{M'_j\}$ es una base del sistema de magnitudes establecido en la teoría física considerada, las magnitudes de la primera base $\{M_i\}$ se podrán expresar mediante sus correspondientes ecuaciones dimensionales:

$$\left[M_1\right] = M'^{a_{11}}_1 * M'^{a_{12}}_2 * \ldots * M'^{a_{1n}}_n$$

$$\left[M_2\right] = M'^{a_{21}}_1 * M'^{a_{22}}_2 * \ldots * M'^{a_{2n}}_n$$

$$\ldots\ldots\ldots\ldots\ldots\ldots\ldots\ldots\ldots\ldots\ldots\ldots\ldots\ldots$$

$$\left[M_m\right] = M'^{a_{m1}}_1 * M'^{a_{m2}}_2 * \ldots * M'^{a_{mn}}_n$$

$$[25.7]$$

Los coeficientes α_{ij} de [25.7], con i de 1 a m y j desde 1 hasta n, indican las dimensiones de la magnitud M_i de la base antigua respecto de las magnitudes de la base nueva $\{M'_j\}$, es decir, que α_{ij} es, por definición, la dimensión de la magnitud M_i respecto de la magnitud M'_j. Así que, sustituyendo las M_i de [25.7] en [25.5], prescindiendo de los corchetes, dado que su significado no es trascendente a estos efectos, tenemos:

$$\left[X\right] = \left(M'^{a_{11}}_1 * M'^{a_{12}}_2 * \ldots * M'^{a_{1n}}_n\right)^{\delta_1} *$$

$$* \left(M'^{a_{21}}_1 * M'^{a_{22}}_2 * \ldots * M'^{a_{2n}}_n\right)^{\delta_2} *$$

$$\ldots\ldots\ldots\ldots\ldots\ldots\ldots\ldots\ldots\ldots\ldots\ldots\ldots\ldots$$

$$* \left(M'^{a_{m1}}_1 * M'^{a_{m2}}_2 * \ldots * M'^{a_{mn}}_n\right)^{\delta_m}$$

Teniendo en cuenta que la potencia de una potencia es otra potencia que tiene por exponente el producto de los exponentes e identificando exponentes con la expresión [25.6] para cada M'_j, resultan las nuevas dimensiones δ'_j de la magnitud X en la base

$\{M'_j\}$ en función de las primeras δ_i respecto de la base $\{M_i\}$ y de las dimensiones α_{ij}, que son las dimensiones de las magnitudes de la primera base respecto de las de la segunda:

$$\delta'_1 = \delta_1 \times \alpha_{11} + \delta_2 \times \alpha_{21} + \ldots + \delta_m \times \alpha_{m1}$$
$$\delta'_2 = \delta_1 \times \alpha_{12} + \delta_2 \times \alpha_{22} + \ldots + \delta_m \times \alpha_{m2}$$

$$\delta'_n = \delta_1 \times \alpha_{1n} + \delta_2 \times \alpha_{2n} + \ldots + \delta_m \times \alpha_{mn}$$

[25.8]

El grupo de ecuaciones [25.8] son relaciones entre números reales, por lo que la adición y la multiplicación que aparecen en ellas son las leyes de composición del cuerpo R. Son semejantes y se deducen con análoga ilación a las que resultan para los cambios de base en los espacios vectoriales[24]. Pueden escribirse en forma abreviada con la forma de sumatorio:

$$\delta'_j = \sum_{i=1}^{m} \delta_i \times \alpha_{ij} \quad ; \quad j = 1, 2, \ldots, n \qquad [25.9]$$

Para simbolizar estas sumas en álgebra es usual utilizar la **notación concentrada** o **indexada**, que no es sino el convenio de eliminar sumatorios, considerando que el índice repetido en los factores de un monomio signifique o sustituya el sumatorio respecto de él, manteniendo constantes los demás subíndices. Con esta simbología que simplifica la escritura las nuevas dimensiones respecto de las antiguas quedarán expresadas con la siguiente fórmula sintética:

$$\delta'_j = \delta_i \times \alpha_{ij}$$
$$i = 1, 2, \ldots, m; \, j = 1, 2, \ldots, n$$

---

[24] El cambio de base en V^3, generalizable a dimensión n, se expone analíticamente en la «Lección 4» del volumen *Matematizar 2*.

Cabe también simbolizar las ecuaciones de cambio de base mediante la **notación matricial**, resultando la expresión siguiente:

$$\begin{bmatrix} \delta'_1 & \delta'_2 & \cdots & \delta'_n \end{bmatrix} = \begin{bmatrix} \delta_1 & \delta_2 & \cdots & \delta_m \end{bmatrix} \begin{bmatrix} \alpha_{11} & \alpha_{12} & \cdots & \alpha_{1n} \\ \alpha_{21} & \alpha_{22} & \cdots & \alpha_{2n} \\ \cdots & \cdots & \cdots & \cdots \\ \alpha_{m1} & \alpha_{m2} & \cdots & \alpha_{mn} \end{bmatrix}$$

Sirva este apartado como prueba de la conexión lógica con el análisis dimensional usual, una vez salvada la laguna del álgebra de magnitudes, dotando de sentido y fundamento a la ecuación [24.1], que suele marcar el comienzo de los textos de Física como principio abstracto e intuitivo, y que aquí hemos inferido en base a las estructuras de los entes diádicos con sus leyes de composición, y así comprobamos que los aspectos dimensionales de las magnitudes y de las ecuaciones físicas quedan descritos por el modelo del álgebra diádica, fundada en el álgebra geométrica.

A su vez, comprobamos con claridad inapelable que toda ecuación dimensional con la forma $[X] = L^{\alpha} \times M^{\beta} \times T^{\gamma}$ no es más que mera simbología vacía de verdadero significado físico-matemático, mientras no se procede a la definición previa y precisa de las **leyes de composición externas generatrices**, tal como se establecen en los apartados XII y siguientes de esta álgebra diádica.

Section XXVI

THE PHYSICAL CONSTANTS

In section XXIV we have analyzed the meaning to be given to physical equations, which must be understood as invariant identities between two dyadic entities, which in the famous case of **Newton's second law,** refers to vector measurements, explicitly written $\overline{F}\, U_F = m\, U_m \odot \overline{a}\, U_a$ (product of section XX) or with symbolic economy $\overline{F}\, U_F = m\, U_m \times \overline{a}\, U_a$ where U_F is the unit of force, U_m the unit of mass and U_a the unit of acceleration. Therefore, for this law, the dyadic meaning must be admitted that the scalar measurement $m\, U_m$ multiplied by the vector measurement $\overline{a}\, U_a$ must be equal to the vector dyad $\overline{F}\, U_F$, where m is a positive scalar of R, called inertial mass, which presents a value specific for each material body.

By virtue of [20.1], which defines the product of a scalar dyad by another vector, and then with [9.3] and [9.4], we can transform Newton's second law in this explicit way:

$$\overline{F}\, U_F = m\, U_m \odot \overline{a}\, U_a = (m \bullet \overline{a})\,(U_m * U_a)] =$$

$$= m \circ [\overline{a}\,(U_m * U_a)] \qquad [26.1]$$

As by the nature of this law m is a scalar, we can apply [11.2] to form the conscious of collinear and homogeneous vector concretes, with which we will have:

$$\frac{\overline{F}\, U_F}{\overline{a}\,(U_m * U_a)} = m \qquad [26.2]$$

So Newton's second law is equivalent to understanding that the primary of the mass, which is a positive real number, is equal to the dyadic quotient between $\overline{F}\, U_F$ y $\overline{a}\,(U_m * U_a)$, and that this number is invariable for each body or point material. So nothing prevents us from considering that m is a **characteristic constant**, without prejudice to the magnitude condition that is attributed to the inertial mass[25]. The scalar dyad that defines the mass of the body will have the form m Um and the unit of mass will not be independent of the units of force and acceleration, because the elements $\overline{F}$ U_F y $\overline{a}$ $(U_m * U_a)$ must be homogeneous, due to the definition of equality [4.2] between vector dyads. Furthermore, nothing prevents setting the units so that the first and second members of the physical equations are not only homogeneous, but uniform, and this is usual for convenience. However, for greater generality we will develop the assumption that these dyads are homogeneous and non-uniform, therefore, the continuity axiom [4.3] will determine that there exists $k \in R$ such that $U_m * U_a = k \circ U_F$. Under these conditions, the doubling of the equality of dyads that integrates the physical equation, described in [24.3], applied to equation [26.1], which can be written:

$$\overline{F} \, U_F = m \circ [\overline{a} \, (U_m * U_a)] = (m \bullet \overline{a}) \, (U_m * U_a)$$

Produces the primary and secondary relationships, appearing a vector equation and an equality between units:

$$\overline{F} = k \times m \bullet \overline{a}$$

$$k \circ U_F = U_m * U_a \Rightarrow U_m = (k \circ U_F) // U_a \qquad [26.3]$$

Remember that the quotient $(k \circ U_F) // U_a$ is not the ordinary one, but the one defined in section XVI. In conclusion, the measurement of the characteristic or specific constant of each

---

[25] It can be argued that the mass of inertia is considered or not a characteristic constant, since we have it for a fundamental physical magnitude; however, since we admit that it is a constant quantity for each material point, it does not seem unreasonable to consider it that way. In any case, it is only a mere convention. On the other hand, regardless of what is accepted in this regard, the interest of studying Newton's second law, one of the essential laws of Physics, is undeniable, so we have decided to include it in this section, if only to observe the ilative consequences of the algebra of magnitudes in the analysis of such a primordial law.

body called mass will be given by a scalar dyad whose primary or measure will be determined by the positive real number given by [26.2] and whose secondary or unit will have the form [26.3], with which it will be that the mass will be expressed by the following homogeneous dyads:

$$m\,U_m = m\ \mathrm{o}\,\frac{k\ \mathrm{o}\,U_F}{U_a}$$

Note that the relationship [26.3] between the units of force U_F, mass U_m, and acceleration U_a means that they cannot be independent. On the other hand, if the units were defined so that that of the first member was uniform with that of the second of the physical equation [26.1], we would simply have the particular and more usual case $k=1$.

Another famous example of constant characteristic is found in the well-known **Hooke's law**. Formulated in common language, this law establishes that stress, defined as force per unit area, is for each body proportional to deformation, understood as variation in length per unit length. Although it is vector in nature, it can be applied in scalar terms, as we justified in section XXIV, and we will do so to vary the reasoning with respect to the previous example. To do this, consider a wire of length $L\,U_L$ and section v, subjected to a force of component $F\,U_F$ in the direction of L, so it will experience a length variation $E\,U_E$, also in the direction of L, and where $E\,U_E$ the proportionality factor, called **Young's modulus**, and thus Hooke's law will have the form of the following dyadic algebra physical equation:

$$\frac{F\,U_F}{S\,U_S} = E\,U_E * \frac{\Delta L\,U_L}{L\,U_L}$$

The dyads $\Delta L\ U_L$ and $L\ U_L$ are homogeneous and uniform by hypothesis, because both represent lengths in the same unit, so that, according to section XI, their dyadic quotient must be a real number, so we can dispense with the U_L unit in numerator and denominator, and Hooke's law can be written in the form:

$$\frac{F U_F}{S U_S} = E U_E \circ \frac{\Delta L}{L}$$

Let $\sigma = F/S$ be the measure of stress or force per unit area in the associated compound unit $U_F /\!/ U_S$; and let $\varepsilon = \Delta L/L$ be the measure of the deformation or variation of length per unit of length, which lacks its own unit. With this notation and by the definition of multiplication by a scalar [9.1] and [9.2], we will have dyadic equality:

$$\sigma \frac{U_F}{U_S} = (E \times \varepsilon)\, U_E \qquad\qquad [26.4]$$

In general, given the definition of equality [4.1] between scalar dyads, in which the identified elements do not have to be uniform, and because of the continuity axiom [4.3], there will be $k \in$ R such that:

$$U_E = k \circ \frac{U_F}{U_S}$$

Under these conditions, the doubling theorem [24.2] for the equality of dyads that make up the physical equation [26.4] can be written, relating primary and secondary:

$$\sigma = k \times E \times \varepsilon$$

$$k \circ \frac{U_F}{U_S} = U_E \qquad\qquad [26.5]$$

On the other hand, according to the definition of division between homogeneous scalar dyads [11.1], or indistinctly with [16.2], and defined the product by a scalar through [9.1] and [9.2], Hooke's law [26.4] will with the form:

$$\frac{\sigma \dfrac{U_F}{U_S}}{\varepsilon\, U_E} = E$$

This means that the ratio between the scalar dyads $\sigma(U_F/\!/U_S)$ and $\mathcal{E}U_E$ will be an invariable real number E for each type of wire. Young's modulus E thus becomes a characteristic constant that specifically relates for each material object the invariant relationship between the applied stress and the deformation produced. With the generic hypothesis that the physical equations are homogeneous dyadic equalities, in this case with the scalar form [4.1], the Young's modulus will be represented by the scalar dyad:

$$E\ U_E = \left(k \times E\right)\ \frac{U_F}{U_S}$$

Also here we have that the relationship [26.5] between the units U_E, U_F and U_S determines the dependency between them, so they cannot all be arbitrarily established. If the units were defined to be uniform, we would have $k=1$.

Apart from the characteristic or specific constants, such as the mass of inertia and Young's modulus, we find others that are independent of the nature of the bodies, remaining invariable in any case, and we will call them **universal constants**. A prominent example is the **mechanical equivalent of heat**. Let $T\ U_T$ be the mechanical work measured with the unit U_T, let us symbolize $Q\,U_Q$ the amount of heat equivalent to the previous work in the unit U_Q and write $M\,U_M$ for the constant of proportionality between these other two quantities with the unit U_M. The universal law that determines the equivalence between heat and work can be written in dyadic algebra with the form of the expression:

$$T\,U_T = (M\,U_M) * (Q\,U_Q)$$

By its nature, it is a scalar physical equation, so we will have to compose the dyads that intervene with the scalar operations that we have defined. The definition of multiplication [12.1] leads us to the expression:

$$T\,U_T = (M \times Q)\,(U_M * U_Q) \qquad\qquad [26.6]$$

In general, the dyads of the first and second member should be homogeneous, without the need for them to be uniform, given the definition of equality [4.1] for specific scalars. The axiom of continuity [4.3] determines that there will be $k \in R$ such that $U_M * U_Q = k \circ U_T$. The doubling theorem [24.2], applied to the scalar physical equation [26.6], allows us to determine the two relations that must be satisfied between its two primary and two secondary ones:

$$T = k \times M \times Q$$

$$k \circ U_T = U_M * U_Q \qquad [26.7]$$

On the other hand, according to the definition of division between homogeneous scalar dyads [11.1], or indistinctly with [16.2], and with the product by a scalar of [9.1] and [9.2], the universal law [26.6] will remain with the shape:

$$\frac{T\,U_T}{Q\left(U_M * U_Q\right)} = M$$

This means that the ratio between the scalar dyads $T\,U_T$ and Q $Q\,(U_M \times U_Q)$ will be a real number M invariable and independent of the nature of the bodies, so it is said to be a universal constant called the mechanical equivalent of heat. With the generic hypothesis that the physical equations equal homogeneous concretes, this constant will be represented by the scalar dyad:

$$M\,U_M = \left(k \times M\right)\,\frac{U_T}{U_Q}$$

Also here we observe that the relationship [26.7] between the units U_T, U_M and U_Q determines the dependency between them, so the three cannot be arbitrarily established. If the units were defined so that they were uniform, which would be the most comfortable, we would have $k = 1$.

Another example of the same type is found in the **constant of universal gravitation**. Although it is considered only one, the law of gravitation is actually two: first, which describes that two material points attract each other with a force, with the direction of the line that joins them, directly proportional to the product of their masses gravitational and inversely proportional to the square of the distance that separates them; and second, that for each material point the ratio between the gravitational mass and the mass of inertia of Newton's second law is constant, and that both are manifestations of the same magnitude called mass.

Although strictly speaking the law of gravitation has a vector formulation, as it can be reduced to another scalar with a single component in the direction of the line that joins the material points, we will follow this path. To distinguish the two masses of each material point, the inertial mass will be symbolized by m and the gravitational μ. Let U_F be the unit of force, U_m the unit of mass, both gravitational and inertial, and U_L the unit of length to measure the distance d between the material points. Under these conditions, the universal law of gravitation can be written in concrete algebra using an expression like the following:

$$F U_F = \frac{(\mu_1 U_m) * (\mu_2 U_m)}{(d U_L)^2}$$

The subscripts 1 and 2 are used to distinguish each of the two material points of gravitational masses μ_1 and μ_2, as well as the inertial masses m_1 and m_2. In turn, being H the ratio of proportionality between the gravitational and inertial masses, they will be written for both material points:

$$\mu_1 U_m = H \circ (m_1 U_m)$$

$$\mu_2 U_m = H \circ (m_2 U_m)$$

Since H is a quotient between homogeneous dyadic numbers, by reason of [11.1], it must be a simple real number, so H lacks unity and is such that $H \in R$. Substituting these two equations in the previous one and, operating with [9.1] and [9.2], the formulation of the law of gravitation with the inertial masses results:

$$F\,U_F = (G\,U_G)* \frac{(m_1\,U_m)*(m_2\,U_m)}{(d\,U_L)^2}$$

Being $G = H^2$ the factor known as the **constant of universal gravitation**, and U_G the corresponding unit. To deduce it, applying the definition [12.1] of multiplication of scalar dyads, we easily arrive at the dyadic algebra equation:

$$F\,U_F = \left(G \times \frac{m_1 \times m_2}{d^2}\right)\left(\frac{U_G * U_m^2}{U_L^2}\right)$$

As in the previous examples, where k is the scalar relationship between the units of both members, which would be the unit $k = 1$ if there was uniformity, in the generic case that the concretes of the equation are not uniform, the doubling of [24.2] brings us the two relationships between primary and secondary:

$$F = k \times G \times \frac{m_1 \times m_2}{d^2}$$

$$k \circ U_F = \frac{U_G * U_m^2}{U_L^2}$$

From which it turns out that the concrete that represents the constant of universal gravitation is:

$$G\,U_G = (k \times G) \circ \frac{U_F * U_L^2}{U_m^2} \qquad [26.8]$$

It might be thought that it would be contradictory that, being H a constant without units, it turns out that $G = H^2$ does have dimensions. However, there is nothing disconcerting in this, since what happens is that while H is indicated only by a real number, given the definition with $\mu U_m = H \circ (m\, U_m)$, which for each material point relates its gravitational mass with that of inertia; instead G is determined by $G = H^2$ and also by [26.8], which means that its nature must be represented by a specific scalar entity, not only by a real number, so that the equality $G = H^2$ only applies refers to its primary element and [26.8] determines its secondary.

In the International System of Units, the value of G has been calculated and established based on the meter m as a unit of length, not to be confused with the same symbol used for mass, the kilogram kg as the unit of mass and the newton N as a unit of force:

$$G = 6{,}67 \times 10^{-11} \circ \frac{N * m^2}{kg^2} = 6{,}67 \times 10^{-11} \circ \frac{m^3}{s^2 * kg}$$

To end this section, we must note that in all the constant analyzes carried out, the constant k has appeared, related to the homogeneity and the uniformity axiom of the dyadic entities identified by the physical equations considered. Hence, this kind of invariants must be considered, which we will call **homogeneity constants**, and they are very important, because they intervene in the splitting of the physical equations of concrete algebra into their corresponding two algebraic and dimensional formulations, which are derived respectively for their primaries and secondary, in accordance with the provisions of [24.2] and [24.3].

In this regard and in light of the preceding analyzes, it is clear that the traditional omission of dyadic or magnitude algebra has forgotten the constants of homogeneity, limiting itself to the case where $k=1$, and thus the doubling of the physical equations in its algebraic and dimensional components are reduced to a specific assumption. And this could contain a critical flaw that could distort the appreciation of the true nature of physical equations. Perhaps the origin of this traditional defect could be found in the importation by Physics of the mathematical method, which in its metric structures operates with abstract and, therefore, uniform units. The doubling theorem evidences and fully justifies that every mathematical equation remains invariant in this process, because, being $k=1$, it happens that the primary and the corresponding dyadic formulation remain identical.

In the preceding analysis we have abstracted from the indirect rule of section XXII, so that to base the logical steps of all the reasoning we have directly used the laws of composition and properties of the algebra of magnitudes, as corresponds to the way of running more precise and typical of strict logic, reserving for this rule the function of simple verification of the conclusions or to examine its own validity a posteriori in the cases described.

We once again verify that the study of physical constants, like dimensional analysis and that all equations and laws, would not make any sense without having previously defined the **laws of external composition generating** sections XXII et seq.

On the other hand, in the preceding analysis we have tacitly assumed the classical **isometric hypothesis,** which assumes that the quantity of magnitude implicit in every unit is always invariable. We leave the opposite option for the second volume of this work, where the «**dysmetric**» **axiom** is considered and the peculiarities of the spaces in which it is fulfilled are studied, with very relevant consequences for the survival of the physical constants.

Apartado XXVI

LAS CONSTANTES FÍSICAS

En el apartado XXIV hemos analizado el significado que debe darse a las ecuaciones físicas, que han de entenderse como identidades invariantes entre dos entes diádicos, que en el caso famoso de la *segunda ley de Newton*, se refiere a mediciones vectoriales, escrita explícitamente $\overline{F}\ U_F = m\ U_m \circledcirc \overline{a}\ U_a$ (producto del apartado XX) o con economía simbólica $\overline{F}\ U_F = m\ U_m \times \overline{a}\ U_a$ donde U_F sea la unidad de fuerza, U_m la unidad de masa y U_a la unidad de aceleración. Debe admitirse, pues, para esta ley el significado diádico de que la medidión escalar $m\ U_m$ multiplicada por la medición vectorial $\overline{a}\ U_a$ debe ser igual a la díada vectorial $\overline{F}\ U_F$, siendo m un escalar positivo de R, llamado masa de inercia, que presenta un valor específico para cada cuerpo material.

En virtud de [20.1], que define el producto de una díada escalar por otra vectorial, y después con [9.3] y [9.4], se podrá transformar la *segunda ley de Newton* de esta manera explícita:

$$\overline{F}\ U_F = m\ U_m \circledcirc \overline{a}\ U_a = (m \bullet \overline{a})\ (U_m * U_a)] =$$
$$= m \circ [\overline{a}\ (U_m * U_a)] \qquad [26.1]$$

Como por la naturaleza de esta ley m es un escalar, podemos aplicar [11.2] para formar el conciente de concretos vectoriales colineales y homogéneos, con lo que tendremos:

$$\frac{\overline{F}\ U_F}{\overline{a}\ (U_m * U_a)} = m \qquad [26.2]$$

De modo que la *segunda ley de Newton* equivale a entender que el primario de la masa, que es un número real positivo, sea igual al cociente diádico entre $\overline{F}\ U_F$ y $\overline{a}\ (U_m * U_a)$, y que este número es invariable para cada cuerpo o punto material. De modo que nada nos impide considerar que m sea una **constante característica**, sin perjuicio de la condición de magnitud que se atribuye a la masa de inercia[26]. La díada escalar que define la masa del cuerpo tendrá la forma m U_m y la unidad de masa no será independiente de las unidades de fuerza y aceleración, porque los elementos $\overline{F}$ U_F y $\overline{a}$ $(U_m * U_a)$ han de ser homogéneos, en razón de la definición de igualdad [4.2] entre concretos vectoriales. Es más, nada impide establecer las unidades de modo que el primero y el segundo miembro de las ecuaciones físicas sean no solo homogéneos, sino uniformes, y esto es lo usual por comodidad. No obstante, para una mayor generalidad desarrollaremos el supuesto de que dichas díadas sean homogéneas y no uniformes, por lo que, el axioma de continuidad [4.3] determinará que exista $k \in \mathbb{R}$ tal que $U_m * U_a = k \circ U_F$. En estas condiciones, el desdoblamiento de la igualdad de díadas que integra la ecuación física, descrito en [24.3], aplicado a la ecuación [26.1], que se puede escribir:

$$\overline{F}\ U_F = m \circ [\overline{a}\ (U_m * U_a)] = (m \bullet \overline{a})\ (U_m * U_a)$$

Produce las relaciones de primarios y secundarios, apareciendo una ecuación vectorial y una igualdad entre unidades:

$$\overline{F} = k \times m \bullet \overline{a}$$

$$k \circ U_F = U_m * U_a \Rightarrow U_m = (k \circ U_F) /\!/ U_a \qquad [26.3]$$

Recuérdese que el cociente $(k \circ U_F) /\!/ U_a$ no es el ordinario, sino el definido en el apartado XVI. En conclusión, la medición de la constante característica o específica de cada cuerpo llamada masa vendrá dada por una díada escalar cuyo primario o medida estará

---

[26] Se puede discutir que la masa de inercia sea considerada o no una constante característica, dado que la tenemos por una magnitud física fundamental; sin embargo, puesto que admitimos que sea una cantidad constante para cada punto material, no parece descabellado considerarlo así. En todo caso, solo se trata de una mera convención. Por otra parte, con independencia de lo que se acepte al respecto, es innegable el interés del estudio de la *segunda ley de Newton*, una de las imprescindibles de la Física, por lo que hemos decidido incluirla en este apartado, aunque solo sea para observar las consecuencias ilativas del álgebra de magnitudes en el análisis de una ley tan primordial.

determinado por el número real positivo dado por [26.2] y cuyo secundario o unidad tendrá la forma [26.3], con lo cual resultará que la masa vendrá expresada por las díadas homogéneas siguientes:

$$m\,U_m = m\; \circ\frac{k \circ U_F}{U_a}$$

Obsérvese que la relación [26.3] entre las unidades de fuerza U_F, de masa U_m y de aceleración U_a significa que no pueden ser independientes. Por otra parte, si las unidades se definiesen de modo que la del primer miembro fuese uniforme con la del segundo de la ecuación física [26.1], se tendría simplemente el caso particular y más usual $k=1$.

Otro ejemplo insigne de constante característica lo encontramos en la conocida *ley de Hooke*. Formulada en lenguaje común, dicha ley establece que la tensión, definida como fuerza por unidad de superficie, es para cada cuerpo proporcional a la deformación, entendida como variación de longitud por unidad de longitud. Aunque tiene naturaleza vectorial, se puede aplicar en términos escalares, como justificamos en el apartado XXIV, y así lo haremos para variar el razonamiento respecto del ejemplo anterior. Para ello, consideremos un alambre de longitud $L\,U_L$ y sección $S\,U_S$, sometido a una fuerza de componente $F\,U_F$ en la dirección de L, por lo que experimentará una variación de longitud $\Delta L\,U_L$, también en la dirección de L, y siendo $E\,U_E$ el factor de proporcionalidad, llamado **módulo de Young**, y así la *ley de Hooke* tendrá la forma de la siguiente ecuación física de álgebra diádica:

$$\frac{F\,U_F}{S\,U_S} = E\,U_E * \frac{\Delta L\,U_L}{L\,U_L}$$

Las díadas $\Delta L\;U_L$ y $L\;U_L$ son homogéneas y uniformes por hipótesis, porque ambas representan longitudes en la misma unidad, de modo que, a tenor del apartado XI, su cociente diádico habrá de ser un número real, por lo que podemos prescindir de la unidad U_L en numerador y denominador, y la *ley de Hooke* se podrá escribir con la forma:

$$\frac{F\,U_F}{S\,U_S} = E\,U_E \circ \frac{\Delta L}{L}$$

Sean $\sigma = F/S$ la medida de la tensión o fuerza por unidad de superficie en la unidad compuesta asociada $U_F/\!/U_S$; y sea $\varepsilon = \Delta L/L$ la medida de la deformación o variación de longitud por unidad de longitud, que carece de unidad propia. Con esta notación y por la definición de multiplicación por un escalar [9.1] y [9.2], tendremos la igualdad diádica:

$$\sigma\,\frac{U_F}{U_S} = \left(E \times \varepsilon\right)\,U_E \qquad [26.4]$$

En general, dada la definición de igualdad [4.1] entre díadas escalares, en la que los elementos identificados no tienen por qué ser uniformes, y en razón del axioma de continuidad [4.3], existirá $k \in R$ tal que:

$$U_E = k \circ \frac{U_F}{U_S}$$

En estas condiciones, el teorema del desdoblamiento [24.2] para la igualdad de díadas que integran la ecuación física [26.4], se puede escribir, relacionando primarios y secundarios:

$$\sigma = k \times E \times \varepsilon$$

$$k \circ \frac{U_F}{U_S} = U_E \qquad [26.5]$$

Por otra parte, de acuerdo con la definición de división entre díadas escalares homogéneas [11.1], o indistintamente con [16.2], y definido el producto por un escalar mediante [9.1] y [9.2], la *ley de Hooke* [26.4] quedará con la forma:

$$\frac{\sigma \dfrac{U_F}{U_S}}{\varepsilon \, U_E} = E$$

Ello significa que la razón entre las díadas escalares $\sigma(U_F /\!/ U_S)$ y $\varepsilon \, U_E$ será un número real E invariable para cada tipo de alambre. El módulo de Young E deviene así en una constante característica que relaciona específicamente para cada objeto material la relación invariante entre la tensión aplicada y la deformación producida. Con la hipótesis genérica de que las ecuaciones físicas sean igualdades diádicas homogéneas, en este caso con la forma escalar [4.1], el módulo de Young quedará representado por la díada escalar:

$$E\,U_E = \left(k \times E\right) \frac{U_F}{U_S}$$

También aquí tenemos que la relación [26.5] entre las unidades U_E, U_F y U_S determina la dependencia entre ellas, por lo que no se pueden establecer todas arbitrariamente. Si se definieran las unidades para que resultasen uniformes, se tendría $k=1$.

Aparte de las constantes características o específicas, como la masa de inercia y el módulo de Young, encontramos otras que son independientes de la naturaleza de los cuerpos, conservándose invariables en cualquier caso, y las llamaremos **constantes universales**. Un ejemplo destacado es el **equivalente mecánico del calor**. Sea $T\,U_T$ el trabajo mecánico medido con la unidad U_T, simbolicemos $Q\,U_Q$ la cantidad de calor equivalente al trabajo anterior en la unidad U_Q y escribamos $M\,U_M$ para la constante de proporcionalidad entre esas otras dos magnitudes con la unidad U_M. La ley universal que determina la equivalencia entre calor y trabajo se puede escribir en álgebra diádica con la forma de la expresión:

$$T\,U_T = (M\,U_M) * (Q\,U_Q)$$

Por su naturaleza, se trata de una ecuación física escalar, por lo que tendremos que componer las díadas que intervienen con las operaciones escalares que tenemos definidas. La definición de multiplicación [12.1] nos lleva a la expresión:

$$T\ U_T = (M \times Q)\ (U_M * U_Q) \qquad [26.6]$$

En general, las díadas del primer y del segundo miembro deberán ser homogéneos, sin necesidad de que sean uniformes, dada la definición de igualdad [4.1] para concretos escalares. El axioma de continuidad [4.3] determina que existirá $k \in R$ tal que $U_M * U_Q = k \circ U_T$. El teorema del desdoblamiento [24.2], aplicado a la ecuación física escalar [26.6], nos permite determinar las dos relaciones que han de satisfacerse entre sus dos primarios y dos secundarios:

$$T = k \times M \times Q$$

$$k \circ U_T = U_M * U_Q \qquad [26.7]$$

Por otra parte, de acuerdo con la definición de división entre díadas escalares homogéneas [11.1], o indistintamente con [16.2], y con el producto por un escalar de [9.1] y [9.2], la ley universal [26.6] quedará con la forma:

$$\frac{T\ U_T}{Q\left(U_M * U_Q\right)} = M$$

Ello significa que la razón entre las díadas escalares $T\ U_T$ y $Q\ (U_M \times U_Q)$ será un número real M invariable e independiente de la naturaleza de los cuerpos, por lo que se dice que es una constante universal llamada equivalente mecánico del calor. Con la hipótesis genérica de que las ecuaciones físicas igualen concretos homogéneos, tal constante quedará representada por la díada escalar:

$$M\ U_M = \left(k \times M\right)\frac{U_T}{U_Q}$$

También aquí observamos que la relación [26.7] entre las unidades U_T, U_M y U_Q determina la dependencia entre ellas, por lo que no se pueden establecer las tres arbitrariamente. Si se definieran las unidades para que resultasen uniformes, que sería lo más cómodo, se tendría $k = 1$.

Otro ejemplo de este mismo tipo lo encontramos en la **constante de la gravitación universal**. Aunque se la tiene por una sola, la *ley de la gravitación* en realidad son dos: primera, la que describe que dos puntos materiales se atraen con una fuerza, con la dirección de la recta que los une, directamente proporcional al producto de sus masas gravitatorias e inversamente proporcional al cuadrado de la distancia que los separa; y segunda, que para cada punto material es constante la razón entre la masa gravitatoria y la masa de inercia de la *segunda ley de Newton*, y que ambas son manifestaciones de la misma magnitud llamada masa.

Aunque en rigor la *ley de la gravitación* tiene formulación vectorial, como se puede reducir a otra escalar con una sola componente en la dirección de la recta que une los puntos materiales, seguiremos este camino. Para distinguir las dos masas de cada punto material, la de inercia la simbolizaremos m y la gravitatoria μ. Sean U_F la unidad de fuerza, U_m la unidad de masa, tanto gravitatoria como de inercia, y U_L la unidad de longitud para medir la distancia d entre los puntos materiales. En estas condiciones, la ley universal de la gravitación se podrá escribir en álgebra concreta mediante una expresión como la siguiente:

$$F\, U_F = \frac{\left(\mu_1\, U_m\right) * \left(\mu_2\, U_m\right)}{\left(d\, U_L\right)^2}$$

Los subíndices 1 y 2 se emplean para distinguir a cada uno de los dos puntos materiales de masas gravitatorias μ_1 y μ_2, así como las masas de inercia m_1 y m_2. A su vez, siendo H la razón de proporcionalidad entre las masas gravitatoria y de inercia, se escribirán para ambos puntos materiales:

$$\mu_1\, U_m = H \circ (m_1\, U_m)$$

$$\mu_2\, U_m = H \circ (m_2\, U_m)$$

Como H es un cociente entre números diádicos homogéneos, en razón de [11.1], ha de ser un simple número real, por lo que H carece de unidad y es tal que $H \in R$. Sustituyendo estas dos ecuaciones en la anterior y, operando con [9.1] y [9.2], resulta la formulación de la *ley de la gravitación* con las masas de inercia:

$$F\, U_F = (G\, U_G) * \frac{(m_1\, U_m) * (m_2\, U_m)}{(d\, U_L)^2}$$

Siendo $G = H^2$ el factor conocido con el nombre de **constante de la gravitación universal**, y U_G la unidad que le corresponda. Para deducirla, aplicando la definición [12.1] de multiplicación de concretos escalares, llegamos con facilidad a la ecuación de álgebra diádica:

$$F\, U_F = \left(G \times \frac{m_1 \times m_2}{d^2} \right) \left(\frac{U_G * U_m^2}{U_L^2} \right)$$

Como en los ejemplos anteriores, siendo k la relación escalar entre las unidades de ambos miembros, que sería la unidad $k = 1$ si hubiera uniformidad, en el caso genérico de que los concretos de la ecuación no sean uniformes, el desdoblamiento de [24.2] nos lleva las dos relaciones entre primarios y secundarios:

$$F = k \times G \times \frac{m_1 \times m_2}{d^2}$$

$$k \circ U_F = \frac{U_G * U_m^2}{U_L^2}$$

De donde resulta que el concreto que representa la constante de la gravitación universal es:

$$G\, U_G = (k \times G) \circ \frac{U_F * U_L^2}{U_m^2} \qquad [26.8]$$

Se podría pensar que fuese contradictorio que, siendo H una constante sin unidades, en cambio resulte que $G=H^2$ sí presente dimensiones. Sin embargo, no hay nada desconcertante en ello, pues lo que sucede es que mientras H queda indicada solo por un número real, dada la definición con $\mu\, U_m = H\circ(m\, U_m)$, que para cada punto material relaciona su masa gravitatoria con la de inercia; en cambio G queda determinada por $G=H^2$ y además por [26.8], lo que supone que su naturaleza deba ser representada por un ente concreto escalar, no únicamente por un número real, de modo que la igualdad $G=H^2$ solo se refiere a su elemento primario y [26.8] determina el secundario.

En el Sistema Internacional de unidades se ha calculado y establecido el valor de G en función del metro m como unidad de longitud, que no hay que confundir con el mismo símbolo utilizado para la masa, del kilogramo kg como unidad de masa y del newton N como unidad de fuerza:

$$ G = 6{,}67 \times 10^{-11} \circ \frac{N*m^2}{kg^2} = 6{,}67 \times 10^{-11} \circ \frac{m^3}{s^2*kg} $$

Para finalizar este apartado debemos advertir que en todos los análisis de constante realizados nos ha aparecido la constante k, relacionada con la homogeneidad y con el axioma de uniformidad de los entes concretos identificados mediante las ecuaciones físicas consideradas. De ahí que deba considerarse esta especie de invariantes a las que llamaremos **constantes de homogeneidad**, y son muy importantes, porque intervienen en el desdoblamiento de las ecuaciones físicas del álgebra concreta en sus correspondientes dos formulaciones algebraica y dimensional, que se derivan respectivamente para sus primarios y secundarios, de acuerdo con lo establecido por [24.2] y [24.3].

A este respecto y a tenor de los análisis precedentes, resulta claro que la tradicional omisión del álgebra diádica o de magnitudes ha olvidado las constantes de homogeneidad, limitándose al caso en que sea $k=1$, y de esta manera el desdoblamiento de las ecuaciones físicas en sus componentes algebraica y dimensional queda reducido a un supuesto específico. Y ello podría contener un vicio crítico que podría distorsionar la apreciación de la verdadera naturaleza de las ecuaciones físicas. Quizá el origen de este defecto tradicional pudiera encontrarse en la importación por la Física del método matemático, que en sus estructuras métricas opera con unidades abstractas y, por tanto, uniformes. El teorema del desdoblamiento evidencia y justifica plenamente que toda ecuación matemática permanezca invariante en este trámite, porque, siendo $k=1$, sucede que el primario y la correspondiente formulación concreta se mantienen idénticos.

En el análisis precedente nos hemos abstraído de la regla indirecta del apartado XXII, de modo que para fundamentar los pasos lógicos de todos los razonamientos nos hemos acogido directamente a las leyes de composición y propiedades del álgebra de magnitudes, como corresponde a la forma de discurrir más precisa y propia de la lógica estricta, reservando para dicha regla la función de simple comprobación de las conclusiones o para examinar a posteriori su propia validez en los casos descritos.

Volvemos a comprobar una vez más que el estudio de las constantes físicas, al igual que el análisis dimensional y que todas las ecuaciones y leyes, no tendría sentido alguno sin haber definido previamente las **leyes de composición externas generatrices** de los apartados XXII y siguientes.

Por otra parte, en el análisis precedente hemos asumido tácitamente la **hipótesis isométrica** clásica, que da por supuesto que la cantidad de magnitud implícita en toda unidad sea siempre invariable. Dejamos la opción contraria y variante general para el segundo volumen de este trabajo, donde se contempla el **axioma «dismétrico»** y se estudian las peculiaridades de los espacios en que se cumpla, con consecuencias muy relevantes para la supervivencia de las constante físicas.

Section XXVII

PHILOSOPHICAL CONSEQUENCES

We have already referred in section XXIV to the meaning of the equations of Physics, whether it is about universal laws or definition formulas, because one of the most important consequences of the algebra of magnitudes is to observe that physical equations do not relate entities Ordinary algebraic, either real numbers or vectors, but establish relationships between dyadic entities, defined in section III. This fact would seem to overturn the way of operating described in all the texts, because in them, although it is striking and even alarming, this unfailing circumstance is not taken into account. However, dyadic algebra turns out to be in such a way that it is reduced to the single rule of section XXII, so that the composition laws defined for these entities, which are inherent to Physics, allow the common algebraic part to be grouped on the one hand and, on the other, the dimensional part, being authorized to operate with the fiction or rather imposture that the symbols of the units of the related magnitudes are vulgar algebraic elements, although this is not really the case at all. And this isomorphic property saves with a lot of luck the negligent and pancist praxis that has prevailed since ancient times from having fallen into a crass error with disastrous consequences, for having forgotten to establish as a principle the necessary algebra of magnitudes that defines the laws of composition with numbers specific to science, which are dyadic entities. Such practice has also been legitimized by the International System of Units and its erroneous hypothesis in order to consider that the quantities of magnitudes present an abelian multiplicative group structure, a gross defect that no one who has followed this work carefully would excuse or allow it to remain in force knowing its falsehood. Anyone who is not a faciliton and has methodically followed the motivations set out here, will have no doubt that whoever claims that unit operations are obvious would not have understood that in science everything must be based on experimentation or definition, nothing has to remain unverified or defined. How, then, could the lack of an algebra of magnitudes be justified? It being the case that Physics deals with parity entities with two ordered elements: first, the abstract algebraic entity; and second, the unitary element of some magnitude, which is by no means a numerical entity in the classical sense.

At the same time, no attentive and responsible reader will understand that others qualify the composite forms of physical quantities as mysterious and even mystical, because they are nothing but the result of generalizing the algebra of geometric segments in the abstract. And in this we do not observe any arcane dressing unattainable for the human understanding, rather the opposite, because, after having defined the algebra of magnitudes, taking advantage of that of the geometry of lengths, there is light and it is verified with inevitable suspicion how much dangerous arbitrariness must hide the composite forms of physical quantities.

Therefore, in view of the First Algebra of Magnitudes of this monograph, we must get used to interpreting physical equations with the meaning of relations between dyadic entities, which are the true elements that science uses, distancing ourselves in this aspect from the pure Mathematics of abstract algebraic structures; although, yes, these are those that operate with the primary element or measure of each magnitude, and this always in accordance with the composition laws defined for the dyads specifically.

To get into the philosophy of the algebra of magnitudes, let us first analyze the case of the work of a force, as a more elementary manifestation of what is understood by energy. By definition, **the work of a force is conceived as the scalar product of the vector concretes that represent the magnitude called**

force by the concrete that refers to the magnitude of the displacement of its point of application, according to the definition of the scalar product of the section XIX. In dimensional terms, such a scalar product can be stated in the abstract as the force magnitude multiplied by the length magnitude. Since every quantity of the force magnitude results from the multiplication of a quantity of the mass magnitude by a quantity of the acceleration magnitude, mechanical work is a derived magnitude, which in the International System will have per unit that given by the following dyadic expression, which composes unit magnitudes:

$$\frac{1\,kg*1\,m^2}{1\,s^2}$$

The result of the previous dyadic operation, formed by a product, two powers and a quotient, all dyadic in nature, of the three fundamental units: kilogram, meter and second, is called, by definition, «joule» and is symbolized by the letter *J*. The compound magnitude called mechanical work, defined as indicated, does not coincide with the common notion of this concept, associated with physical effort. For example, to hold a heavy object hanging from a pulley it is necessary to counteract its weight by holding firmly one end of the rope; whoever resists the action of the weight in this way will feel that they are making a great effort; but from a mechanical point of view, if the body is not moving, no work will have been done. Therefore, for Physics, work is derived from the movement of the point of application of all force and leads to the ambiguous notion of **energy**, defined as the **ability to produce work**, or perhaps better said, the **ability to transform itself into work**.

We observe in nature a multitude of phenomena that reveal an infinity of what we could understand as energy deposits, such as the wind, capable of moving a wind turbine and producing electricity; or bodies in motion, which when colliding with others

313

move them and produce work; or the gasoline of a vehicle that, consumed by an engine, will propel it from one place to another, producing work; and many other similar cases easy to imagine. All these energy stores have in common that the stored energy does not manifest itself until, by means of some suitable artifice, it is released or transformed into work. Thus, work can be considered as one of the many manifestations of energy that exist for Physics. Well, even if the essence of what we call energy is completely ignored with such joy, we understand that we can measure it through some of its manifestations, and specifically through work, so **we have to admit that energy is a measurable magnitude in the same units as work.**

Let's take some examples. Let's think about a weight clock, we observe that, when they are in the highest position, the clock's machinery will work by the action of the descent due to the weight to the lowest position of its career, overcoming the resistance of the various internal mechanisms; at the lowest position the watch will stop and at an intermediate point it will only run for a fraction of the time from the highest position; and this would mean that the clock weights contain different energy simply because of the position in height they occupy at each moment. From which it is inferred that bodies can be depositories of energy simply by reason of the situation in a gravitational field. This magnitude is usually indicated by the product of three dyads $(M \ kg)*(g \ m/\!\!s^2)*(z \ m)$, where M is a measure of mass in kilograms, g is the measure of acceleration of gravity in meters per second squared dyadic and z the vertical elevation measure in meters; it is a derived magnitude with dimensions of the work magnitude; however, if z were constant, the mass would not move and there would be no work. Now, if it were allowed to fall freely from a given position, the mass would set in motion and do work, as in the case of the weight clock. This means that the magnitude $(M \ kg)*(g \ m/\!\!s^2)*(z \ m)$ must be recognized as having the capacity to generate work or to transform into it, so it is admitted that it is a manifestation of energy called potential energy[27], name which suggestively alludes to the quality of possibility, in the sense that it does not exist in action, but will certainly manifest itself if sufficient conditions are present.

On the other hand, rational mechanics considers that a mass M kg moving at speed v m//s carries an amount of energy by reason of its movement that has been established in the well-known dyadic formula ($\frac{1}{2}m{\times}v^2$ kg$*$m^2//s^2), considering it another manifestation of energy, called kinetic energy, since its dimensional expression is the same as that of mechanical work. This form of energy arises from the observation that a mass in motion is such that, if it is opposed by a force, for example, by arranging a weight hanging from a pulley and attached to the moving mass, the force will slow down its movement and make raise the weight, performing work that is equivalent to the loss of kinetic energy of the moving mass. Therefore, it seems that a moving mass behaves like a deposit of energy that can be transformed into work by means of suitable devices[28].

Mechanical theory deduces the well-known conservation of energy theorem, concluding that the sum of kinetic and potential energy remains constant, and this justifies the general principle of conservation of energy, with the well-known statement that **in nature energy is neither created nor destroyed, but is transformed**[29]. A somewhat different but equivalent wording of this principle is the **impossibility of the perpetual mobile**, that is,

---

[27] The mathematical model of the function of forces that supports the concept of potential energy is exposed in the author's syllabus, Mathematize 3, pp. 219 and following.

[28] Rational mechanics provides the mathematical justification for this physical fact, through the well-known theorem of living forces or kinetic energy, found in Mathematize 3, pp. 224, 242 and following.

[29] The mechanical justification of the principle of conservation of energy is exposed in Mathematize 3, pp. 225 and following.

there is no mechanism that produces energy without altering itself and without taking an equivalent amount from the outside. This means that no device will be capable of producing, without external input, not even the energy necessary to overcome the inevitable internal friction between its various elements. Repeated proofs of this principle are the innumerable failures of so many chimerical inventors of all times and places who have unsuccessfully pursued the continuous movement, the ideal seductive solution to the energy needs of humanity.

Notwithstanding the above, experience shows countless situations in which it seems that the conservation of energy is not fulfilled. If we drop a stone from a certain height, we observe that when it reaches the ground it stops and remains at rest, having lost its potential energy and without showing the least kinetic energy. Or a cyclist descending a hill at a constant speed, applying the brakes, will lose potential energy and, however, will not increase his speed and, therefore, his kinetic energy. In such cases it would appear that energy deviates from the law that marks its conservation. However, in these and in all the cases that can be imagined, it is observed that there has not been the slightest loss of energy, but that it has been transformed into another kind of it that we call heat: the stone and the ground will have been heated in In the first example, as in the case of the cyclist, the brakes and the rim will heat up.

At present we define **heat** as the **energy that passes from one body to another and that causes its dilation and changes of state**. However, the recognition of heat as another way of manifesting energy was not easy for Physics. Until 1780 the magnitudes of heat and temperature were considered similar, which was a serious obstacle to understanding thermal phenomena. Instead, today we distinguish them clearly, understanding by **temperature** the **magnitude that expresses the degree or level of heat of the bodies**. Until the end of the 18th century, the prevailing theory to explain the nature of heat assumed that it was an imponderable fluid called caloric. The

316

current thermodynamics has abandoned the caloric and measures the amount of heat that passes from one body to another by establishing a criterion of equality, defining the addition and choosing a unit. Thus, it is said that two quantities of heat are equal when, absorbed by the same body under the same conditions of pressure and temperature, it turns out that the changes in it are identical. The criterion of addition is conceived with the hypothesis that the amount of heat necessary to produce a certain transformation in a given body is proportional to its mass. Finally, as a unit of heat, the **calorie**, or amount of heat necessary for a gram of water at the normal pressure of an atmosphere to raise its temperature from 14,5 °C to 15,5 °C, has been established.

The experiment of James Prescott Joule (1818-1889) connected thermodynamics with mechanics and showed that heat should be recognized as a main manifestation of energy, because experience strongly shows that for the transformation of mechanical work into heat, or vice versa , the conservation principle is always verified, so it has been possible to observe the mechanical equivalent of heat and establish that the amount of heat in the unit called calorie is equal to 4,186 joules. This connection is perhaps the direct link that elevates the derived magnitude called mechanical work to the category of energetic magnitude, because before finding it, work was rather a merely abstract definition of this mathematical manifestation of energy.

The concept of heat is not enough, however, to formulate a complete statement of the **principle of conservation of energy**, because we appreciate cases that require something more, such as, for example, it is not explained that a coal locomotive starts up, increasing its mechanical energy, without receiving heat from the outside. And this must occur because the coal that you house and burn in your home must be a depository of energy in some way, an energy class that is called **internal energy**. The same phenomenon is observed, for example, if a certain amount of water is heated, supplying it with a little energy, the increase in volume produced by the thermal increase will be very small, so the work done by the expansion against the The forces of atmospheric pressure will be negligible, and if a lower-temperature body is subsequently immersed in the water, it will heat up and the water will lose the energy stored in the initial heating, suggesting that it would have been somehow stored in its interior. And, although this internal energy cannot be known in absolute terms, its variations can be observed. And so, on this basis, in 1847 Helmholtz enunciated the principle of conservation in its most general form, admitting that **the amount of heat transferred to a body is used to increase its internal energy and to produce external work,** a statement known as the **first law of thermodynamics.**

This brief disquisition will serve to appreciate two things: first, that we are referring to energy and its manifestations without having precise knowledge or perhaps without having the least idea of the essence of that magnitude; and second, that the observation of physical magnitudes is not only a matter of analyzing their mathematical or dimensional form, but also requires experimentation and theorizing.

Let's see an illustrative example: let's consider a pair of forces, remember that they are two equal and opposite coplanar forces separated by a certain distance; let M be its moment, the dimensions of the moment are the same as those of a job, because it is the product of a force and a length; the question is, can it be concluded that the moment of the couple is a manifestation of the energy magnitude, given that its dimensional expression is that of a work? The elementary work dT of a pair of forces is given by the product of its moment M and the rotated differential angle $d\theta$, expressed in radians[30], according to the primary differential formula between measures $dT = M \times d\theta$. Recall that the radian is a way to measure angles that

---

[30] The detailed analysis of this case can be found in the volume of applications of the syllabus, Mathematize 3, p. 220.

is defined as the length of the circumference arc equal to a radius, so any measurement of this magnitude is the quotient between two lengths, the length of the arc between the length of the radius, Thus, according to section XI, an abstract real number will result for the measurement of the angles with radians, without any unit; and this motivates that the work of the moment has the same dimensional shape as the moment itself. So what is the difference between the moment magnitudes of a pair and the work of a pair?; we will have to seek the answer in the analysis of the physical fact, and thus, we must understand that diversity must be similar to the distinction between a force and its work, because it seems appreciable that a force cannot be transformed into any form of energy, since it is by the movement of its point of application that a job is developed. So, similarly, it is not the moment of the couple that produces a job, but its rotation; and thus, we will have to conclude that the moment of the pair cannot be characterized as a form or manifestation of energy; while the work of the pair should be. This reflection alerts us to the prudence with which derived magnitudes must be judged, since only with their dimensional equation it is not possible to establish what their nature is, something else is required, and that plus must be sought in the direct physical observation of phenomena, combined with precise reflection on what is observed.

On the other hand, we must warn in this section about certain insidious notations, which induce confusion and which are born from the traditional forgetfulness of dyadic algebra. In the texts and in the International System of Units we can find isolated symbologies such as s^{-1}, m^{-1} or similar, which would seem to suggest that they refer to the inverse units of the second or the meter or any other pattern. However, we discovered in section XIV that the unit and inverse elements do not exist for the multiplication of dyads or scalar measurements, so that the notation U^{-1}, U being any unit, cannot mean the inverse unit of U, but which is another way, analogous to numerical powers, of writing a divisor or denominator of a dyadic fractional notation.

Let's see an example: the International System indicates with the isolated notation s^{-1} the compound unit with a divisor or denominator equal to one second s and a dividend or numerator without dimension, such as the number of cycles of a wave or the number of radians or the number of revolutions; therefore, although the dividend or numerator is not explicit, it must be understood as present depending on the context of the equation in which it appears; so, for example, if you refer to a wave frequency, s^{-1} will indicate cycles per second; or if it is related to an angular velocity, it will mean radians per second or revolutions per second.

Recapitulating: in the First algebra of magnitudes we have tried to justify the traditional way of operating with physical measurements, in accordance with the unique rule of section XXII, deduced in this monograph based on coherent laws of composition between dyadic mathematical entities, considered the representatives ideal quantities of magnitudes. To arrive at this rule we have had to admit certain axioms and make hypotheses or assumptions such as that the quantity of every physical magnitude is represented by an abstract geometric segment and that thus the geometric algebra of segments is applicable to any magnitude. This shows that the aforementioned rule, subliminally infused into the intelligences, because it is not even mentioned by texts in any field, and thus not by teachers, who tacitly assume it without the least explanation or motivation, going directly to operate with the symbols of the units as if they were ordinary algebraic elements, infecting the intellects with this unconscious vice and degrading the teaching quality, it is by no means evident that it is a correct way of composing magnitudes, not even after having founded the laws of composition that they justify, that rather they contribute to put it in quarantine; because, after having revealed the illusions that it hides and, having born such a habit of a crass outrageous forgetfulness, as it is that the operations with measurements have remained until now epistemically undefined, depriving the operational symbols of the physical equations of the exact meaning that corresponds to them, it is inevitable to feel distrust about the dubious suitability of the algebraic simulations that have been accepted to justify the traditional operation and out of a mere sense of responsibility to produce the minimum intellectual disorder to the careless current of thought prevailing in this matter.

On the contrary, if these antecedents were questioned, the doors would be opened wide to new research on the appropriate ways of composing magnitudes, an area that to this day remains unexplored by the sclerosis of tradition. And this, because it could well happen, and perhaps it is the most probable thing, that the composite magnitudes do not respond at all well to the algebra of Euclidean geometry, which would be causing an insidious stagnation in the development of Physics, which could be saved by laws composition of magnitudes that more accurately reflect natural reality. And this will undoubtedly be the way to go towards new dyadic algebras, which follow the course of this first one and allow us to carve out the forgotten pillar of science, perhaps leading us to new horizons in the search for laws and definitions that better represent that the current phenomena of nature.

For our part, we believe we have contributed to this with the development of dyadic algebra and especially with the precise formulation of the **generating external composition laws** of sections XII et seq., which give meaning to all the laws and equations of physics, Therefore, once presented, they become essential to us and it seems incredible to us that they have not been established before and that we have accepted without altering this symbolic pseudo-algebra that means nothing.

And with this reflection, which alerts us to this primordial pending and forgotten question of traditional Physics, this First Algebra of Magnitudes is concluded, which will undoubtedly admit improvements, extensions and substantial changes in future editions; therefore, in the case of a virgin and pioneering subject, we apologize for the inevitable imperfections or omissions that we may have incurred, typical of any original work, which in its first edition aims above all to prevent the tenacious and toxic forgetting of a support primordial of science, as is without question the algebra of magnitudes. After all, the possible faults that our explanations suffer will always be milder and less harmful than the silent prejudices of the current carelessness. We hope that the possible shortcomings inherent in all innovation will be compensated by the honest attempt to unveil and carve this foundation of science, and with this we trust that we have humbly contributed to recycling the physical knowledge of faithful readers, adding a fundamental foundation to its intellectual baggage and allowing them to gain insight into all their prior knowledge, because for them the best intention has been put without sparing efforts.

Not surprisingly, a first innovative and fascinating contribution of this First Algebra of Magnitudes are the **«dysmetric» spaces**, which arise naturally from the simple observation of the dyadic elements, in which it is not prohibited at all to consider that the secondary, or unitary element of the pair, contains different quantities of the associated magnitude depending on its position in space and time, even without materially varying the body or phenomenon taken as a physical unit, and due to whatever causes. This observation, which we could call the **«dysmetric» axiom forecast**, is a new, unappealable mathematical tool, which must be capable of explaining and describing an infinity of natural phenomena, as outlined in the second volume of this work.

Apartado XXVII

CONSECUENTES FILOSÓFICOS

Ya nos hemos referido en el apartado XXIV al significado de las ecuaciones de la Física, tanto si se trata de leyes universales como de fórmulas de definición, porque uno de los consiguientes más trascendentes del álgebra de magnitudes es observar que las ecuaciones físicas no relacionan entes algebraicos ordinarios, ya sean números reales o vectores, sino que establecen relaciones entre entes diádicos, definidos en el apartado III. Este hecho parecería derribar el modo de operar descrito en todos los textos, porque en ellos, aunque resulte llamativo y hasta alarmante, no se tiene en cuenta dicha circunstancia indefectible. Sin embargo, el álgebra diádica resulta ser de tal modo que se reduce a la regla única del apartado XXII, de forma que las leyes de composición definidas para estos entes, que son inherentes a la Física, permiten agrupar por un lado la parte algebraica común y, por otro, la parte dimensional, quedando autorizado operar con la ficción o más bien impostura de que los símbolos de las unidades de las magnitudes relacionadas sean vulgares elementos algebraicos, aunque realmente no sea así en absoluto. Y esta propiedad isomórfica salva con mucha dosis de suerte a la negligente y pancista praxis imperante desde antiguo de haber caído en un craso error de consecuencias desastrosas, por haber olvidado instituir como principio la necesaria álgebra de magnitudes que defina las leyes de composición con los números específicos de la ciencia, que son los entes diádicos. Tal praxis ha sido además legitimada por el Sistema Internacional de Unidades y su hipótesis errónea en orden a considerar que las cantidades de magnitudes presenten estructura de grupo multiplicativo abeliano, craso defecto que nadie que haya seguido con atención este trabajo disculparía ni permitiría que se mantuviese vigente a sabiendas de su falsedad. Cualquiera que no sea un facilitón y

323

haya seguido metódicamente las motivaciones expuestas aquí, no tendrá duda de que quien alegase que las operaciones con unidades sean obvias, no habría entendido que en ciencia todo debe asentarse en la experimentación o en la definición, nada ha de quedar sin verificar ni definir. ¿Cómo se podría justificar, entonces, la falta de un álgebra de magnitudes?, siendo el caso que la Física maneja entes paritarios con dos elementos ordenados: primero, el ente algebraico abstracto; y segundo, el elemento unitario de alguna magnitud, que no es ni mucho menos un ente numérico en sentido clásico.

A su vez, tampoco entenderá ningún lector atento y responsable que otros califiquen de misteriosas y hasta de místicas las formas compuestas de las magnitudes físicas, porque no son sino el resultado de generalizar en abstracto el álgebra de los segmentos geométricos. Y en ello no se observa ningún aderezo arcano inalcanzable para el entendimiento humano, más bien todo lo contrario, porque, tras haber definido el álgebra de magnitudes, aprovechando la de la geometría de longitudes, se hace la luz y se constata con suspicacia inevitable cuánta peligrosa arbitrariedad han de esconder las formas compuestas de las magnitudes físicas.

Por tanto, a la vista de la *Primera álgebra de magnitudes* de esta monografía, debemos acostumbrarnos a interpretar las ecuaciones físicas con el significado de relaciones entre entes diádicos, que son los verdaderos elementos de que se sirve la ciencia, distanciándose en este aspecto de la Matemática pura de las estructuras algebraicas abstractas; aunque, eso sí, estas son las que operan con el elemento primario o medida de cada magnitud, y ello siempre de conformidad con las leyes de composición definidas para las díadas específicamente.

Para adentrarnos en la filosofía del álgebra de magnitudes, analicemos en primer lugar el caso del trabajo de una fuerza, como manifestación más elemental de lo que se entiende por energía. Por definición, **el trabajo de una fuerza se concibe como el producto escalar de los concretos vectoriales que representen la magnitud llamada fuerza por el concreto que se refiera a la**

magnitud del desplazamiento de su punto de aplicación, de acuerdo con la definición del producto escalar del apartado XIX. En términos dimensionales, tal producto escalar se puede indicar en abstracto como la magnitud fuerza multiplicada por la magnitud longitud. Como toda cantidad de la magnitud fuerza resulta de la multiplicación de una cantidad de la magnitud masa por una cantidad de la magnitud aceleración, el trabajo mecánico es una magnitud derivada, que en el Sistema Internacional tendrá por unidad la dada por la siguiente expresión diádica, que compone magnitudes unitarias:

$$\frac{1\,kg * 1\,m^2}{1\,s^2}$$

El resultado de la operación diádica anterior, formada por un producto, dos potencias y un cociente, todos de naturaleza diádica, de las tres unidades fundamentales: kilogramo, metro y segundo, se denomina, por definición, «julio» y se simboliza con la letra J. La magnitud compuesta denominada trabajo mecánico, definida como se ha indicado, no coincide con la noción común de este concepto, asociada al esfuerzo físico. Por ejemplo, para sujetar un objeto pesado colgado de una polea es preciso contrarrestar su peso sujetando con firmeza un extremo de la cuerda; quien resista de este modo la acción del peso sentirá que realiza un gran esfuerzo; pero desde el punto de vista mecánico, si el cuerpo no se mueve, no se habrá realizado ningún trabajo. Por tanto, para la Física, el trabajo es derivado del movimiento del punto de aplicación de toda fuerza y conduce a la noción ambigua de la **energía**, definida como la **capacidad para producir un trabajo**, o quizá mejor dicho, la **facultad de transformarse en trabajo**.

Observamos en la naturaleza multitud de fenómenos que revelan infinidad de los que podríamos entender como depósitos de energía, tales como el viento, capaz de mover un aerogenerador y producir electricidad; o los cuerpos en movimiento, que al chocar con otros los trasladan y producen un trabajo; o la gasolina de un vehículo que, consumida por un motor, lo impulsará de un lugar a otro, produciendo un trabajo; y muchísimos otros casos semejantes fáciles de imaginar. Todos estos depósitos energéticos tienen en común que la energía almacenada no se manifiesta hasta que, por medio de algún artificio adecuado, se libera o se transforma en trabajo. Así, pues, el trabajo puede ser considerado como una de las múltiples manifestaciones de la energía que existen para la Física. Pues bien, aunque se ignore por completo la esencia de lo que llamamos energía con tanta alegría, entendemos que podemos medirla a través de algunas de sus manifestaciones, y en concreto por medio del trabajo, por lo que **hemos de admitir que la energía sea una magnitud medible en las mismas unidades que el trabajo.**

Pongamos algunos ejemplos. Pensemos en un reloj de pesas, observamos que, cuando estas se encuentren en la posición más elevada, la maquinaria del reloj funcionará por la acción del descenso debido al peso hasta la posición más baja de su carrera, venciendo la resistencia de los diversos mecanismos internos; en la posición más baja el reloj se parará y en un punto intermedio solo podrá funcionar durante una fracción de tiempo respecto de la posición más elevada; y ello significaría que las pesas del reloj contengan diferente energía simplemente por la posición en altura que ocupen en cada momento. De donde se infiere que los cuerpos pueden ser depositarios de energía simplemente por razón de la situación en un campo gravitatorio. Esta magnitud suele indicarse por el producto de tres díadas $(M\,kg) * (g\,m\,/\!\!/s^2) * (z\,m)$, donde M sea una medida de masa en kilogramos, g la medida de aceleración de la gravedad en metros por segundo al cuadrado diádico y z la medida de cota vertical en metros; se trata de una magnitud derivada con dimensiones de la magnitud trabajo; sin embargo, si z fuese constante, la masa no se movería y no habría trabajo alguno. Ahora bien, si se le dejase caer libremente desde una posición dada, la masa se pondría en movimiento y desarrollaría un trabajo, como en el caso del reloj de pesas. Ello supone que a la magnitud $(M\,kg) * (g\,m\,/\!\!/s^2) * (z\,m)$ deba reconocérsele la capacidad de generar trabajo o de transformarse en él, por lo que se admite que sea una manifestación de la energía denominada energía potencial[31], nombre que alude sugerentemente a la calidad de posibilidad, en el sentido de que no se da en acto, pero con certeza se manifestará si se presentasen las condiciones suficientes.

Por otra parte, la mecánica racional considera que una masa M kg en movimiento a velocidad v $m/\!/s$ es portadora de una cantidad de energía por razón de su movimiento que se ha establecido en la conocida fórmula diádica ($\frac{1}{2}m \times v^2$ $kg * m^2/\!/s^2$), considerándola otra manifestación de la energía, llamada energía cinética, dado que su expresión dimensional es la misma que la del trabajo mecánico. Esta forma de energía brota de la observación de que una masa en movimiento es tal que, si a ella se opone una fuerza, por ejemplo, disponiendo un peso colgado de una polea y unido a la masa móvil, la fuerza frenará su movimiento y hará ascender el peso, desarrollando un trabajo que equivale a la pérdida de energía cinética de la masa móvil. Por tanto, parece que una masa en movimiento se comporte como un depósito de energía que pueda transformarse en trabajo mediante adecuados artificios[32].

La teoría mecánica deduce el conocido teorema de conservación de la energía, concluyendo que la suma de la energía cinética y de la potencial se mantiene constante, y ello justifica el principio general de conservación de la energía, con el conocido enunciado de que **en la naturaleza la energía ni se crea ni se destruye, sino que se transforma**[33]. Una redacción algo diferente pero equivalente de este principio es la **imposibilidad del móvil perpetuo**, es decir, que

---

[31] El modelo matemático de la función de fuerzas que soporta el concepto de energía potencial se expone en el temario del autor, *Matematizar 3*, pp. 219 y siguientes.

[32] La mecánica racional proporciona la justificación matemática de este hecho físico, mediante el conocido teorema de las fuerzas vivas o de la energía cinética, que se encuentra en *Matematizar 3*, pp. 224, 242 y siguientes.

[33] La justificación mecánica del principio de conservación de la energía se expone en *Matematizar 3*, pp. 225 y siguientes.

no existe ningún mecanismo que produzca energía sin alterarse y sin tomar del exterior una cantidad equivalente. Ello supone que ningún artificio será capaz de producir sin aportación exterior ni siquiera la energía necesaria para vencer los inevitables rozamientos internos entre sus diversos elementos. Son pruebas reiteradas de este principio los innumerables fracasos de tantos inventores quiméricos de todos los tiempos y lugares que infructuosamente han perseguido el movimiento continuo, seductora solución ideal a las necesidades energéticas de la humanidad.

No obstante lo anterior, la experiencia exhibe incontables situaciones en las que parece que la conservación de la energía no se cumpla. Si dejamos caer una piedra desde cierta altura, observamos que al llegar al suelo se detiene y queda en reposo, habiendo perdido su energía potencial y sin mostrar la menor energía cinética. O un ciclista que descienda una pendiente a velocidad constante, accionando los frenos, perderá energía potencial y, sin embargo, no aumentará su velocidad ni, por tanto, su energía cinética. Aparentaría en tales supuestos que la energía se apartase de la ley que marca su conservación. Sin embargo, en estos y en todos los casos que se imaginen, se observa que no ha habido la menor pérdida de energía, sino que esta se ha transformado en otra especie de ella que llamamos calor: se habrán calentado la piedra y el suelo en el primer ejemplo, así como en el caso del ciclista se calentarán los frenos y la llanta.

En la actualidad el **calor** lo definimos como la **energía que pasa de un cuerpo a otro y que causa su dilatación y sus cambios de estado**. Sin embargo, el reconocimiento del calor como otra forma de manifestarse la energía no le resultó fácil a la Física. Hasta 1780 las magnitudes de calor y de temperatura se consideraban semejantes, lo cual era un obstáculo serio para comprender los fenómenos térmicos. En cambio, hoy las distinguimos claramente, entendiendo por **temperatura** la **magnitud que expresa el grado o nivel de calor de los cuerpos**. Hasta finales del siglo XVIII la teoría predominante para explicar la naturaleza del calor suponía que se trataba de un fluido imponderable llamado calórico. La termodinámica actual ha abandonado el calórico y mide la cantidad de calor que pasa de un cuerpo a otro estableciendo un criterio de igualdad, definiendo la adición y eligiendo una unidad. Así, se dice que dos cantidades de calor son iguales cuando, absorbidas por un mismo cuerpo en las mismas condiciones de presión y temperatura, resulte que los cambios en él sean idénticos. El criterio de la adición se concibe con la hipótesis de que la cantidad de calor necesaria para producir en un cuerpo determinada transformación sea proporcional a su masa. Finalmente, como unidad de calor se ha establecido la **caloría**, o cantidad de calor necesario para que un gramo de agua a la presión normal de una atmósfera eleve su temperatura de 14,5 °C a 15,5 °C.

El experimento de James Prescott Joule (1818-1889) conectó la termodinámica con la mecánica y evidenció que el calor deba ser reconocido como una manifestación principal de la energía, porque la experiencia demuestra tenazmente que para la transformación de trabajo mecánico en calor, o viceversa, se verifica siempre el principio de conservación, por lo que se ha podido observar el equivalente mecánico del calor y establecer que la cantidad de calor de la unidad llamada caloría es igual a 4,186 julios. Esta conexión sea quizá el enlace directo que encumbra a la magnitud derivada llamada trabajo mecánico a la categoría de magnitud energética, pues antes de encontrarla, el trabajo era más bien una definición meramente abstracta de esta manifestación matemática de la energía.

El concepto de calor no basta, sin embargo, para formular un enunciado completo del **principio de conservación de la energía**, porque apreciamos casos que requieren algo más, como por ejemplo, no se explica el que una locomotora de carbón se ponga en marcha, aumentando su energía mecánica, sin recibir calor del exterior. Y ello ha de producirse debido a que el carbón que alberga y quema en su hogar debe ser depositario de energía de alguna manera, clase energética que se denomina **energía interna**. El mismo fenómeno se observa, por ejemplo, si se calienta una cierta cantidad de agua, suministrándole un poco de energía, el aumento de volumen producido por el incremento térmico será muy pequeño, por lo que el trabajo realizado por la expansión en contra de las fuerzas de la presión atmosférica será despreciable, y si a continuación se sumerge en el agua un cuerpo a menor temperatura, este se calentará y el agua perderá la energía almacenada en el calentamiento inicial, lo que sugiere que antes habría quedado almacenada de algún modo en su interior. Y, si bien esta energía interna no se puede conocer en términos absolutos, sí que se pueden observar sus variaciones. Y así, con esta base, en 1847 Helmholtz enunció el principio de conservación en su forma más general, admitiendo que **la cantidad de calor transferida a un cuerpo se emplea en aumentar su energía interna y en producir un trabajo exterior**, enunciado conocido como la *primera ley de la termodinámica*.

Sirva esta breve disquisición para apreciar dos cosas: primera, que estamos refiriéndonos a la energía y sus manifestaciones sin tener conocimiento preciso o quizá sin tener la menor idea de la esencia de esa magnitud; y segunda, que la observación de las magnitudes físicas no solo es cuestión de analizar su forma matemática o dimensional, sino que requiere experimentar y teorizar.

Veamos un ejemplo ilustrador: consideremos un par de fuerzas, recordemos que se trata de dos fuerzas coplanarias iguales y opuestas separadas una cierta distancia; sea M su momento, las dimensiones del momento son las mismas que las de un trabajo, porque se trata del producto de una fuerza por una longitud; la cuestión es, ¿se puede concluir que el momento del par sea una manifestación de la magnitud energía, dado que su expresión dimensional es la de un trabajo? El trabajo elemental dT de un par de fuerzas viene dado por el producto de su momento M por el ángulo diferencial girado $d\theta$, expresado en radianes[34], de acuerdo con la fórmula diferencial primaria entre medidas

---

[34] El análisis detallado de este caso se puede encontrar en el volumen de aplicaciones del temario, *Matematizar 3*, p. 220.

$dT = M \times d\theta$. Recordemos que el radián es una manera de medir ángulos que se define como la longitud de arco de circunferencia igual a un radio, por lo que toda medición de esta magnitud es el cociente entre dos longitudes, la longitud del arco entre la longitud del radio, así que, de acuerdo con el apartado XI, resultará para la medición de los ángulos con radianes un número real abstracto, sin ninguna unidad; y ello motiva que el trabajo del momento tenga la misma forma dimensional que el propio momento. Entonces, ¿cual es la diferencia entre las magnitudes momento de un par y trabajo de un par?; la respuesta habremos de buscarla en el análisis del hecho físico, y así, hemos de entender que la diversidad ha de ser parecida a la distinción entre una fuerza y su trabajo, porque parece apreciable que una fuerza no pueda transformarse en ninguna forma de energía, ya que es por el movimiento de su punto de aplicación por lo que se desarrolla un trabajo. De modo que, análogamente, no es el momento del par lo que produce un trabajo, sino su rotación; y así, habremos de concluir que el momento del par no pueda ser caracterizado como una forma o manifestación de la energía; mientras que el trabajo del par sí debería serlo. Esta reflexión nos alerta sobre la prudencia con que deben juzgarse las magnitudes derivadas, pues solo con su ecuación dimensional no se puede llegar con fundamento a establecer cuál sea su naturaleza, se requiere algo más, y ese plus debe buscarse en la observación física directa de los fenómenos, combinada con la reflexión precisa acerca de lo observado.

Por otra parte, hemos de prevenir en este apartado sobre ciertas notaciones insidiosas, que inducen a confusión y que nacen del tradicional olvido del álgebra diádica. En los textos y en el sistema Internacional de Unidades podemos encontrar simbologías aisladas como s^{-1}, m^{-1} o similares, que parecerían sugerir que se refieran a las unidades inversas del segundo o del metro o de cualquier otro patrón. Sin embargo, descubrimos en el apartado XIV que no existen los elementos unidad ni inverso para la multiplicación de díadas o mediciones escalares, de modo que la notación U^{-1}, siendo U una unidad cualquiera, no puede significar la unidad inversa de U, sino que es otra forma, análoga a la de las potencias numéricas, de escribir un divisor o el denominador de una notación fraccionaria diádica. Veamos un ejemplo: el Sistema Internacional indica con la notación aislada s^{-1} la unidad compuesta con un divisor o denominador igual a un segundo s y un dividendo o numerador sin dimensión, tal como el número de ciclos de una onda o el número de radianes o el número de revoluciones; por ello, aunque no se explicite el dividendo o numerador, debe entenderse presente en función del contexto de la ecuación en que aparezca; así que, por ejemplo, si se refiere a una frecuencia de onda, s^{-1} indicará ciclos por segundo; o si se relaciona con una velocidad angular, significará radianes por segundo o revoluciones por segundo.

Recapitulando: en la *Primera álgebra de magnitudes* hemos procurado justificar el modo tradicional de operar con mediciones físicas, de acuerdo con la regla única del apartado XXII, deducida en esta monografía en función de unas leyes de composición coherentes entre entes matemáticos diádicos, considerados los representantes idóneos de las cantidades de magnitudes. Para llegar a esa regla hemos debido admitir determinados axiomas y hacer hipótesis o suposiciones tales como que la cantidad de toda magnitud física sea representada por un segmento geométrico abstracto y que así el álgebra geométrica de segmentos sea aplicable a cualesquiera magnitudes. Ello evidencia que la citada regla, insuflada subliminalmente en las inteligencias, porque ni siquiera es mencionada por los textos de cualquier ámbito, y con ello tampoco por los profesores, que la asumen tácitamente sin la menor explicación ni motivación, pasando directamente a operar con los símbolos de las unidades como si fuesen elementos algebraicos ordinarios, infectando los intelectos con este vicio inconsciente y envileciendo la calidad docente, no es ni mucho menos evidente que sea una forma correcta de componer magnitudes, ni siquiera tras haber fundamentado las leyes de composición que la justifican, que más bien contribuyen a ponerla en cuarentena; porque, tras haber revelado las ilusiones que oculta y, habiendo nacido tal costumbre de un craso olvido garrafal, como lo es que las operaciones con mediciones han permanecido hasta ahora sin definir epistémicamente, privando a los símbolos operacionales de las ecuaciones físicas del significado exacto que les corresponda, resulta inevitable sentir desconfianza sobre la dudosa idoneidad de las simulaciones algebraicas que se han admitido para justificar la operativa tradicional y por mero sentido de responsabilidad producir el mínimo trastorno intelectual a la descuidada corriente de pensamiento imperante en esta materia.

Por el contrario, si se cuestionasen esos antecedentes, se abrirían de par en par las puertas a nuevas investigaciones sobre las formas adecuadas de componer magnitudes, ámbito que a día de hoy permanece inexplorado por la esclerosis de la tradición. Y ello, porque bien pudiera suceder, y quizá sea lo más probable, que las magnitudes compuestas no respondiesen del todo bien al álgebra de la geometría euclidiana, lo que estaría provocando un insidioso marasmo en el desarrollo de la Física, que podría salvarse mediante leyes de composición de magnitudes que reflejen con mayor precisión la realidad natural. Y este habrá de ser sin duda el camino a recorrer hacia nuevas álgebras diádicas, que sigan el rumbo de esta primera y que permitan labrar el pilar olvidado de la ciencia, llevándonos quizá hacia nuevos horizontes en la búsqueda de leyes y definiciones que representen mejor que las actuales los fenómenos de la naturaleza.

Por nuestra parte, creemos haber contribuido a ello con el desarrollo del álgebra diádica y en especial con la formulación precisa de las **leyes de composición externas generatrices** de los apartados XII y siguientes, que dotan de significado a todas las leyes y ecuaciones de la física, por lo que, una vez presentadas, se nos hacen imprescindibles y nos parece increíble que no se hayan establecido antes y que hayamos aceptado sin alterarnos esa pseudoálgebra simbólica que no significa nada.

Y con esta reflexión, que alerta sobre esa cuestión primordial pendiente y olvidada de la Física tradicional, se da por finalizada esta *Primera álgebra de magnitudes*, que habrá de admitir sin duda mejoras, ampliaciones y cambios sustanciales en futuras ediciones; por lo que, tratándose de un tema virgen y pionero, pedimos disculpas por las inevitables imperfecciones u omisiones en las que podamos haber incurrido, propias de toda obra original, que en su primera edición pretende ante todo prevenir del tenaz y tóxico olvido de un soporte primordial de la ciencia, como lo es sin discusión el álgebra de magnitudes. Después de todo, las eventuales faltas de que adolezcan nuestras explicaciones siempre serán más leves y menos nocivas que los prejuicios silenciosos de los descuidos vigentes. Esperamos que las posibles carencias inherentes a toda innovación sean compensadas por el intento honesto de desvelar y labrar este puntal de la ciencia, y con ello confiamos haber contribuido con toda humildad a reciclar los conocimientos físicos de los fieles lectores, añadiendo un cimiento fundamental a su bagaje intelectual y propiciando que gane en lucidez el entendimiento de todo su saber previo, pues para ellos se ha puesto la mejor intención sin escatimar esfuerzos.

No en vano, una primera aportación innovadora y fascinante de esta *Primera álgebra de magnitudes* son los **espacios «dismétricos»**, que surgen con naturalidad de la sencilla observación de los elementos diádicos, en lo que no está prohibido en absoluto considerar que el secundario, o elemento unitario del par, contenga diferentes cantidades de la magnitud asociada en función de su posición en el espacio y en el tiempo, aun sin variar materialmente el cuerpo o fenómeto tomado como unidad física, y debido a las causas que sean. Esta observación, que podría llamarse **previsión del axioma «dismétrico»**, es una nueva herramienta matemática inapelable, que ha de ser capaz de explicar y describir infinidad de fenómenos naturales, como se esboza en el segundo volumen de este trabajo.

Section XXVIII

DYADIC OR PHYSICAL ALGEBRA COMPENDIUM
NEED AND BASIC DEVELOPMENT

This section summarizes the matter developed in the First Algebra of Magnitudes with a course aimed at university professors and students. One more didactic effort to convince that it is necessary to correct the deplorable situation of Physics in its operational foundations. The research is inspired by the philosophical legacy of the fathers of modern Physics, whose testimony warns of the harmful **underlying presuppositions in the application of common algebraic operations to physical magnitudes**, summarized in the current false hypothesis of the International System of Units, which attributes to the quantities of magnitudes the abelian multiplicative group structure, with the practical consequence that it is assumed arbitrarily and with a good deal of absurdity that the operations with magnitudes correspond to those of rational numbers or ordinary fractions. Crass and harmful error of the highest scientific institutions, which contaminates everything, for which they should hurry to remedy this insidious contamination of Physics, unbecoming of modern times. Here the nature of this congenital malformation is revealed, which insidiously intoxicates teaching, severely curtailing educational quality and scientific excellence with confusing principles, depriving everyone of their right to receive complete information, free from latent or tacit assumptions.

This work aims to shed light on how to overcome this unworthy deficiency. For this, an epistemic physical algebra based on Euclidean geometry is developed as a remedy, which can be extended to other more complex ones. The compendium collects the fundamental principles of this incredibly absent matter, depriving Physics of a fundamental pillar. Something similar to that Mathematics would have been built without arithmetic. The compendium is organized into the following articles:

Number		Page
1	Introduction, theoretical framework and background	337
2	Methodology and results	345
3	Definition of concrete entity or physical dyad and equality	351
4	The dyadic addition	357
5	Deduction of dyadic subtraction	361
6	Dyadic multiplication by a scalar	365
7	Deduction of the uniform dyadic division	371
8	Geometric multiplication of lengths	379
9	Dyadic multiplication of scalar magnitudes	389
10	Deduction of the scalar dyadic division	393
11	Multiplication between scalar and vector dyads	397
12	Definition of the scalar and vector products of vector dyads	401
13	Algebraic structure of dyadic sets	405
14	Dyadic analysis of the Thales' Theorem	411
15	The Pythagorean Theorem: The first diaidic form of Mathematics	421
16	Dyadic form of physical equations	433
17	Effects of the principle of symbolic economy	439
18	Discussion and Conclusions	443

Article 1

INTRODUCTION, THEORETICAL FRAMEWORK
AND BACKGROUND

From the beginning in the elementary study of Physics, it is customary to use operations with entities that indicate concrete quantities of magnitudes and, by suggestion of arithmetic operations with abstract numbers such as real ones, it is naturally believed that concrete operations should follow the same slide rules and it is normal not to question tradition.

So what unscrupulously is learned and taught to do from a young age unconsciously, seeming so natural, in reality, it is not only not obvious, but it is totally incorrect, since it dispenses with something capital: epistemic definitions of the laws of composition between entities that represent measurements or quantities of magnitudes and their units. So it is not strange that the effects of this omission worried and continue to disturb the sages of Physics of all times; what is more, what is striking is that it must be the prominent ones who philosophize about it and that no one else discusses the tradition, because the root of the problem is very elementary. This is what R. M. Cooke and J. Hilgevoord, The Algebra of Physical Magnitudes, refer to, summarizing the debates of the classics as follows:

> Philosophers have long been interested in the question of the physical presuppositions underlying the application of algebraic operations to physical magnitudes, and this interest has quickened as a result of the existence of hidden variables underlying quantum mechanics. (p. 363)

These presuppositions allude to the gap raised by the distinguished Spanish physicist, Professor Julio Palacios, reflected in the prologue of his Dimensional Analysis (second edition, 1964, Espasa Calpe). This is how he describes the current traditional unknown:

> A widely held opinion, which goes back to Clerk Maxwell, and in which many physicists of my generation have participated, is that

these symbols —refers to the right parentheses that enclose the names of the different quantities— and, therefore, the formulas Dimensional refers to units, and is written like this, for example:

$$1\,erg = \frac{1\,g \times 1\,cm^2}{1\,s^2}$$

without realizing that we would be in a bind if an inquisitive student asked us how to multiply a square centimeter by a gram and divide the product by a second squared. (p. 12)

This incongruous omission pending to resolve in this matter of the operations with quantities of physical magnitudes is disturbing for scientific logic, and for this reason it has caused the proliferation of diverse and contradictory opinions regarding its nature and formulation, discussions that would simply be put to an end defining the necessary composition laws. A group of authors such as R. C. Tolman ascribe to the symbols of dimensional expressions a certain impenetrable or mystical character and consider that «The true essence of magnitudes, from the physical point of view, is represented by their dimensional formula» (Physics Review, p. 25, 1917). This hypothesis does not seem to be true, because it would suppose that such disparate magnitudes as the moment of a force and its work, which can both be expressed in «newton *meter», were essentially manifestations of the same magnitude, energy, which seems clearly an unacceptable delirium. Great authors such as Planck indicate that «It is as meaningless to speak of the "real" dimension of a magnitude as it is of the "real" name of an object», which would mean that physical magnitudes should be hidden from the understanding. Planck seems to indicate to us that we must not forget that physical magnitudes are mental entities and that, like any other name that indicates an extramental object, they are the result of the arbitrariness of thought. The positivist faction of the Vienna Circle, headed by Bridgman, states that «Dimensions do not have absolute value at all, but must be defined precisely from the process used to measure the respective magnitude» (Dimensional Analysis, Yale,

University Press). Bridgman seems to suggest again that in the realm of magnitudes there must be a good deal of arbitrariness; which made Planck so uncomfortable that he criticized positivism in the famous conference entitled Religion und Naturwissenschaft:

> The views of the positivists cannot be fought from a purely logical point of view. And yet a careful examination of them reveals that they are inadequate and sterile, because they dispense with a circumstance that is of decisive importance for scientific progress. As much as positivism boasts of being free from prejudice, it has to start from a fundamental premise if it is not to degenerate into an unintelligible solipsism. This premise is that every physical measurement can be reproduced in such a way that the result is independent of the observer's personality, the place and time in which the measurement is made, and any other circumstance. All of this simply reveals that the decisive factor for the measurement result lies outside the observer and that, consequently, the measurements pose problems involving causal connections in an objective reality independent of the observer..

The philosophical pandemonium that results in this matter pushes conformism with the usual way of operating with quantities of magnitudes without even wondering if it is compatible with scientific logic and without becoming aware of the incongruity that the lack of epistemic definition of its laws of composition, so that for the majority this capital vice, which is real, does not exist, seriously affecting the teaching quality and the complete training of the students who, at least, have the right to a curriculum that explains the gap and to decide their position intellectual about it. And, for their part, physicists and scientists have the responsibility to base their work on solid and coherent bases. What kind of physicist is he who does not know what he is doing when operating with magnitudes and is satisfied with it?

Apartado XXVIII

COMPENDIO DE ÁLGEBRA DIÁDICA O FÍSICA
NECESIDAD Y DESARROLLO BÁSICO

En este apartado se compendia la materia desarrollada en la *Primera álgebra de magnitudes* con una ilación dirigida a profesores y estudiantes universitarios. Un esfuerzo didáctico más para convencer de que hay que corregir la deplorable situación de la Física en sus fundamentos operacionales. La investigación se inspira en el legado filosófico de los padres de la Física moderna, cuyo testimonio advierte sobre las nocivas **presuposiciones subyacentes en la aplicación de operaciones algebraicas comunes a las magnitudes físicas**, resumidas en la hipótesis falsa actual del sistema Internacional de Unidades, que atribuye a las cantidades de magnitudes la estructura de grupo multiplicativo abeliano, con la consecuencia práctica de que se supone arbitrariamente y con buena dosis de absurdo que las operaciones con magnitudes se correspondan con las de los números racionales o fracciones ordinarias. Craso y nocivo error de las más altas instituciones científicas, que lo contamina todo, por lo que deberían apresurarse en remediar esta contaminación insidiosa de la Física, impropia de los tiempos modernos. Aquí se desvela la naturaleza de esta malformación congénita, que intoxica insidiosamente la docencia, cercenando gravemente la calidad educativa y la excelencia científica con principios confusos, privando a todos de su derecho a recibir información cabal, libre de suposiciones latentes o tácitas.

Este trabajo pretende arrojar luz para salvar esta indigna deficiencia. Para ello se desarrolla como remedio un **álgebra física epistémica** basada en la geometría euclidiana, ampliable a otras más complejas. El compendio recoge los principios fundamentales de esta materia increíblemente ausente, privando a la Física de un pilar fundamental. Algo parecido a que las Matemáticas se hubieran construido sin la aritmética. El compendio se organiza en los siguientes artículos:

Número		Página
1	Introducción, marco teórico y antecedentes	342
2	Metodología y resultados	348
3	Definición de ente concreto o díada física e igualdad	354
4	La adición diádica	359
5	Deducción de la sustracción diádica	363
6	Multiplicación diádica por un escalar	368
7	Deducción de la división diádica uniforme	375
8	Multiplicación geométrica de longitudes	384
9	Multiplicación diádica de magnitudes escalares	391
10	Deducción de la división diádica escalar	395
11	Multiplicación entre díadas escalares y vectoriales	399
12	Definición de los productos escalar y vectorial de díadas vectoriales	403
13	Estructura algebraica de los conjuntos diádicos	408
14	Análisis diádico del *Teorema de Tales*	416
15	El *Teorema de Pitágoras*: La primera forma diádica de la Matemática	427
16	Forma diádica de las ecuaciones físicas	436
17	Efectos del principio de economía simbólica	441
18	Discusión y conclusiones	446

Artículo 1

INTRODUCCIÓN, MARCO TEÓRICO
Y ANTECEDENTES

Desde la iniciación en el estudio elemental de la Física es costumbre servirse de las operaciones con entes que indican cantidades concretas de magnitudes y, por sugestión de las operaciones aritméticas con números abstractos como los reales, se cree con naturalidad que las operaciones concretas deban seguir las mismas reglas de cálculo y es normal no cuestionarse la tradición.

Así que lo que sin escrúpulos se aprende y se enseña a hacer desde pequeños inconscientemente, pareciendo tan natural, en realidad, no solo no es nada evidente, sino que es totalmente incorrecto, puesto que se prescinde de algo capital: de las definiciones epistémicas de las leyes de composición entre entes que representen mediciones o cantidades de magnitudes y de sus unidades. Así que no es extraño que los efectos de esta omisión preocupasen y sigan inquietando a los sabios de la Física de todos los tiempos; es más, lo llamativo es que hayan de ser los prominentes quienes filosofen al respecto y que nadie más discuta la tradición, porque la raíz del problema es muy elemental. A ello se refieren R. M. Cooke y J. Hilgevoord, *The Algebra of Physical Magnitudes*, que resumen los debates de los clásicos de la siguiente manera:

> Los filósofos han estado interesados por mucho tiempo en la cuestión de las presuposiciones subyacentes a la aplicación de operaciones algebraicas a magnitudes físicas, y este interés se ha acelerado como resultado del papel aparente que estas presuposiciones desempeñan en relación con la existencia de variables ocultas subyacentes a la mecánica cuántica. (p. 363)

A estas presuposiciones alude la laguna planteada por el insigne físico español, profesor Julio Palacios, reflejada en el prólogo de su *Análisis dimensional* (segunda edición, 1964, Espasa Calpe). Así describe la vigente incógnita tradicional:

Una opinión muy extendida, que se remonta a Clerk Maxwell, y de la que hemos participado muchos físicos de mi generación, es que dichos símbolos —alude a los paréntesis rectos que encierran los nombres de las distintas magnitudes— y, por tanto, las fórmulas dimensionales se refieren a las unidades, y así se escribe, por ejemplo:

$$1\,ergio = \frac{1\,g \times 1\,cm^2}{1\,s^2}$$

sin caer en la cuenta de que nos veríamos en un aprieto si un alumno inquisitivo nos preguntase cómo se hace para multiplicar un centímetro cuadrado por un gramo y dividir el producto por un segundo elevado a cuadrado. (p. 12)

Dicha omisión incongruente pendiente de resolver en esto de las operaciones con cantidades de magnitudes físicas es perturbadora para la lógica científica, y por ello ha provocado la proliferación de opiniones diversas y contradictorias respecto a su naturaleza y formulación, discusiones a las que se pondría fin simplemente definiendo las leyes de composición necesarias. Un grupo de autores como R. C. Tolman atribuyen a los símbolos de las expresiones dimensionales cierto carácter impenetrable o místico y consideran que «La verdadera esencia de las magnitudes, desde el punto de vista físico, está representada por su fórmula dimensional» (*Physics Review*, p. 25, 1917). Esta hipótesis no parece que pueda ser cierta, porque supondría que magnitudes tan dispares como el momento de una fuerza y su trabajo, que pueden expresarse ambas en «newton∗metro», fuesen esencialmente manifestaciones de la misma magnitud, la energía, lo cual parece a todas luces un desvarío inaceptable. Grandes autores como Planck indican que «Tan falto de sentido es hablar de la dimensión "real" de una magnitud como del nombre "real" de un objeto», lo que supondría que las magnitudes físicas habrían de ocultarse al entendimiento. Planck parece indicarnos que no hemos de olvidar que las magnitudes físicas son entes mentales y que, como cualquier otro nombre que señale a un objeto extramental, son fruto de la arbitrariedad del pensamiento. La facción positivista del Círculo de Viena, encabezada por

Bridgman, dispone que «Las dimensiones no tienen en modo alguno valor absoluto, sino que han de definirse, precisamente, a partir del proceso que se utilice para medir la magnitud respectiva» (*Dimensional Analysis*, Yale, University Press). Bridgman parece sugerir nuevamente que en el ámbito de las magnitudes debe de haber una buena dosis de arbitrariedad; lo que incomodaba tanto a Planck, que criticó el positivismo en la famosa conferencia titulada *Religion und Naturwissenschaft*:

> Las opiniones de los positivistas no pueden ser combatidas desde un punto de vista puramente lógico. Y, sin embargo, un examen detenido de las mismas revela que son inadecuadas y estériles, porque prescinden de una circunstancia que tiene importancia decisiva para el progreso científico. Por mucho que alardee el positivismo de estar exento de prejuicios, tiene que partir de una premisa fundamental si no quiere degenerar en un solipsismo ininteligible. Tal premisa consiste en que toda medida física puede ser reproducida de tal modo que el resultado es independiente de la personalidad del observador, del lugar y tiempo en que se efectúa la medición, y de cualquier otra circunstancia. Todo esto revela simplemente que el factor decisivo para el resultado de la medición está fuera del observador y que, en consecuencia, las medidas plantean problemas que implican conexiones causales en una realidad objetiva independiente del observador.

El pandemónium filosófico que resulta en esta materia empuja al conformismo con la manera usual de operar con cantidades de magnitudes sin tan siquiera preguntarse si es compatible con la lógica científica y sin tomar conciencia de la incongruencia que supone la falta de definición epistémica de sus leyes de composición, por lo que para la mayoría este vicio capital, que es real, no existe, afectando gravemente a la calidad docente y a la formación cabal de los alumnos que, al menos, tienen derecho a un currículo que explique la laguna y a decidir su posición intelectual al respecto. Y, por su parte, los físicos y científicos tienen la responsabilidad de fundamentar sus trabajos en bases sólidas y coherentes. ¿Qué clase de físico es aquel que no sabe lo que hace al operar con magnitudes y se conforma con ello?

Article 2

In the same way that there are algebras for numbers and abstract vectors, universally accepted, an algebra of dyadic entities, representatives of quantities of magnitudes, should be established, because only in this way would the prevailing confusion and ignorance and the **current educational dogma** be ended, being better clarified the meanings of the different composite magnitudes, as humbly outlined in this compendium.

There is a well-known and intuitive magnitude, which is **geometric length**, and this should be the starting point to base a generic algebra of physical magnitudes. To do this, it is necessary to first establish the laws of composition of the segments, insofar as they are the most elementary figures, and define a **geometric or graphical algebra** to compose them before moving on to their corresponding **analytical algebra**. Geometric addition does not offer any difficulty, it is enough to conceive it as the graphic juxtaposition of segments, which analytically requires measuring them with the same unit of length. On the other hand, geometric multiplication can be conceived in such a way as to produce new magnitudes from length, area with two factors, volume with three, or hypervolumes with more than three; and the multiplied segments can be expressed in any length units, which do not have to coincide, as the addition requires. The length is therefore said to be a fundamental magnitude and the others are called derivatives or compounds through multiplication, and it should be noted that this operation, in principle graphical, differs substantially from the notion of the arithmetic product, given by the addition of a multiplying as many times as the multiplier indicates.

When segments are multiplied in this way, it is not possible to attribute the arithmetic multiplier function to any of them, which shows that this operation must be clearly differentiated from ordinary multiplication.

Once the algebra of segments is established, nothing prevents associating any quantity of any magnitude with the quantity of length of a segment. To do this, it would be enough to identify the empirical unit of the magnitude considered with an arbitrary or abstract unit of length. In this way, a one-to-one correspondence could be established between the set of all quantities of the given magnitude and that of all abstract lengths, that is, without real scale. The **affinity postulate** consists in admitting the previous operation and handling the quantities of magnitudes as if they were abstract geometric segments, which is to suppose that, although the quantities are different by nature, their quantities are affine to those of the length quantity. This postulate, in combination with the composition of areas and volumes in Euclidean space, allows defining the dyadic multiplication between any scalar magnitudes.

Starting as the foundation of the algebra of geometric segments, the algebra of lengths is developed and, by reasonable generalization, the precise definition of the laws of composition for any magnitude is reached with relative ease. This reveals the hidden frameworks of the derived units and the meanings that can be attributed to them can be judged more accurately. So the notion of dimension of all magnitude has to be considered after and not before having conceived an algebra of magnitudes, whose analytical mathematical expression is the concrete entities or dyadic elements.

Hence, the method followed in this exposition should be presented according to the following sequence: first, to establish the basic concepts of physical magnitudes in general; then, assign them a mathematical entity and create the concrete or dyadic entities; then define an algebra for such special entities, precise representatives of the quantities of measurable natural magnitudes; then investigate the meaning of definition equations, universal laws and other physical entities; and, finally, explain the principles of dimensional analysis. Well, this is how it is done in the preceding nuclear part of this text.

However, in a didactic compendium such as the one presented here, it is necessary to limit the length and select the substantial, so that the possibilities of the resulting complex algebra are glimpsed together without diminishing the clarity of the exposition. Hence, here only the fundamental aspects can be exposed in the most concise way possible: the definition of the concrete or dyadic entities of Physics, the elementary criterion of equality, the dyadic sets and the basic operations of addition, subtraction, and some multiplication and division of magnitudes.

The results of what **should be taught to science students**, in the degree that corresponds to their level of studies, are included in the following articles, replacing the **traditional toxic dogma** by an **algebraic logic for Physics, reasoned and grounded with coherence as science demands**.

Artículo 2

METODOLOGÍA Y RESULTADOS

De la misma manera que existen álgebras para los números y vectores abstractos, aceptadas universalmente, debería asentarse un álgebra de los entes diádicos, representantes de las cantidades de magnitudes, porque solo así se acabaría con la confusión e ignorancia imperantes y con el **dogma educativo vigente**, quedando mejor aclarados los significados de las distintas magnitudes compuestas, tal como humildemente se esboza en este compendio.

Existe una magnitud muy conocida e intuitiva, que es la **longitud geométrica**, y esta debe ser el punto de partida para fundamentar un álgebra genérica de magnitudes físicas. Para ello, es preciso establecer primero las leyes de composición de los segmentos, en tanto que son las figuras más elementales, y definir un **álgebra geométrica o gráfica** para componerlos antes de pasar a su correspondiente **álgebra analítica**. La adición geométrica no ofrece ninguna dificultad, basta concebirla como la yuxtaposición gráfica de segmentos, lo que analíticamente exige medirlos con la misma unidad de longitud. En cambio la multiplicación geométrica puede concebirse de modo que produzca nuevas magnitudes a partir de la longitud, el área con dos factores, el volumen con tres, o los hipervolúmenes con más de tres; y los segmentos multiplicados pueden expresarse en unidades de longitud cualesquiera, que no tienen por qué coincidir, como exige la adición. La longitud se dice por ello que sea una magnitud fundamental y las demás se llaman derivadas o compuestas por medio de la multiplicación, debiéndose advertir que esta operación, en principio gráfica, se diferencia sustancialmente con la noción del producto aritmético, dado por la adición de un multiplicando tantas veces como indique el multiplicador.

Cuando se multiplican segmentos de esta forma no es posible atribuir a ninguno de ellos la función de multiplicador aritmético, lo que evidencia que debe diferenciarse claramente esta operación de la multiplicación ordinaria.

348

Una vez establecida el álgebra de segmentos, nada impide asociar toda cantidad de cualquier magnitud a la cantidad de longitud de un segmento. Para ello, bastaría con identificar la unidad empírica de la magnitud considerada con una unidad de longitud arbitraria o abstracta. De este modo se podría establecer una correspondencia biunívoca entre el conjunto de todas las cantidades de la magnitud dada y el de todas las longitudes abstractas, es decir, sin escala real. El **postulado de afinidad** consiste en admitir la operativa anterior y manejar las cantidades de magnitudes como si fuesen segmentos geométricos abstractos, lo que equivale a suponer que, si bien las magnitudes son diferentes por naturaleza, sus cantidades son afines a las de la magnitud longitud. Este postulado, en combinación con la composición de áreas y volúmenes en el espacio euclídeo, permite definir la multiplicación diádica entre cualesquiera magnitudes escalares.

Partiendo como fundamento del álgebra de los segmentos geométricos, se desarrolla la de longitudes y, por razonable generalización, se llega con relativa facilidad a la definición precisa de las leyes de composición para cualesquiera magnitudes. Con ello quedan al descubierto los entramados ocultos de las unidades derivadas y pueden juzgarse con más acierto los significados que se les pueda atribuir. De modo que la noción de dimensión de toda magnitud ha de considerarse después y no antes de haber concebido un álgebra de magnitudes, cuya expresión matemática analítica son los entes concretos o elementos diádicos.

De ahí que el método seguido en esta exposición debería presentarse según la siguiente secuencia: primero, asentar los conceptos básicos propios de las magnitudes físicas en general; luego, asignarles entidad matemática y crear los entes concretos o diádicos; a continuación definir un álgebra para tales entes especiales, representantes precisos de las cantidades de magnitudes naturales medibles; investigar después el significado de las ecuaciones de definición, de las leyes universales y de otros entes físicos; y, finalmente, explicar los principios del análisis dimensional. Pues bien, así se hace en la parte nuclear precedente de este texto.

No obstante, en un compendio didáctico como el que aquí se realiza, es preciso limitar la extensión y seleccionar lo sustancial, de modo que se vislumbren en conjunto las posibilidades de la compleja álgebra resultante sin mermar la claridad expositiva. De ahí que aquí solo se puedan exponer de la forma más concisa posible los aspectos fundamentales: la definición de los entes concretos o diádicos de la Física, el criterio elemental de igualdad, los conjuntos diádicos y las operaciones básicas de la adición, sustracción, y algunas multiplicaciones y divisiones de magnitudes.

Los resultados de **lo que debería impartirse a los alumnos de ciencias**, en el grado que corresponda a su nivel de estudios, se incluyen en los artículos siguientes, sustituyendo el tóxico **dogma tradicional** por una **lógica algebraica para la Física, razonada y fundamentada con coherencia como exige la ciencia.**

Article 3

It is convenient to call **measurement** the quantity, extension or portion of a magnitud expressed in the form $q\,U$, as a symbol of the times q, a real number, that a unit quantity U is present in a phenomenon, calling q measured with the unit U of the magnitude included in the observed event. And similarly if the measure were a vector $\overline{q}$ of R^3.

The magnitudes whose measurements are such that $q \in R$ or that $|\overline{q}| \in R$ and that can take any value are called **continuous**, on the other hand, those in which the measurements can only be whole numbers, with $q \in Z$ or $|\overline{q}| \in Z$, they are called **discrete**. It is observed that the operations with discrete magnitudes are included in the continuous ones, since their measurements will be represented by whole numbers, a subset of the real numbers, so that the continuous ones present greater generality than the discrete ones; and the continuous ones will be explained in the abstract in many cases by means of the length, which fictitiously represents them all, because any of them can be assimilated to the real line, resulting in any case the same reasoning scheme.

In this way, every pair formed by a real number or a vector, followed by a unit that reflects a certain quantity of some magnitude is, by definition, a **concrete entity or physical dyad**.

If the primary q is a real number, the concrete will be called **scalar** and it will mean the quantity of a magnitude equal to q times the quantity of the same contained in the unit, which can only be described in the abstract by means of some symbol empirically associated with some phenomenon; so that the notation $(q\,U)$ or (q,U) or $q\,U$, without superfluous parentheses, to represent the measurement of a magnitude by the unit U with the real number q has parity in nature, hence the name of physical dyad.

Similarly, if the primary is characterized by a vector $\overline{q}$, the dyad will be said to be **vector**.

The scalar parity entities associated with each unit U form sets that can be symbolized $\{R,U\}$, and are capable of being composed together by internal composition laws, establishing applications of the Cartesian product $\{R,U\}\times\{R,U\}$ in $\{R,U\}$; and they can also be composed with the elements of other sets, such as R, through laws of external composition, with applications of $R\times\{R,U\}$ in $\{R,U\}$, so the task of establishing for them an adequate algebra, which must be tried to be as isomorphic as possible with the structure of the field of real numbers, because these are the universal model. We will frame these types of operations in the so-called **additive scalars**, indicating them with the sign «⊕».

On the other hand, when there are diverse dyadic sets associated to different units U_1 and U_2, such as $\{R,U_1\}$ and $\{R,U_2\}$, external composition laws can be defined, which we will call **multiplicative scalars** and symbolize with the asterisk «∗», through applications of $\{R,U_1\}\times\{R,U_2\}$ in $\{R,U_1 *U_2\}$. **They are laws that generate new magnitudes.**

In turn, vector dyads form sets with each U unit that can be symbolized $\{R^3,U\}$ and are capable of being composed of each other by internal composition laws, with applications of the Cartesian product $\{R^3,U\}\times\{R^3,U\}$ in $\{R^3,U\}$; and they can also be composed with the elements of other sets, such as R, by means of external composition laws, with applications of $R\times\{R^3,U\}$ in $\{R^3,U\}$, ensuring that it is isomorphic with the structure of the vector space R^3 on R. We will frame these types of operations in the **additive vector** calls, marking them with the sign «⊕».

In the same way as with scalar dyadic sets, for the vectors associated with different units U_1 and U_2, such as $\{R^3,U_1\}$ and $\{R^3,U_2\}$, we can define external composition laws, called **multiplicative vectors**, which we will symbolize with the asterisk «∗», by applying $\{R^3,U_1\}\times\{R^3,U_2\}$ in $\{R^3,U_1 *U_2\}$. Like scalars, these external laws are also **generators of new magnitudes.**

Since dyadic entities are made up of pairs of elements linked together and inseparable, a **mathematical primary** of ordinary algebra and a unitary **physical secondary**, their specific algebra must obey operational criteria with **dyads**, so proper composition laws must be established that allow the construction of a sui generis structure similar to the forms of classical dyadic algebra, a precursor of tensor algebra. On the other hand, the dyadic nature of the concrete entity would justify its being indicated with terms such as concrete dyad, physical dyad, dimensional dyad or other similar nomenclature.

Likewise, a **criterion of dyadic equality** is required, which will have to identify two concrete elements when the quantity of the magnitude to which they refer is the same, which in analytical terms will mean that, if they correspond to the same unit of measurement, the primaries must match. Thus, given two scalar dyads (q_1,U_1) and (q_2,U_2), it will be said that they are equal if and only if they represent the same quantity of the magnitude to which the units U_1 and U_2 belong, indicating said equality with the form $(q_1,U_1)=(q_2,U_2)$. Similarly, if the dyads were vector, the equality $(\overline{q}_1,U_1)=(\overline{q}_2,U_2)$ would be written. Note that equality can be expressed with any of the forms admitted for the pair notation: $(q_1,U_1)=(q_2,U_2)$ or $(q_1\ U_1)=(q_2\ U_2)$ or $q_1\ U_1=q_2\ U_2$, and so similar for vector dyads: $(\overline{q}_1,U_1)=(\overline{q}_2,U_2)$ or $(\overline{q}_1\ U_1)=(\overline{q}_2\ U_2)$ or $\overline{q}_1\ U_1=\overline{q}_2\ U_2$.

At the end of article 7 of this same section, this criterion of equality is generalized, once the multiplication of a scalar by a dyad and the dyadic division are defined, which are necessary to give it complete and precise meaning.

Artículo 3

DEFINICIÓN DE ENTE CONCRETO O DÍADA FÍSICA E IGUALDAD

Se conviene en llamar **medición** a la cantidad, extensión o porción de una magnitud expresada con la forma $q\ U$, como símbolo de las veces q, número real, que una cantidad unitaria U esté presente en un fenómeno, denominando a q medida con la unidad U de la magnitud incluida en el hecho observado. Y análogamente si la medida fuese un vector $\overline{q}$ de R^3.

Las magnitudes cuyas medidas sean tales que $q \in \mathrm{R}$ o que $|\overline{q}| \in \mathrm{R}$ y que puedan tomar cualquier valor se denominan **continuas**, en cambio, aquellas en que las medidas solo puedan ser números enteros, con $q \in \mathrm{Z}$ o $|\overline{q}| \in \mathrm{Z}$, se llaman **discretas**. Se observa que las operaciones con magnitudes discretas quedan comprendidas en las continuas, pues sus medidas vendrán representadas por números enteros, subconjunto de los números reales, por lo que las continuas presentan mayor generalidad que las discretas; y las continuas quedarán explicadas en abstracto en muchos casos por medio de la longitud, que las representa ficticiamente a todas, porque cualquiera de ellas se puede asimilar a la recta real, resultando en todo caso el mismo esquema de razonamiento.

De este modo, todo par formado por un número real o un vector, seguido de una unidad que refleje cierta cantidad de alguna magnitud es, por definición, un **ente concreto** o **díada física**.

Si el primario q es un número real, el concreto se llamará **escalar** y significará la cantidad de una magnitud igual a q veces la cantidad de la misma contenida en la unidad, que solo podrá describirse en abstracto mediante algún símbolo asociado empíricamente a algún fenómeno; de modo que la notación $(q\ U)$ o (q,U) o $q\ U$, sin paréntesis superfluos, para representar la medición de una magnitud mediante la unidad U con el número real q tiene naturaleza paritaria, es un par, de ahí el nombre de díada física.

Análogamente, si el primario quedase caracterizado por un vector $\overline{q}$, la díada se dirá que es **vectorial**.

Los entes paritarios escalares asociados a cada unidad U forman conjuntos que se pueden simbolizar $\{R,U\}$, y son susceptibles de componerse entre sí mediante leyes de composición interna, estableciendo aplicaciones del producto cartesiano $\{R,U\}\times\{R,U\}$ en $\{R,U\}$; y también pueden componerse con los elementos de otros conjuntos, como por ejemplo R, mediante leyes de composición externa, con aplicaciones de $R\times\{R,U\}$ en $\{R,U\}$, por lo que hay que abordar la tarea de establecer para ellos un álgebra adecuada, que habrá de procurarse sea lo más isomorfa posible con la estructura del cuerpo de los números reales, porque estos son el modelo universal. Este tipo de operaciones las encuadraremos en las llamadas **escalares aditivas**, señalándolas con el signo «$\oplus$».

Por otra parte, cuando se tengan conjuntos diádicos diversos, asociados a unidades diferentes U_1 y U_2, tales como $\{R,U_1\}$ y $\{R,U_2\}$, se podrán definir leyes de composición externas, que llamaremos **escalares multiplicativas** y simbolizaremos con el asterisco «$*$», mediante aplicaciones de $\{R,U_1\}\times\{R,U_2\}$ en $\{R,U_1*U_2\}$. **Son leyes generadoras de nuevas magnitudes.**

A su vez, las díadas vectoriales forman conjuntos con cada unidad U que se pueden simbolizar $\{R^3,U\}$ y son susceptibles de componerse entre sí mediante leyes de composición interna, con aplicaciones del producto cartesiano $\{R^3,U\}\times\{R^3,U\}$ en $\{R^3,U\}$; y también pueden componerse con los elementos de otros conjuntos, como por ejemplo R, mediante leyes de composición externa, con aplicaciones de $R\times\{R^3,U\}$ en $\{R^3,U\}$, procurando que sea isomorfa con la estructura del espacio vectorial R^3 sobre R. Este tipo de operaciones las encuadraremos en las llamadas **vectoriales aditivas**, señalándolas con el signo «$\oplus$».

De la misma forma que con los conjuntos diádicos escalares, para los vectoriales asociados a unidades diferentes U_1 y U_2, tales como $\{R^3,U_1\}$ y $\{R^3,U_2\}$, se podrán definir leyes de composición externas, llamadas **vectoriales multiplicativas**, que simbolizaremos con el asterisco «*», mediante aplicaciones de $\{R^3, U_1\} \times \{R^3, U_2\}$ en $\{R^3, U_1 * U_2\}$. Como las escalares, estas leyes externas son también **generadoras de nuevas magnitudes**.

Estando formados los entes diádicos por parejas de elementos enlazados entre sí e inseparables, un **primario matemático** del álgebra ordinaria y un **secundario físico** unitario, su álgebra específica deberá obedecer a criterios operacionales con **díadas**, por lo que deben establecerse leyes de composición propias que permitan construir una estructura sui géneris similar a las formas del álgebra diádica clásica, precursora a su vez de la tensorial. Por otra parte, la naturaleza diádica del ente concreto justificaría que fuese indicado con términos como díada concreta, díada física, díada dimensional u otra nomenclatura similar.

Asimismo, se precisa de un **criterio de igualdad diádica**, que habrá de identificar dos elementos concretos cuando la cantidad de la magnitud a que se refieran sea la misma, lo que en términos analíticos significará que, si corresponden a la misma unidad de medida, los primarios deban coincidir. Así, dadas dos díadas escalares (q_1, U_1) y (q_2, U_2), se dirá que son iguales si y solo si representan la misma cantidad de la magnitud a que pertenecen las unidades U_1 y U_2, indicándose dicha igualdad con la forma $(q_1, U_1) = (q_2, U_2)$. Análogamente si las díadas fuesen vectoriales, se escribiría la igualdad $(\overline{q}_1, U_1) = (\overline{q}_2, U_2)$. Nótese que la igualdad se puede expresar con cualquiera de las formas admitidas para la notación de pares: $(q_1, U_1) = (q_2, U_2)$ o $(q_1 \ U_1) = (q_2 \ U_2)$ o $q_1 \ U_1 = q_2 \ U_2$, y de modo similar para las díadas vectoriales: $(\overline{q}_1, U_1) = (\overline{q}_2, U_2)$ o $(\overline{q}_1 \ U_1) = (\overline{q}_2 \ U_2)$ o $\overline{q}_1 \ U_1 = \overline{q}_2 \ U_2$.

Al final del artículo 7 de este mismo apartado se generaliza este criterio de igualdad, una vez definidas la multiplicación de un escalar por una díada y la división diádica, que son necesarias para darle sentido completo y preciso.

Article 4

There is a physical magnitude that can be inspiring of how it should be operated with all the others, it is the longitude. And it is precisely Euclidean geometry that has resolved the way of composing lengths, through the **geometric algebra of segments**, which establishes the addition and subtraction of lengths by means of the adequate graphic juxtaposition of the segments to be composed, analytically by means of the prior requirement that the measurements of the components are referred to the same unit or **axiom of uniformity**, simply so that they can be counted as equal elements. Thus, given two quantities of length expressed, for example, in centimeters, (q_1 cm) and (q_2 cm), the juxtaposition of these lengths will have a length equal to [(q_1+q_2) cm]. Identifying this operation with the addition of segments, the dyadic sum of lengths could be symbolized with the proper sign «$\oplus$», and the analytical definition of segment addition would be:

$$(q_1\, cm)\oplus(q_2\, cm)=[(q_1+q_2)\, cm]$$

This definition can be generalized to any magnitude, **idealizing every physical quantity with the fiction that it is a quantity of length,** what we have called the **affinity postulate**. And so it is easy to arrive at the generic dyadic addition as a **law of internal composition** or application of the Cartesian product $\{R,U\}\times\{R,U\}$ in $\{R,U\}$, for scalar magnitudes, or application of $\{R^3,U\}\times\{R^3,U\}$ in $\{R^3,U\}$, for vectors. Thus the two respective analytical definitions result, which, expressed with superfluous parentheses, for greater significance and distinction of their elements, are:

$$(q_1\, U)\oplus(q_2\, U)=[(q_1+q_2)\, U]$$
$$(\overline{q}_1\, U)\oplus(\overline{q}_2\, U)=[(\overline{q}_1+\overline{q}_2)\, U]$$

Note that, although in both the same «$\oplus$» sign appears for the dyadic addition, in the first equation it represents the scalar dyadic sum and in the second the vector one, in the same way that the « + » sign of the first is the addition of R and that of the second is the vector sum of R^3.

All of which is nothing more than a reflection of the existence of an isomorphism between the structures of the dyadic sets $\{R,U\}$ and $\{R^3,U\}$.

Artículo 4

LA ADICIÓN DIÁDICA

Existe una magnitud física que puede ser inspiradora de cómo deba operarse con todas las demás, es la longitud. Y precisamente la geometría euclidiana tiene resuelta la manera de componer longitudes, mediante el **álgebra geométrica de segmentos**, que asienta la adición y sustracción de longitudes mediante la yuxtaposición gráfica adecuada de los segmentos a componer, analíticamente mediante la exigencia previa de que las mediciones de los componentes estén referidas a la misma unidad o **axioma de uniformidad**, simplemente para que puedan contarse como elementos iguales. Así, dadas dos cantidades de longitud expresadas, por ejemplo, en centímetros, $(q_1 \ cm)$ y $(q_2 \ cm)$, la yuxtaposición de estas longitudes tendrá una longitud igual a $[(q_1 + q_2) \ cm]$. Identificando esta operación con la adición de segmentos, la suma diádica de longitudes se podría simbolizar con el signo propio «$\oplus$», y la definición analítica de adición de segmentos sería:

$$(q_1 \ cm) \oplus (q_2 \ cm) = [(q_1 + q_2) \ cm]$$

Esta definición se puede generalizar a cualquier magnitud, **idealizando toda cantidad física con la ficción de que sea una cantidad de longitud**, lo que hemos llamado **postulado de afinidad**. Y así se llega con facilidad a la adición diádica genérica como **ley de composición interna** o aplicación del producto cartesiano $\{R, U\} \times \{R, U\}$ en $\{R, U\}$, para las magnitudes escalares, o aplicación de $\{R^3, U\} \times \{R^3, U\}$ en $\{R^3, U\}$, para las vectoriales. Resultan así las dos definiciones analíticas respectivas, que expresadas con paréntesis superfluos, para mayor significación y distinción de sus elementos, son:

$$(q_1 \ U) \oplus (q_2 \ U) = [(q_1 + q_2) \ U]$$

$$(\overline{q}_1 \ U) \oplus (\overline{q}_2 \ U) = [(\overline{q}_1 + \overline{q}_2) \ U]$$

Nótese que, aunque en ambas aparece el mismo signo «$\oplus$» para la adición diádica, en la primera ecuación representa la suma diádica escalar y en la segunda la vectorial, de la misma forma que el signo «+» de la primera es la adición de R y el de la segunda es la suma vectorial de R^3.

Todo lo cual no es más que el reflejo de la existencia de un isomorfismo entre las estructuras de los conjuntos diádicos $\{R, U\}$ y $\{R^3, U\}$.

Article 5

DEDUCTION OF DYADIC SUBTRACTION

The **dyadic subtraction** can be deduced from the addition and based on the generic subtraction criterion, which establishes the difference between a minuend and a subtrahend as that remainder or difference such that added to the subtrahend gives the minuend. To do this, let the scalar dyadic sum $(dU) \oplus (sU) = (mU)$. The parentheses are dispensable, but they are kept to mark the dyads well, and the symbology of the dyadic addition has simply been adapted to indicate with the letters m a minuend, s a subtrahend and d for the difference that corresponds to them. The usual subtraction criterion, as an operation that, given an addition, allows one of the addends to be obtained as a function of the sum and the other by adding, is just another way of writing the initial sum, which allows establishing the uniform dyadic difference , distinguished with the sign «$\ominus$», by the equation:

$$(m\,U) \ominus (s\,U) = (d\,U)$$

The definition of dyadic addition applied to the initial sum allows it to be written in the form $[(d+s)\,U] = (m\,U)$. The simple criterion of equality of dyads consists in considering them equal when their primaries and secondaries coincide. The equality of primaries leads to the relation $(d+s) = m$. For its part, the definition of subtraction in R leads to $d = m-s$. So, substituting d in $(m\,U) \ominus (s\,U) = (d\,U)$, we finally have:

$$(m\,U) \ominus (s\,U) = [(m-s)\,U]$$

And this is the analytic definition of dyadic subtraction, and it means that the dyadic difference between two uniform scalar dyads, called minuend and subtrahend, is a dyad called difference whose primary is the subtraction in R of the primaries and with the same secondary as them. It is a law of internal composition of $\{R,U\} \times \{R,U\}$ in $\{R,U\}$.

The subtraction of uniform vector dyads is isomorphic with the scalar, given the abelian and additive group structure of R^3, which presents the same formal properties for the sum of vectors that occur with the real numbers.

Artículo 5

DEDUCCIÓN DE LA SUSTRACCIÓN DIÁDICA

La **resta diádica** se puede deducir a partir de la adición y en función del criterio genérico de sustracción, que establece la diferencia entre un minuendo y un sustraendo como aquel resto o diferencia tal que sumado al sustraendo dé el minuendo. Para ello, sea la suma diádica escalar $(d\ U) \oplus (s\ U) = (m\ U)$. Los paréntesis son prescindibles, pero se mantienen para marcar bien la díadas, y simplemente se ha adaptado la simbología de la adición diádica para indicar con las letras m un minuendo, s un sustraendo y d para la diferencia que les corresponda. El criterio usual de sustracción, como operación que, dada una adición, permite obtener uno de los sumandos en función de la suma y del otro sumando, no es más que otra forma de escribir la suma inicial, lo que permite establecer la diferencia diádica uniforme, distinguida con el signo «$\ominus$», mediante la ecuación:

$$(m\ U) \ominus (s\ U) = (d\ U)$$

La definición de adición diádica aplicada a la suma inicial permite escribirla con la forma $[(d+s)\ U] = (m\ U)$. El criterio simple de igualdad de díadas consiste en considerarlas iguales cuando coincidan sus primarios y secundarios. La igualdad de primarios lleva a la relación $(d+s) = m$. Por su parte, la definición de sustracción en R conduce a $d = m - s$. De modo que, sustituyendo d en $(m\ U) \ominus (s\ U) = (d\ U)$, se tiene finalmente:

$$(m\ U) \ominus (s\ U) = [(m-s)\ U]$$

Y esta es la definición analítica de sustracción diádica, y significa que la diferencia diádica entre dos díadas escalares uniformes, llamadas minuendo y sustraendo, es una díada llamada diferencia cuyo primario es la sustracción en R de los primarios y con el mismo secundario que ellos. Se trata de una ley de composición interna de $\{R, U\} \times \{R, U\}$ en $\{R, U\}$.

La sustracción de díadas vectoriales uniformes es isomorfa con la escalar, dada la estructura de grupo aditivo y abeliano de R^3, que presenta las mismas propiedades formales para la suma de vectores que se dan con los números reales.

Article 6

DYADIC MULTIPLICATION BY A SCALAR

The dyadic addition allows to conceive the case in which all the addends of a sum are equal, that is to say that, referring to the same magnitude or being **homogeneous**, in addition, they are referred to the same unit, that is, they are **uniform**. So, being p a real number and, given a scalar physical dyad $(q\,U)$, with q belonging to R, the dyadic addition $(q\,U) \oplus (q\,U) \oplus \ldots \oplus (q\,U)$ can be formed with p adding us. Well, the result of this sum, which means p times the quantity represented by the physical pair $(q\,U)$, can be briefly symbolized as a multiplication indicated by $p \circ (q\,U)$, where the symbol «$\circ$» indicates the operation of adding equal dyads a certain number of times. The similarity of this operation with the arithmetic multiplication «$\times$» is evident, because the factor p acts as a multiplier; however, their difference is notable, because this refers to operations with real numbers, while the present one operates with real numbers and physical dyads, then, the sets that relate these forms of multiplication do not coincide, hence, strictly speaking, they must be assigned different operational symbols, so as not to favor the algebraic confusion of disparate laws of composition.

It is clear that the same sum of identical dyadic elements can be represented with the form $(q\,U) \circ p$, without more than attributing p the multiplier function on both sides, which is equivalent to axiomatizing the **commutative property** of this kind of multiplication, and with it its analytical definition can be written like this:

$$p \circ (q\,U) = (q\,U) \circ p = (q\,U) \oplus (q\,U) \oplus \ldots \oplus (q\,U) \text{ with } p \text{ addends}$$

The definition of dyadic addition allows you to write the second member of the definition equation above like this:

$$(q\,U) \oplus (q\,U) \oplus \ldots \oplus (q\,U) = [(q + q + \ldots + q)\,U] \text{ with } p \text{ addends}$$

The definition of multiplication «$\times$» in R allows us to write the abbreviated sum $q + q + \ldots + q$ with p addends, with the form $p \times q$

or $q \times p$, because the multiplication of R is commutative. Therefore, in conclusion, the **multiplication of scalar dyads with real elements of R** is established by the equation:

$$p \circ (q\,U) = (q\,U) \circ p = [(p \times q)\,U] = [(q \times p)\,U]$$

If the dyad were of a vector nature $(\overline{q}\,U)$, being $\overline{q}$ an element of R^3, the development is analogous. To do this, using the same sign «∘» for this composition law, even knowing that it is a different operation, the sum to describe in this case would be $(\overline{q}\,U) \oplus (\overline{q}\,U) \oplus \ldots \oplus (\overline{q}\,U)$ with p addends; but here the symbol «⊕» indicates the vector dyadic addition, which allows us to write the following:

$$(\overline{q}\,U) \oplus (\overline{q}\,U) \oplus \ldots \oplus (\overline{q}\,U) = [(\overline{q} + \overline{q} + \ldots + \overline{q})\,U] \text{ with } p \text{ addends}$$

Note that the «+» that appears in this equation is not the addition of R, but the addition of R^3, even though the same sign is used. And, as R^3 has a vector space structure over R, designating the product of a scalar by a vector with the sign «•», we arrive at the formulation described in the final definition equation:

$$p \circ (\overline{q}\,U) = (\overline{q}\,U) \circ p = [(p \bullet \overline{q})\,U] = [(\overline{q} \bullet p)\,U]$$

The **multiplication of real numbers by scalar dyads** is nothing but a law of external composition or application of the Cartesian product $R \times \{R,U\}$ in $\{R,U\}$ on the left and the symmetric on the right of $\{R,U\} \times R$ in $\{R,U\}$.

In turn, the **multiplication of real numbers by vector dyads** defines an external composition law or application of the Cartesian product $R \times \{R^3,U\}$ in $\{R^3,U\}$ on the left and the symmetric on the right of $\{R^3,U\} \times R$ in $\{R^3,U\}$.

It is observed that the forms deduced for this kind of multiplication justify that it can be operated symbolically as if all the symbols were elements of R, without actually being it, commuting and grouping them with the usual rules, which is due to the isomorphism that is established between the different algebraic structures that participate in reasoning. However, it would be wrong to understand such a way of operating as something immediate, because it is not, it requires justification, otherwise it would be to contaminate logical development with inadmissible presuppositions.

Note that in the preceding arguments it has been tacitly assumed that the multiplying number p belonging to R is also an integer, since we always speak of p addends. However, the logical deductions can be repeated considering any rational number p/h, with p and h integers, simply by forming the sums $q/h+q/h+ \dots +q/h$ with p as addends, that is, also with p as multiplier. And thus the reason for the definition is completed with any number of R, leaving aside the incommensurables, for which limit concepts would have to be introduced, which would also include these, covering the entire spectrum of real numbers.

As we already did at the end of section IX and we repeat here due to its importance and inclusion in this alternative formulation, we must ask ourselves what happens to a dyad $(q\,U)$ when its unit is multiplied by a number p. If the dyad $(q\,U)$ indicates the quantity of magnitude equal to q times the quantity indicated by U, the new dyad $(q\,p{\circ}U)$ must represent the quantity of p times the quantity $(q\,U)$, which allows us to complete the definition of the multiplicative operation in this section with the following equivalent analytic forms:

$$(q\,p{\circ}U)=p{\circ}(q\,U)=(p{\times}q\,U)=(p{\times}q){\circ}U$$

$$(p\,U)=p{\circ}(1\,U)=p{\circ}U=(1\,p{\circ}U)$$

Another very obvious property that can be useful in some deductions is obtaining the multiplicative form of every dyadic expression. In this regard, given any dyad $(q\,U)$, the following reasoning can be put together:

$$(q\,U)=q{\circ}(1\,U)=q{\circ}U$$

Therefore, the quantity of magnitude that any dyad symbolizes is the product of its primary by its secondary.

Artículo 6

MULTIPLICACIÓN DIÁDICA POR UN ESCALAR

La adición diádica permite concebir el caso en que todos los sumandos de una suma sean iguales, es decir que, refiriéndose a la misma magnitud o siendo **homogéneos**, además, estén referidos a la misma unidad, o sea, que sean **uniformes**. Así que, siendo p un número real y, dada una díada física escalar $(q\ U)$, con q perteneciente a R, se puede formar la adición diádica $(q\ U) \oplus (q\ U) \oplus \ldots \oplus (q\ U)$ con p sumandos. Pues bien, el resultado de esta suma, que significa p veces la cantidad que represente el par físico $(q\ U)$, se puede simbolizar abreviadamente como una multiplicación indicada por $p \circ (q\ U)$, donde el símbolo «∘» indica la operación de sumar díadas iguales un determinado número de veces. Es evidente la similitud de esta operación con la multiplicación aritmética «×», porque el factor p actúa como multiplicador; sin embargo, su diferencia es notable, porque esta se refiere a operaciones con números reales, mientras que la presente opera con números reales y díadas físicas, luego, los conjuntos que relacionan dichas formas de multiplicación no coinciden, de ahí que en rigor deban ser asignados símbolos operacionales distintos, para no favorecer la confusión algebraica de leyes de composición dispares.

Está claro que la misma suma de elementos diádicos idénticos se puede representar con la forma $(q\ U) \circ p$, sin más que atribuir a p la función de multiplicador por ambos lados, lo que equivale a axiomatizar la **propiedad conmutativa** de esta especie de multiplicación, y con ello su definición analítica se puede escribir así:

$$p \circ (q\ U) = (q\ U) \circ p = (q\ U) \oplus (q\ U) \oplus \ldots \oplus (q\ U) \text{ con } p \text{ sumandos}$$

La definición de adición diádica permite escribir el segundo miembro de la ecuación de definición anterior de esta manera:

$$(q\ U) \oplus (q\ U) \oplus \ldots \oplus (q\ U) = [(q + q + \ldots + q)\ U] \text{ con } p \text{ sumandos}$$

La definición de multiplicación «×» en R permite escribir la suma abreviada $q + q + \ldots + q$ con p sumandos, con la forma $p \times q$ o

$q \times p$, porque la multiplicación de R es conmutativa. Por lo que, en conclusión, la **multiplicación de díadas escalares con elementos reales de R** queda establecida por la ecuación:

$$p \circ (q\ U) = (q\ U) \circ p = [(p \times q)\ U] = [(q \times p)\ U]$$

Si la díada fuese de índole vectorial $(\overline{q}\ U)$, siendo $\overline{q}$ un elemento de R^3, el desarrollo es análogo. Para ello, empleando el mismo signo «$\circ$» para esta ley de composición, aun sabiendo que se trata de una operación diferente, la suma a describir en este caso sería $(\overline{q}\ U) \oplus (\overline{q}\ U) \oplus \ldots \oplus (\overline{q}\ U)$ con p sumandos; pero aquí el símbolo «$\oplus$» indica la adición diádica vectorial, que permite escribir lo siguiente:

$$(\overline{q}\ U) \oplus (\overline{q}\ U) \oplus \ldots \oplus (\overline{q}\ U) = [(\overline{q} + \overline{q} + \ldots + \overline{q})\ U] \text{ con } p \text{ sumandos}$$

Nótese que el «$+$» que aparece en esta ecuación no es la adición de R, sino la de R^3, aunque se utilice el mismo signo. Y, como R^3 tiene estructura de espacio vectorial sobre R, designando el producto de un escalar por un vector con el signo «$\bullet$», se llega a la formulación descrita en la ecuación final de definición:

$$p \circ (\overline{q}\ U) = (\overline{q}\ U) \circ p = [(p \bullet \overline{q})\ U] = [(\overline{q} \bullet p)\ U]$$

La **multiplicación de números reales por díadas escalares** no es sino una ley de composición externa o aplicación del producto cartesiano $R \times \{R, U\}$ en $\{R, U\}$ por la izquierda y la simétrica por la derecha de $\{R, U\} \times R$ en $\{R, U\}$.

A su vez, la **multiplicación de números reales por díadas vectoriales** define una ley de composición externa o aplicación del producto cartesiano $R \times \{R^3, U\}$ en $\{R^3, U\}$ por la izquierda y la simétrica por la derecha de $\{R^3, U\} \times R$ en $\{R^3, U\}$.

Se observa que las formas deducidas para esta especie de multiplicación justifican que pueda operarse simbólicamente como si todos los símbolos fuesen elementos de R, sin serlo realmente, conmutándolos y agrupándolos con las reglas usuales, lo que se debe al isomorfismo que se establece entre las distintas estructuras algebraicas que participan en los razonamientos. Sin embargo, sería erróneo entender tal forma de operar como algo inmediato, porque no lo es, requiere justificación, otra cosa sería contaminar el desarrollo lógico con presuposiciones inadmisibles.

Nótese que en los razonamientos precedentes se ha supuesto tácitamente que el número multiplicador p perteneciente a R sea además entero, puesto que se habla siempre de p sumandos. Sin embargo, pueden repetirse las deducciones lógicas considerando cualquier número racional p/h, con p y h enteros, sin más que formar las sumas $q/h + q/h + \ldots + q/h$ con p sumandos, es decir, también con p como multiplicador. Y así se completa el porqué de la definición con cualquier número de R, dejando aparte los incomensurables, para los que habría que introducir los conceptos de límite, con lo que también quedarían estos incluidos, cubriendo todo el espectro de los números reales.

Como ya hicimos al final del apartado IX y repetimos aquí por su importancia e inclusión en esta formulación alternativa, debemos preguntarnos qué le ocurre a una díada $(q\ U)$ cuando su unidad se multiplica por un número p. Si la díada $(q\ U)$ señala la cantidad de magnitud igual a q veces la cantidad indicada por U, la nueva díada $(q\ p{\circ}\,U)$ debe representar la cantidad de p veces la cantidad $(q\ U)$, lo cual nos permite completar la definición de la operación multiplicativa de este apartado con las siguientes formas analíticas equivalentes:

$$(q\ p{\circ}\,U) = p{\circ}(q\ U) = (p{\times}q\ U) = (p{\times}q){\circ}\,U$$

$$(p\ U) = p{\circ}(1\ U) = p{\circ}\,U = (1\ p{\circ}\,U)$$

Otra propiedad muy evidente que puede resultar de utilidad en algunas deducciones es la obtención de la forma multiplicativa de toda expresión diádica. A este respecto, dada cualquier díada $(q\ U)$, se puede hilar el siguiente razonamiento:

$$(q\ U) = q{\circ}(1\ U) = q{\circ}\,U$$

Por tanto, la cantidad de magnitud que simboliza una díada cualquiera es el producto de su primario por su secundario.

Article 7

DEDUCTION OF THE UNIFORM DYADIC DIVISION

The dyadic multiplication by a scalar, which relates concrete entities referred to the same unit, taking advantage of the generic criterion of dividing, allows us to deduce the analytical form of the kind of division in which dividend and divisor are uniform, scalar or vector dyads. In the first case, given two dyads $(q_1 U)$ and $(q_2 U)$, by the properties of R, there will always be a real number p such that $q_1 = p \times q_2$. This allows us to affirm the dyadic equality $[(p \times q_2) U] = (q_1 U)$. So, considering the definition of multiplication of a scalar by a concrete, the first member can be written in the form $[(p \times q_2) U] = p \circ (q_2 U)$. Combining both equations, we have $p \circ (q_2 U) = (q_1 U)$. And this expression, with respect to the multiplicative form «$\circ$», can be interpreted as a division between the dividend $(q_1 U)$ and the divisor $(q_2 U)$, resulting in the quotient $p = q_1 / q_2$.

Distinguishing this operation with a **double bar**, inclined or horizontal, we arrive at the result that the **dyadic division of uniform scalar elements**, that is, referred to the same unit, must give as a result a real number, which will be precisely the quotient in R of the primaries between dividend and divisor. Analytically we will have:

$$\frac{(q_1 U)}{(q_2 U)} = \frac{q_1}{q_2} = p$$

The observation of this equation proves that the form of division established here operates like that of R, allowing us to simplify the equal terms U that appear both in the numerator and in the denominator. However, such a substantial property is not due to the dogmatic application of the algebra of real numbers, but to the dyadic definitions that are being formulated. What happens is that the resulting isomorphism allows us to operate with the various symbols as if they were all elements of R, without being it, as the units of magnitudes are not.

For vector dyads, it should be noted that the algebra of R^3 is such that multiplication by a scalar relates **collinear vectors**, so that uniform dyadic division will only be possible when the dividend and divisor primaries are in turn collinear. So, let us now be the vector dyads $(\overline{q}_1 U)$ and $(\overline{q}_2 U)$, such that vectors $\overline{q}_1$ y $\overline{q}_2$ are collinear. Given the algebra of the vector space R^3, the existence of a scalar p of R such that $\overline{q}_1 = p \bullet \overline{q}_2$ is certain. Operating with vector algebra and with the definition of a dyadic product by a scalar, the following reasoning can be written with full justification: in R^3, the initial equality $\overline{q}_1 = p \bullet \overline{q}_2$ allows establishing the identity of the dyads $(\overline{q}_1 U) = [(p \bullet \overline{q}_2) U]$; the product «∘» converts the second member into $[(p \bullet \overline{q}_2) U] = p \circ (\overline{q}_2 U)$, from which it follows that $(\overline{q}_1 U) = p \circ (\overline{q}_2 U)$, and in this equation, in relation to the multiplicative operation «∘», we can apply the generic criterion of division and consider the factor $(\overline{q}_1 U)$ as a dividend, the $(\overline{q}_2 U)$ as a divisor and p as a quotient, with which it is concluded that the quotient between two vector dyads with collinear and uniform primaries, that is, referred to the same unit U, is a real number p such that $\overline{q}_1 = p \bullet \overline{q}_2$. Expressing this analytically, we have the definition equation of this uniform **vector dyadic quotient:**

$$\frac{\left(\overline{q}_1 U\right)}{\left(\overline{q}_2 U\right)} = p \ \text{ of } \ R \ \text{ and such that } \ \overline{q}_1 = p \bullet \overline{q}_2$$

Here we also observe the permissiveness of simplification of the identical symbols U that appear both in the numerator and in the denominator, but with the difference that the quotient of the second member is not that of R, but that of R^3, and **only for collinear vectors**. With this, the presuppositions that are admitted without any rigor are made manifest when the symbolic tradition is applied without further ado, based on the principle of economy of operational signs, which handles all the elements that relate quantities of physical magnitudes with the unfounded tacit assumption that they are behave as elements of R.

On the contrary, with the algebra of magnitudes, the different transformations allowed for the dyadic equations are justified one by one, establishing them with the **quality due to scientific, logical and didactic rigor.**

In the event that the dividend and divisor units are not uniform but homogeneous, there will be a dividend $(q_1 U_1)$ and a divisor $(q_1 U_2)$. The axiom of continuity guarantees that there exists a real number k of R such that $U_1 = k \circ U_2$. Now it can be formed and operated dyadic with the following quotient:

$$\frac{(q_1 U_1)}{(q_2 U_2)} = \frac{\left[q_1 (k \circ U_2)\right]}{(q_2 U_2)} = \frac{\left[(q_1 \times k) U_2\right]}{(q_2 U_2)}$$

Once the numerator and denominator have been reduced to uniform units, that is, equal, we can apply the property of the uniform quotient and eliminate the unit U_2 from the numerator and denominator, resulting in:

$$\frac{\left[(q_1 \times k) U_2\right]}{(q_2 U_2)} = \frac{q_1 \times k}{q_2} = p \times k$$

We have, then, that the dyadic quotient of two homogeneous elements $(q_1 U_1)$ and $(q_2 U_2)$, is a real number, without dimension, which is given by the ordinary quotient of the second member of the previous expression.

For homogeneous and non-uniform vector dyads $(\overline{q_1} U_1)$ and $(\overline{q_2} U_2)$, with $U_1 = k \circ U_2$, following the same reasoning as above, with vectors $\overline{q_1}$ and $\overline{q_2}$ being collinear, we will verify $\overline{q_1} = p \bullet \overline{q_2}$, with $p \in R$, and the quotient of collinear vectors $\overline{q_1} / \overline{q_2} = p$ can be formed. Under these conditions, the following reasoning can be made:

$$\frac{(\overline{q_1} U_1)}{(\overline{q_2} U_2)} = \frac{\left[\overline{q_1} (k \circ U_2)\right]}{(\overline{q_2} U_2)} = \frac{(\overline{q_1} U_2)}{(\overline{q_2} U_2)} \times k = \frac{\overline{q_1}}{\overline{q_2}} \times k = p \times k$$

In conclusion, the dyadic quotient of two homogeneous, non-uniform and collinear vector measurements is given by the factor $k{\times}p$, where k is the real number that represents the dyadic ratio between the homogeneous units U_1 and U_2 of the secondaries, and p the real number which indicates the ratio between the collinear vectors $\overline{q}_1$ and $\overline{q}_2$ of the dyadic primaries.

Thus we are now in a position to complete the **dyadic equality criterion** in section IV. It is enough to take into account that the axiom of continuity guarantees the existence of the real number k such that $U_1 = k{\circ}U_2$. So, if two dyads $(q_1\, U_1)$ and $(q_2\, U_2)$ are equal, which is denoted $(q_1\, U_1) = (q_2\, U_2)$, their dyadic ratio must obviously be the unit of real numbers. Therefore, we can conclude the following:

$$\frac{\left(q_1\, U_1\right)}{\left(q_2\, U_2\right)} = p \times k = 1 \;\Rightarrow\; p = \frac{1}{k}$$

We have verified that exactly the same relationship between p and k holds for vector dyads with collinear measures. Consequently, for both scalar and vector magnitudes, it can be stated that, **if two dyads are equal, the algebraic ratio p of their primaries is the inverse of the dyadic ratio k of their secondaries.**

These properties of the definition of equality of quantities of magnitudes, which we remember only make sense for homogeneous dyads, that is, representative of the same magnitude, although the units to which they refer are different, are essential for the construction and interpretation of the laws and equations of Physics.

Artículo 7

DEDUCCIÓN DE LA DIVISIÓN DIÁDICA UNIFORME

La multiplicación diádica por un escalar, que relaciona entes concretos referidos a la misma unidad, aprovechando el criterio genérico de dividir, permite deducir la forma analítica de la especie de división en que dividendo y divisor sean díadas uniformes, escalares o vectoriales. En el primer caso, dadas dos díadas $(q_1\ U)$ y $(q_2\ U)$, por las propiedades de R, siempre existirá un número real p tal que $q_1 = p \times q_2$. Ello permite afirmar la igualdad diádica $[(p \times q_2)\ U] = (q_1\ U)$. Así que, considerando la definición de multiplicación de un escalar por un concreto, el primer miembro se puede escribir con la forma $[(p \times q_2)\ U] = p \circ (q_2\ U)$. Combinando ambas ecuaciones, se tiene $p \circ (q_2\ U) = (q_1\ U)$. Y esta expresión, respecto de la forma multiplicativa «$\circ$», se puede interpretar como una división entre el dividendo $(q_1\ U)$ y el divisor $(q_2\ U)$, dando como resultado el cociente $p = q_1 / q_2$.

Distinguiendo esta operación con una **doble barra**, inclinada u horizontal, se llega al resultado de que la **división diádica de elementos escalares uniformes**, es decir, referidos a la misma unidad, ha de dar como resultado un número real, que será precisamente el cociente en R de los primarios entre dividendo y divisor. Analíticamente se tendrá:

$$\frac{\left(q_1\ U\right)}{\left(q_2\ U\right)} = \frac{q_1}{q_2} = p$$

La observación de esta ecuación prueba que la forma de división aquí establecida opera como la de R, permitiendo simplificar los términos iguales U que aparezcan a la vez en el numerador y en el denominador. Sin embargo, tan sustancial propiedad no se debe a que se aplique dogmáticamente el álgebra de los números reales, sino a las definiciones diádicas que se están formulando. Lo que pasa es que el isomorfismo resultante permite operar con los diversos símbolos como si fueran todos elementos de R, sin serlo, como no lo son las unidades de magnitudes.

Para concretos vectoriales hay que advertir que el álgebra de R^3 es tal que la multiplicación por un escalar relaciona **vectores colineales**, por lo que la división diádica uniforme solo será posible cuando los primarios de dividendo y divisor sean a su vez colineales. De modo que, sean ahora los concretos vectoriales $(\overline{q}_1\ U)$ y $(\overline{q}_2\ U)$, tales que los vectores $\overline{q}_1$ y $\overline{q}_2$ sean colineales. Dada el álgebra del espacio vectorial R^3, es segura la existencia de un escalar p de R tal que $\overline{q}_1 = p \bullet \overline{q}_2$. Operando con el álgebra vectorial y con la definición de producto diádico por un escalar, se podrá escribir con plena justificación el siguiente razonamiento: en R^3, la igualdad inicial $\overline{q}_1 = p \bullet \overline{q}_2$ permite establecer la identidad de las díadas $(\overline{q}_1\ U) = [(p \bullet \overline{q}_2)\ U]$; el producto «$\circ$» convierte el segundo miembro en $[(p \bullet \overline{q}_2)\ U] = p \circ (\overline{q}_2\ U)$, de donde resulta que $(\overline{q}_1\ U) = p \circ (\overline{q}_2\ U)$, y en esta ecuación, en relación con la operación multiplicativa «$\circ$», se puede aplicar el criterio genérico de la division y considerar el factor $(\overline{q}_1\ U)$ como un dividendo, el $(\overline{q}_2\ U)$ como un divisor y p como un cociente, con lo que se concluye que el cociente entre dos díadas vectoriales con primarios colineales y uniformes, es decir, referidos a la misma unidad U, es un número real p tal que $\overline{q}_1 = p \bullet \overline{q}_2$. Expresando esto analíticamente, se tiene la ecuación de definición de este **cociente diádico vectorial uniforme**:

$$\frac{\left(\overline{q}_1\ U\right)}{\left(\overline{q}_2\ U\right)} = p \ \text{ de } \ R \ \text{ y } \ \text{tal} \ \text{ que } \ \overline{q}_1 = p \bullet \overline{q}_2$$

Se observa también aquí la permisividad de simplificación de los símbolos idénticos U que aparezcan a la vez en el numerador y en el denominador, pero con la diferencia de que el cociente del segundo miembro no es el de R, sino el de R^3, y **solo para vectores colineales**. Con ello quedan manifiestas las presuposiciones que sin ningún rigor se admiten cuando se aplica sin más la tradición simbólica, basada en el principio de economía de signos operacionales, que maneja todos los elementos que relacionan cantidades de magnitudes físicas con la suposición tácita infundada de que se comporten como elementos de R.

Por el contrario, con el álgebra de magnitudes se van justificando una a una las distintas transformaciones permitidas para las ecuaciones diádicas, asentándolas con la **calidad debida al rigor científico, lógico y didáctico.**

En el caso de que las unidades de dividendo y divisor no sean uniformes pero sí homogéneas, se tendrá un dividendo $(q_1\,U_1)$ y un divisor $(q_2\,U_2)$. El axioma de continuidad garantiza que exista un número real k de R tal que $U_1 = k \circ U_2$. Ahora se puede formar y operar diádicamente con el siguiente cociente:

$$\frac{\left(q_1\,U_1\right)}{\left(q_2\,U_2\right)} = \frac{\left[q_1\left(k \circ U_2\right)\right]}{\left(q_2\,U_2\right)} = \frac{\left[\left(q_1 \times k\right)U_2\right]}{\left(q_2\,U_2\right)}$$

Una vez reducidos el numerador y el denominador a unidades uniformes, es decir, iguales, se puede aplicar la propiedad del cociente uniforme y eliminar la unidad U_2 de numerador y denominador, resultando:

$$\frac{\left[\left(q_1 \times k\right)U_2\right]}{\left(q_2\,U\right)_2} = \frac{q_1 \times k}{q_2} = p \times k$$

Se tiene, entonces, que el cociente diádico de dos elementos homogéneos $(q_1\,U_1)$ y $(q_2\,U_2)$, es un número real, sin dimensión, que viene dado por el cociente ordinario del segundo miembro de la expresión anterior.

Para díadas vectoriales homogéneas y no uniformes $(\overline{q}_1\,U_1)$ y $(\overline{q}_2\,U_2)$, con $U_1 = k \circ U_2$, siguiendo el mismo razonamiento anterior, siendo los vectores $\overline{q}_1$ y $\overline{q}_2$ colineales, se verificará $\overline{q}_1 = p \bullet \overline{q}_2$, con $p \in \mathrm{R}$, y se podrá formar el cociente de vectores colineales $\overline{q}_1 / \overline{q}_2 = p$. En estas condiciones, se puede hilar el siguiente razonamiento:

$$\frac{\left(\overline{q}_1\,U_1\right)}{\left(\overline{q}_2\,U_2\right)} = \frac{\left[\overline{q}_1\left(k \circ U_2\right)\right]}{\left(\overline{q}_2\,U_2\right)} = \frac{\left(\overline{q}_1\,U_2\right)}{\left(\overline{q}_2\,U_2\right)} \times k = \frac{\overline{q}_1}{\overline{q}_2} \times k = p \times k$$

En conclusión, el cociente diádico de dos mediciones vectoriales homogéneas, no uniformes y colineales viene dado por el factor $k \times p$, siendo k el número real que representa la razón diádica entre las unidades homogéneas U_1 y U_2 de los secundarios, y p el número real que indica la razón entre los vectores colineales $\overline{q}_1$ y $\overline{q}_2$ de los primarios diádicos.

Así ya estamos en condiciones de completar el **criterio de igualdad diádica** del apartado IV. Basta tener en cuenta que el axioma de continuidad garantiza la existencia del número real k tal que $U_1 = k \circ U_2$. De modo que, si dos díadas $(q_1 \ U_1)$ y $(q_2 \ U_2)$ son iguales, lo que se denota $(q_1 \ U_1) = (q_2 \ U_2)$, su razón diádica ha de ser obviamente la unidad de los números reales. Por tanto, podemos concluir lo siguiente:

$$\frac{\left(q_1 \ U_1 \right)}{\left(q_2 \ U_2 \right)} = p \times k = 1 \ \Rightarrow \ p = \frac{1}{k}$$

Hemos comprobado que exactamente la misma relación entre p y k se tiene para díadas vectoriales con medidas colineales. Por consiguiente, tanto para magnitudes escalares como vectoriales se puede afirmar que, **si dos díadas son iguales, la razón algebraica p de sus primarios es la inversa de la razón diádica k de sus secundarios.**

Estas propiedades de la definición de igualdad de cantidades de magnitudes, que recordemos solo tiene sentido para díadas homogéneas, es decir, representativas de la misma magnitud, aunque las unidades a que se refieran sean distintas, son esenciales para la construcción e interpretación de las leyes y ecuaciones de la Física.

In the same way that dyadic addition based on the geometric sum of segments has become generalized, the dyadic multiplication of magnitudes, pending definition, must be inspired by the **geometric product of lengths**. To do this, the first thing to do is to observe how this operation works, which will be denoted by the mathematical asterisk « $*$ », to differentiate it from the arithmetic product and break the erroneous illusion caused by the traditional identity of symbols with which different composition laws are represented. Thus, given two segments S_1 and S_2, geometric multiplication consists, by definition, in forming with them a rectangle whose dimensions are the segments themselves. The product of two segments is thus not another length, but a different magnitude called area or surface, whose analytical form can be expressed with a specific symbol such as the product $S_1 * S_2$. In this way, the factors of geometric multiplication are two lengths, those that correspond to the segments S_1 and S_2, while the product $S_1 * S_2$ is a determined quantity of the new magnitude derived from the length and called surface or area.

It is observed in the previous definition that geometric multiplication does not require that the factors be expressed in uniform units, as occurs with addition, because whatever the segments S_1 and S_2 are, the corresponding rectangle can always be formed with them. In turn, since it is a geometric operation with figures, which is not expressed numerically, none of the factors can act as a multiplier, which highlights the «arithmetization» that is arbitrarily assumed, revealing the incoherence unacceptable of current formulations with magnitudes.

This operation has a remarkable property that will be the basis of the generalization adopted for the multiplication of any magnitudes. It is the following **geometric fact**: let the lengths of the segments S_1 and S_2 be given by the dyads $(L_1 U_{L_1})$ and $(L_2 U_{L_2})$, where $L_1 \in R$ and indicates the measure of S_1 in the unit of length U_{L_1} and $L_2 \in R$ and indicates the measurement of S_2 in the unit of length U_{L_2}. It is recalled that **the parentheses are superfluous, but are expressed to mark the dyadic factors.** Under these conditions, it is materially verified that the surface of the rectangle symbolized by the product $S_1 * S_2$ is such that it can be measured in units of area equal to the area of the unit rectangle formed by the units of length when multiplying them geometrically, that is, the indicated surface by the geometric product $U_{L_1} * U_{L_2}$, and such measurement is equal to the arithmetic product of the dyadic primaries $L_1 \times L_2$. Expressed this property analytically, we have the capital fundamental **equation of geometric multiplication:**

$$(L_1 U_{L_1}) * (L_2 U_{L_2}) = [(L_1 \times L_2) (U_{L_1} * U_{L_2})]$$

It is necessary to observe the difference and the relationship that this epistemic equation establishes between the geometric product of lengths, symbolized by a mathematical asterisk « * », and the ordinary arithmetic product, indicated by the typical cross «×», as well as the correspondence between the factor lengths and the resulting area by multiplying them, that is, by composing them with this operation. This relationship is what justifies that this composition law is considered a multiplicative operation, but this does not mean much less that it is ordinary multiplication. On the contrary, it is a law of composition clearly different from the arithmetic product of R. In figure 10 the examples that graphically clarify the crucial property described above are visualized and developed.

In the case that there are three segments to be multiplied, S_1, S_2 and S_3, the logical development is totally analogous, with the difference that its geometric product generates, by definition, instead of a surface, a symbolically designated straight parallelepipedic volume $S_1 * S_2 * S_3$ such that its dimensions are precisely the three multiplied segments. Also here the

Geometric experiment of the areas

Given two lengths expressed in the same unit U_l, if an **abstract rectangle without scale** is formed with its numerical parts, it is observed that, dividing it into ideal squares with sides equal to one, the number of these is equal to the product of the measures of the lengths given relative to the unit. This observation of geometry allows defining the product of two lengths a $a\,U_l$ and $b\,U_l$ or two concrete numbers with the same unit, interpreting it as an area that is symbolized.

$$a\,U_l * b\,U_l = [(a \times b)\,(U_l * U_l)] = [(a \times b)\,U_l^2\,]$$

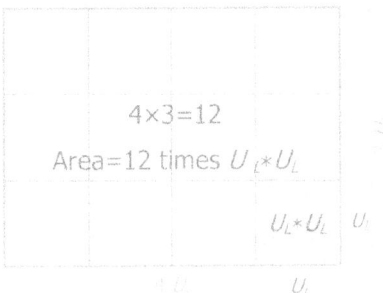

On the left, the case in which the lengths or dyadic are not expressed in the same unit as U_{l1} and $b\,U_{l2}$, in the abstract rectangle built with them, it is observed that their product can be associated with the quantity called area, which is measured by means of rectangles equal to the unit of area symbolized $U_{l1} \cdot U_{l2}$, justifying the same product definition:

$$a\,U_{l1} * b\,U_{l2} = [(a \times b)\,(U_{l1} * U_{l2})]$$

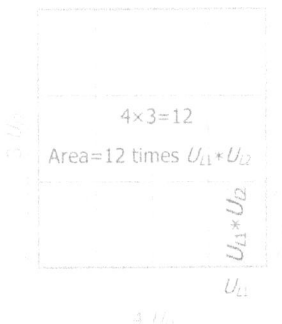

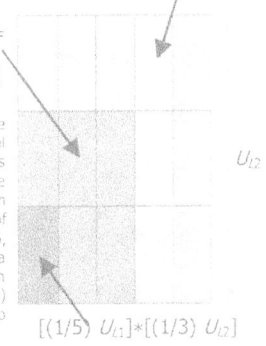

$$(3/5)\,U_{l1} * (2/3)\,U_{l2} =$$
$$= [(6/15)\,(U_{l1} * U_{l2})]$$

On the right the product of two lengths with fractional measure $[(3/5)\,U_{l1}] \cdot [(2/3)\,U_{l2}]$. Dividing one of the dimensions into five equal segments and the other into three, results in a set of equal rectangles whose sides measure 1/5 of U_{l1} and 1/3 of U_{l2}, the number of these equal elements that make up the unit is equal to $5 \times 3 = 15$, which coincides with the product of the denominators, and the number of equal elements that fit in the assumed fractional measure is $3 \times 2 = 6$, which coincides with the product of the numerators; the fractional area will be 3×2 elements of the 5×3 total rectangles, which is the fraction $(2 \times 3)/(3 \times 5)$, which is equal to the product of fractions $(3/5) \times (2/3) = 6/15$, so here the form of the definition of dyadic multiplication also holds.

Figure 10

fundamental equation of geometric multiplication with three factors is verified:

$$(L_1\,U_{L1}) * (L_2\,U_{L2})) * (L_3\,U_{L3}) = [(L_1 \times L_2 \times L_3)\,(U_{L1} * U_{L2} * U_{L3})]$$

In figure 11 the definition of this composition law and its fundamental geometric property are clarified with an example,

Experimental significance of the product three-length geometric

$$4 \times 2 \times 3 = 24$$

Volume = 24 times $(U_{L1} * U_{L2} * U_{L3})$

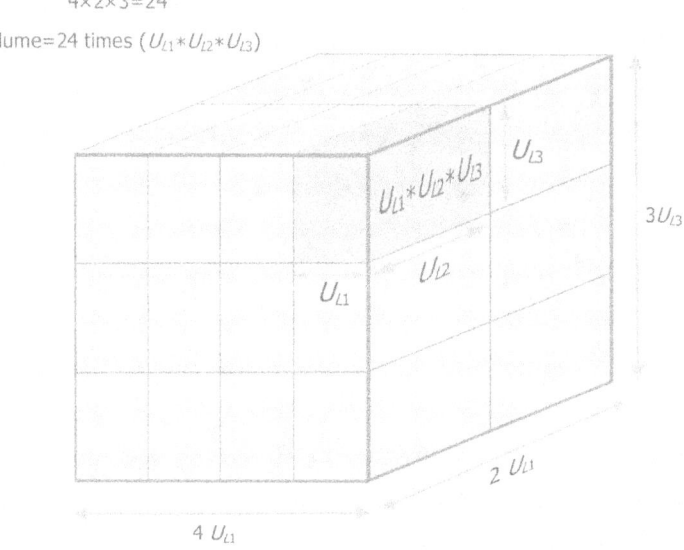

Given three lengths 4 U_{L1}, 2 U_{L2} and 3 U_{L3}, an **abstract straight parallelepiped without scale** can be formed with them and ideally decomposed by delimiting the corresponding symbolic length on each edge. Thus, they result in a series of parallelepipeds with the same ideal unit measurements, so they are congruent and equal. The new magnitude that results from composing three lengths is called volume, and the fact that the number of elementary parallelepipeds is equal to 24 makes it possible to refer to the quantity of volume indicating that one of these elements measures 24 times, which nothing prevents symbolizing with the Notation similar to the algebraic $U_{L1} \cdot U_{L2} \cdot U_{L3}$, writing this result [24 ($U_{L1} \cdot U_{L2} \cdot U_{L3}$)]. With this, the operation of composing three lengths consisting of forming a straight parallelepiped with them can be called multiplication of the concrete numbers or initial dyads given by three lengths, and this operation is symbolized (4 U_{L1})×(2 U_{L2})×(3 U_{L3})=[(4×2×3) ($U_{L1} \cdot U_{L2} \cdot U_{L3}$)], resulting in that the numerical part is equal to 4×2×3=24. So it can be defined that multiplying lengths is to obtain another quantity of the magnitude called volume whose measure is the arithmetic product of the numerical parts of the factors and whose unit of volume is expressed as the geometric product of the units of the factors. Since the unit elements are composed in the same way regardless of the order in which the factor units are composed, the commutative and associative properties of geometric multiplication must be axiomatized..

Figure 11

which are the basis of the generalization that will define the dyadic product of any physical magnitudes.

Given the current erroneous «arithmetization» of the composition of lengths and other physical quantities, which is tolerated by the false hypothesis of the International System of Units, which attributes the Abelian multiplicative group structure to magnitudes, it is necessary to insist once again that the multiplication of lengths does not correspond to the ordinary product in R, but rather it is a graphical operation, it is a law of composition with geometric segments, as straight figures of one dimension, in such a way that with two segments it produces a surface and with three a volume. It is, therefore, a **law of generating external composition**.

We are, then, before a misconception in the physical foundations, which is saved with full coherence through the algebra that is established here.

Artículo 8

MULTIPLICACIÓN GEOMÉTRICA DE LONGITUDES

De la misma manera que se ha generalizado la adición diádica con base en la suma geométrica de segmentos, la multiplicación diádica de magnitudes, pendiente de definición, debe inspirarse en el **producto geométrico de longitudes**. Para ello, lo primero es observar cómo funciona esta operación, que se denotará con el asterisco matemático «$*$», para diferenciarla del producto aritmético y romper la ilusión errónea que provoca la tradicional identidad de símbolos con que se representan leyes de composición diferentes. De este modo, dados dos segmentos S_1 y S_2, la multiplicación geométrica consiste, por definición, en formar con ellos un rectángulo que tenga por dimensiones los propios segmentos. El producto de dos segmentos no es, pues, otra longitud, sino una magnitud diferente que se denomina área o superficie, cuya forma analítica se puede expresar con un símbolo específico como el producto $S_1 * S_2$. De este modo, los factores de la multiplicación geométrica son dos longitudes, las que correspondan a los segmentos S_1 y S_2, mientras que el producto $S_1 * S_2$ es una cantidad determinada de la nueva magnitud derivada de la longitud y denominada superficie o área.

Se observa en la definición anterior que la multiplicación geométrica no exige que los factores se expresen en unidades uniformes, como ocurre con la adición, porque cualesquiera que sean los segmentos S_1 y S_2 siempre se podrá formar con ellos el correspondiente rectángulo. A su vez, puesto que se trata de una operación geométrica con figuras, que no se expresa numéricamente, ninguno de los factores puede hacer la función de multiplicador, lo que pone en evidencia la «aritmetización» que se le supone arbitrariamente, revelando la incoherencia inaceptable de las actuales formulaciones con magnitudes.

Esta operación presenta una propiedad notable que será el fundamento de la generalización adoptada para la multiplicación de cualesquiera magnitudes. Se trata del siguiente **hecho geométrico**: sean las longitudes de los segmentos S_1 y S_2 dadas por las díadas $(L_1\ U_{L1})$ y $(L_2\ U_{L2})$, donde $L_1 \in R$ e indica la medida de S_1 en la unidad de longitud U_{L1} y $L_2 \in R$ e indica la medida de S_2 en la unidad de longitud U_{L2}. Se recuerda que **los paréntesis son superfluos, pero se expresan para marcar los factores diádicos**. En estas condiciones, se comprueba materialmente que la superficie del rectángulo simbolizado por el producto $S_1 * S_2$ es tal que puede medirse en unidades de área iguales al área del rectángulo unitario que forman las unidades de longitud al multiplicarlas geométricamente, es decir, la superficie indicada por el producto geométrico $U_{L1} * U_{L2}$, y tal medida resulta igual al producto aritmético de los primarios diádicos $L_1 \times L_2$. Expresada esta propiedad analíticamente, se tiene la capital **ecuación fundamental de la multiplicación geométrica**:

$$(L_1\ U_{L1}) * (L_2\ U_{L2}) = [(L_1 \times L_2)\ (U_{L1} * U_{L2})]$$

Hay que observar la diferencia y la relación que esta ecuación epistémica establece entre el producto geométrico de longitudes, simbolizado con un asterisco matemático «$*$», y el producto aritmético ordinario, señalado por la típica aspa «$\times$», así como la correspondencia entre las longitudes de los factores y la superficie resultante al multiplicarlas, es decir, al componerlas con esta operación. Dicha relación es lo que justifica que esta ley de composición sea considerada una operación multiplicativa, pero sin que ello quiera decir ni mucho menos que se trate de la multiplicación ordinaria. Por el contrario, es una ley de composición manifiestamente diferente del producto aritmético de R. En la figura 10 se visualizan y desarrollan los ejemplos que aclaran gráficamente la crucial propiedad descrita en lo que precede.

En el caso de que sean tres los segmentos a multiplicar, S_1, S_2 y S_3, el desarrollo lógico es totalmente análogo, con la diferencia de que su producto geométrico engendra, por definición, en vez de una superficie, un volumen paralelepipédico recto designado simbólicamente $S_1 * S_2 * S_3$ tal que sus dimensiones sean precisamente los tres segmentos multiplicados. También aquí se

Experimento geométrico de las áreas

Dadas dos longitudes expresadas en la misma unidad U_L, si se forma un **rectángulo abstracto sin escala** con sus partes numéricas, se observa que, dividiéndolo en cuadrados ideales de lado igual a la unidad, el número de éstos resulta igual al producto de las medidas de las longitudes dadas respecto de la unidad. Esta observación de la geometría permite definir el producto de dos longitudes a U_L y b U_L o dos números concretos con la misma unidad, interpretándola como un área que se simboliza:

$$a\ U_L * b\ U_L = [(a \times b)\ (U_L * U_L)] = [(a \times b)\ U_L^2]$$

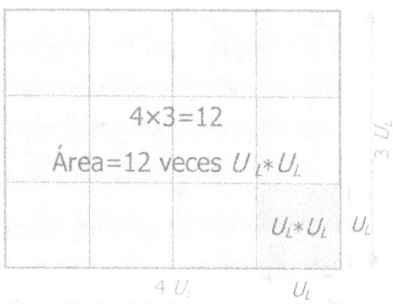

$$4 \times 3 = 12$$

Área $= 12$ veces $U_L * U_L$

$$U_L * U_L \quad U_L$$

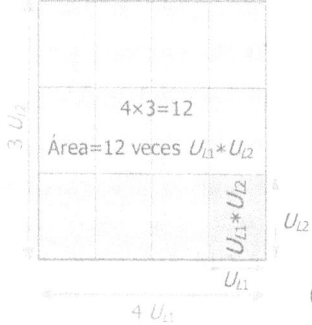

$$4 \times 3 = 12$$

Área $= 12$ veces $U_{L1} * U_{L2}$

$$U_{L1} * U_{L2}$$

$$U_{L2}$$

$$U_{L1}$$

A la izquierda el caso en que las longitudes o concretos no se expresan en la misma unidad a U_{L1} y b U_{L2}, en el rectángulo abstracto construido con ellas se observa que su producto se puede asociar a la magnitud denominada área, que queda medida por medio de rectángulos iguales a la unidad de área simbolizada $U_{L1} * U_{L2}$, justificándose la misma definición de producto:

$$a\ U_{L1} * b\ U_{L2} = [(a \times b)\ (U_{L1} * U_{L2})]$$

$$(3/5)\ U_{L1} * (2/3)\ U_{L2} =$$

$$= [(6/15)\ (U_{L1} * U_{L2})]$$

$$U_{L1} * U_{L2}$$

$$U_{L1}$$

A la derecha el producto de dos longitudes con medida fraccionaria $[(3/5)\ U_{L1}] * [(2/3)\ U_{L2}]$. Dividiendo una de las dimensiones en cinco segmentos iguales y en tres la otra, resulta un conjunto de rectángulos iguales cuyos lados miden 1/5 de U_{L1} y 1/3 de U_{L2}, el número de estos elementos iguales que componen la unidad es igual a 5×3=15, que coincide con el producto de los denominadores, y el número de elementos iguales que caben en la medida fraccionaria supuesta es de 3×2=6, que coincide con el producto de los numeradores; el área fraccionaria será 3×2 elementos de los 5×3 rectángulos totales, que es la fracción (2×3)/(3×5), que resulta igual al producto de fracciones (3/5)×(2/3)=6/15, conque aquí también se cumple la forma de la definición de la multiplicación concreta.

$$U_{L2}$$

$$[(1/5)\ U_{L1}] * [(1/3)\ U_{L2}]$$

Figura 10

verifica la **ecuación fundamental de la multiplicación geométrica con tres factores:**

$$(L_1\ U_{L1}) * (L_2\ U_{L2})) * (L_3\ U_{L3}) = [(L_1 \times L_2 \times L_3)\ (U_{L1} * U_{L2} * U_{L3})]$$

En la figura 11 se aclara con un ejemplo la definición de esta ley de composición y su propiedad geométrica fundamental, que son

Significado experimental del producto geométrico de tres longitudes

$4 \times 2 \times 3 = 24$

Volumen=24 veces ($U_{L1} * U_{L2} * U_{L3}$)

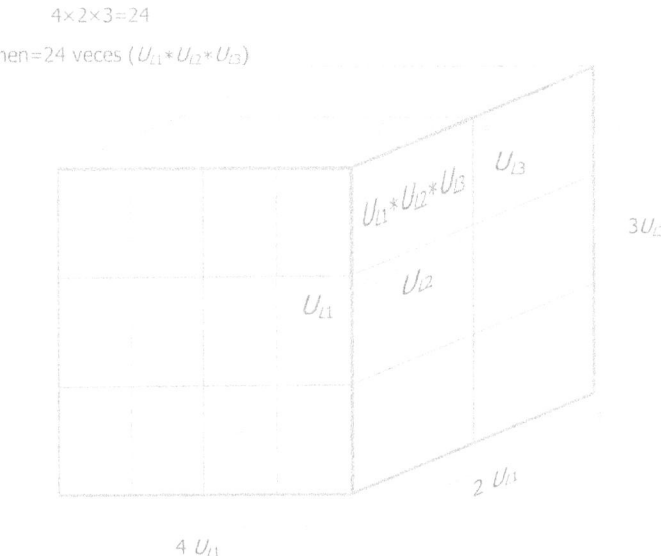

4 U_{L1}

Dadas tres longitudes 4 U_{L1}, 2 U_{L2} y 3 U_{L3}, se puede formar con ellas un **paralelepípedo recto abstracto sin escala** y descomponerlo idealmente delimitando en cada arista la longitud simbólica que corresponda. Resultan así una serie de paralelepípedos con las mismas medidas unitarias ideales, por lo que son congruentes e iguales. La nueva magnitud que resulta de componer tres longitudes se denomina volumen, y el hecho de que el número de paralelepípedos elementales resulte igual a 24 permite referirse a la cantidad de volumen indicando que mide 24 veces uno de esos elementos, que nada impide simbolizar con la notación semejante a la algebraica $U_{L1} * U_{L2} * U_{L3}$, escribiendo este resultado [24 ($U_{L1} * U_{L2} * U_{L3}$)]. Con ello, la operación de componer tres longitudes consistente en formar con ellas un paralelepípedo recto se puede denominar multiplicación de los números concretos o diadas iniciales dados por tres longitudes, y esta operación se simboliza (4 U_{L1})*(2 U_{L2})*(3 U_{L3})=[(4×2×3)·($U_{L1} * U_{L2} * U_{L3}$)], resultando que la parte numérica es igual a 4×2×3=24. De modo que se puede definir que multiplicar longitudes es obtener otra cantidad de la magnitud llamada volumen cuya medida sea el producto aritmético de las partes numéricas de los factores y cuya unidad de volumen se exprese como producto geométrico de las unidades de los factores. Como los elementos unitarios quedan compuestos de la misma manera con independencia del orden en que se compongan las unidades de los factores, deben axiomatizarse las propiedades conmutativa y asociativa de la multiplicación geométrica.

Figura 11

la base de la generalización que definirá el producto diádico de cualesquiera magnitudes físicas.

Dada la errónea «aritmetización» actual de la composición de longitudes y demás magnitudes físicas, que está tolerada por la hipótesis falsa del Sistema Internacional de Unidades, que atribuye a las magnitudes la estructura de grupo multiplicativo abeliano, es preciso volver a insistir en que la multiplicación de longitudes no se corresponde con el producto ordinario en R, sino que es una operación gráfica, es una ley de composición con segmentos geométricos, en tanto que figuras rectas de una dimensión, de forma tal que con dos segmentos produce una superficie y con tres un volumen. Se trata, por tanto, de una **ley de composición externa generatriz**.

Estamos, pues, ante una concepción errónea en los fundamentos físicos, que es salvada con plena coherencia mediante el álgebra que aquí se establece.

Article 9

The geometric multiplication of segments allows and justifies the generation of the product between all the scalar magnitudes. It is enough to simulate that the quantity of a certain magnitude is indicated by a segment or a quantity referred to an **abstract unit of length** that resembles the unit considered as a reference quantity of that magnitude, which is equivalent to the observation that, given a unit of length U_L and another U_M of any magnitude M, between the dyadic sets $\{R,U_L\}$ and $\{R,U_M\}$ it is possible to establish one-to-one correspondence without more than imaginatively identifying the U_M unit with the U_L, since formally any element $(q\,U_M)$ of the second set may be associated biunivocally with the element $(q\,U_L)$ of the first. We will say in this case that the magnitude M is affine to the length, and the attribution of this quality to the physical magnitudes we will call it the **affinity postulate**.

Since we have defined physical magnitudes as those properties of nature that can be measured, and we have seen that dyadic sets arise from measurement, this definition is equivalent to having as physical quantities those natural properties affine to length.

In this way, given two scalar physical dyads $(q_1\,U_1)$ and $(q_2\,U_2)$, the dyadic multiplication is defined, by generalization of the geometric one, as the abstract rectangle of surface indicated by $(q_1\,U_1)*(q_2\,U_2)$ such that the measure of its area expressed in the **abstract unit** U_1*U_2 is fixed by the following epistemic definition equation:

$$(q_1\,U_1)*(q_2\,U_2)=[(q_1{\times}q_2)\,(U_1*U_2)]$$

This equation determines a composition law that applies the Cartesian product $\{R,U_1\}{\times}\{R,U_2\}$ on $\{R,U_1*U_2\}$. It is an **external generating law**, which creates a new magnitude whose unit is U_1*U_2.

If the dyadic factors to multiply were three, the dyadic multiplication would form an **abstract parallelepipedic volume,** whose measure would be given by the definition equation:

$$(q_1 U_1) * (q_2 U_2) * (q_3 U_3) = [(q_1 \times q_2 \times q_3)(U_1 * U_2 * U_3)]$$

In this case the external composition law applies the Cartesian product set $\{R,U_1\} \times \{R,U_2\} \times \{R,U_3\}$ in $\{R,U_1 * U_2 * U_3\}$.

And, in general, for any number of factors n, the dyadic multiplication of quantities of scalar physical magnitudes will be defined by a **hypervolume,** whose measure would be determined by the following analytical equation:

$$(q_1 U_1) * (q_2 U_2) * \ldots * (q_n U_n) =$$
$$= [(q_1 \times q_2 \times \ldots \times q_n)(U_1 * U_2 * \ldots * U_n)]$$

The external composition law thus defined represents an application or function of the Cartesian product set indicated by $\{R,U_1\} \times \{R,U_2\} \times \ldots \times \{R,U_n\}$ in $\{R,U_1 * U_2 * \ldots * U_n\}$.

The external nature of these multiplicative composition laws, as explained in section XIV, irrefutably denies the existence of unit or inverse elements of dyadic entities, so the negative exponents that result in algebraic expressions should not be interpreted as inverses of other measurements, but as denominators of dyadic fractions. Thus, the noxious and scandalous gap in the International System of Units in this matter is overcome, which admits notations such as m^{-1}, kg^{-1} or s^{-1}, without defining at all what kind of entities these symbologies refer to, taking for granted childishly that they obey vulgar algebra, an erroneous and garrafal presupposition that vitiates all physical content from the root. Crass mistake produced by not asking what is the meaning of the inverse of a meter, a kilogram or a second, as would correspond to any physicist responsible for his science.

390

Artículo 9

MULTIPLICACIÓN DIÁDICA DE MAGNITUDES ESCALARES

La multiplicación geométrica de segmentos permite y justifica la generación del producto entre todas las magnitudes escalares. Basta con simular que la cantidad de cierta magnitud quede indicada por un segmento o una cantidad referida a una **unidad de longitud abstracta** que semeje la unidad considerada como cantidad de referencia de esa magnitud, lo que equivale a la observación de que, dadas una unidad de longitud U_L y otra U_M de una magnitud cualquiera M, entre los conjuntos diádicos $\{R, U_L\}$ y $\{R, U_M\}$ es posible establecer correspondencias biunívocas sin más que indentificar imaginariamente la unidad U_M con la U_L, ya que formalmente cualquier elemento $(q\ U_M)$ del segundo conjunto podrá asociarse biunívocamente con el elemento $(q\ U_L)$ del primero. Diremos en este caso que la magnitud M es afín a la longitud, y la atribución de esta cualidad a las magnitudes físicas lo denominaremos **postulado de afinidad**.

Puesto que hemos definido las magnitudes físicas como aquellas propiedades de la naturaleza susceptibles de medición, y hemos visto que de la medición surgen los conjuntos diádicos, esta definición es equivalente a tener por magnitudes físicas aquellas propiedades naturales afines a la longitud.

De este modo, dadas dos díadas físicas escalares $(q_1\ U_1)$ y $(q_2\ U_2)$, se define la multiplicación diádica, por generalización de la geométrica, como el **rectángulo abstracto** de superficie indicada por $(q_1\ U_1) * (q_2\ U_2)$ tal que la medida de su área expresada en la **unidad abstracta** $U_1 * U_2$ queda fijada por la ecuación epistémica de definición siguiente:

$$(q_1\ U_1) * (q_2\ U_2) = [(q_1 \times q_2)\ (U_1 * U_2)]$$

Esta ecuación determina una ley de composición que aplica el producto cartesiano $\{R, U_1\} \times \{R, U_2\}$ en $\{R, U_1 * U_2\}$. Se trata de una **ley externa generatriz**, que crea una nueva magnitud cuya unidad es $U_1 * U_2$.

Si los factores diádicos a multiplicar fuesen tres, la multiplicación diádica formaría un **volumen paralelepipédico abstracto**, cuya medida vendría dada por la ecuación de definición:

$$(q_1 \ U_1) * (q_2 \ U_2) * (q_3 \ U_3) = [(q_1 \times q_2 \times q_3) \ (U_1 * U_2 * U_3)]$$

En este caso la ley de composición externa aplica el conjunto producto cartesiano $\{R, U_1\} \times \{R, U_2\} \times \{R, U_3\}$ en $\{R, U_1 * U_2 * U_3\}$.

Y, en general, para un número cualquiera de factores n, la multiplicación diádica de cantidades de magnitudes físicas escalares vendrá definida por un **hipervolumen**, cuya medida estaría determinada por la ecuación analítica siguiente:

$$(q_1 \ U_1) * (q_2 \ U_2) * \ \ldots \ * (q_n \ U_n) =$$

$$= [(q_1 \times q_2 \times \ \ldots \ \times q_n) \ (U_1 * U_2 * \ \ldots \ * U_n)]$$

La ley de composición externa así definida representa una aplicación o función del conjunto producto cartesiano indicado por $\{R, U_1\} \times \{R, U_2\} \times \ \ldots \ \times \{R, U_n\}$ en $\{R, U_1 * U_2 * \ \ldots \ * U_n\}$.

La naturaleza externa de estas leyes de composición multiplicativas, como se expuso en el apartado XIV, niegan irrefutablemente la existencia de elementos unidad ni inversos de los entes diádicos, por lo que los exponentes negativos que resulten en las expresiones algebraicas no deben interpretarse como inversos de otras mediciones, sino como denominadores de fracciones diádicas. Se salva así la nociva y escandalosa laguna del sistema Internacional de Unidades en esta materia, que admite notaciones como m^{-1}, kg^{-1} o s^{-1}, sin definir en absoluto a qué clase de entes se refieran estas simbologías, dando por sentado puerilmente que obedezcan al álgebra vulgar, presuposición errónea y garrafal que vicia todo el contenido físico desde la raíz. Craso yerro producido por no preguntarse cuál es el significado del inverso de un metro, de un kilogramo o de un segundo, como correspondería a cualquier físico responsable de su ciencia.

Article 10

DEDUCTION OF THE DYADIC SCALAR DIVISION

It is possible to deduce the analytical definition of the dyadic quotient without more than attending to the generic concept of division. To do this, just imagine an abstract rectangle whose surface is identified with a dyadic dividend $(a\,U_1)$, one of its dimensions with the divisor $(b\,U_2)$ and the other with the dyadic quotient $[c\,(U_1/\!/U_2)]$. The unit associated with c must be identified with the dyadic ratio of units $[c\,(U_1/\!/U_2)]$, because the unit rectangle must have the unit U_1 by area and by dimensions U_2 and $U_1/\!/U_2$. In the same way, the three indicated dyads cannot be independent, but must satisfy the division condition, that is, the quotient multiplied by the divisor must equal the dividend; or, in other words, the dyadic product of the two dimensions of the abstract rectangle must be equal to its surface; and it will be written analytically like this:

$$\left(a\,U_1\right) = \left(b\,U_2\right)*\left(c\,\frac{U_1}{U_2}\right)$$

The parentheses are superfluous, but as usual they are kept to mark well the dyads involved in the formula, which can be interpreted according to the generic division criterion, which leads to consider the factor $[c\,(U_1/\!/U_2)]$ as the quotient between the total surface of the abstract rectangle $(a\,U_1)$ and the other of its two dimensions $(b\,U_2)$. And this analytically can be described as follows:

$$\left(c\,\frac{U_1}{U_2}\right) = \frac{\left(a\,U_1\right)}{\left(b\,U_2\right)}$$

The geometry of the abstract rectangle is such that $a=b\times c$, given the fundamental property of geometric multiplication, so that $c=a/b$ with the algebra of R. So, substituting $c=a/b$ in the first member of the last equality, we will finally have this other dyadic expression:

393

$$\left(c\,\frac{U_1}{U_2}\right) = \frac{(a\,U_1)}{(b\,U_2)} = \left(\frac{a}{b}\,\frac{U_1}{U_2}\right)$$

And between the second and third terms of this equation the **analytical definition of the dyadic division** between the scalar dyadic concretes $(a\,U_1)$ and $(b\,U_2)$, deduced by the preceding reasoning, is already observed. So it can be concluded that the quotient of these two dyads is equal to a dyadic element whose primary is the quotient of the primaries of the factors and whose secondary is the dyadic division of the units of the dividend and the divisor. Expressed analytically:

$$\frac{(a\,U_1)}{(b\,U_2)} = \left(\frac{a}{b}\,\frac{U_1}{U_2}\right)$$

In this way, we verify, as for the rest of the operations previously analyzed, that the symbols of the units behave ideally like the other elements of R, but this consequence is not due to the traditional symbolic logic, and it is insisted, it would be an error It is crass and inadmissible to consider it this way, because it has been irrefutably justified that this formal behavior is due to the concept of dyadic multiplication through abstract rectangles, it is not the result of the properties of the operations in R. It is simply an isomorphism, not an identity in absolute.

On the other hand, it is noted that with the **double bar** the division of scalar concrete analyzed in this article has been symbolized, an operation different from the quotient of homogeneous dyads, which it has been agreed to represent with the same sign. And it is that the diversity of algebraic laws is such that, although symbolic exhaustiveness is sought for pedagogical clarity, it is inevitable and even sometimes convenient to resort to a certain degree to the principle of symbolic economy, but this without it being permissible to confuse the different operations indicated with the same signs.

Artículo 10

DEDUCCIÓN DE LA DIVISIÓN DIÁDICA ESCALAR

Es posible deducir la definición analítica del cociente diádico sin más que atender al concepto genérico de división. Para ello, basta con imaginar un rectángulo abstracto cuya superficie quede identificada con un dividendo diádico ($a\ U_1$), una de sus dimensiones con el divisor ($b\ U_2$) y la otra con el cociente concreto [$c\ (U_1/\!/U_2)$]. La unidad asociada a c debe identificarse con el cociente diádico de unidades $U_1/\!/U_2$, porque el rectángulo unitario ha de tener por área la unidad U_1 y por dimensiones U_2 y $U_1/\!/U_2$. De la misma manera, las tres díadas indicadas no pueden ser independientes, sino que deben satisfacer la condición de la división, es decir, que el cociente multiplicado por el divisor debe ser igual el dividendo; o, dicho de otro modo, el producto diádico de las dos dimensiones del rectángulo abstracto debe ser igual a su superficie; y ello se escribirá analíticamente así:

$$\left(a\,U_1\right) = \left(b\,U_2\right) * \left(c\,\frac{U_1}{U_2}\right)$$

Los paréntesis son superfluos, pero como de costumbre se mantienen para marcar bien las díadas que intervienen en la fórmula, que se puede interpretar en función del criterio genérico de división, lo que lleva a considerar el factor [$c\ (U_1/\!/U_2)$] como el cociente entre la superficie total del rectángulo abstracto ($a\ U_1$) y la otra de sus dos dimensiones ($b\ U_2$). Y ello analíticamente se podrá describir así:

$$\left(c\,\frac{U_1}{U_2}\right) = \frac{\left(a\,U_1\right)}{\left(b\,U_2\right)}$$

La geometría del rectángulo abstracto es tal que $a = b \times c$, dada la propiedad fundamental de la multiplicación geométrica, por lo que $c = a/b$ con el álgebra de R. De modo que, sustituyendo $c = a/b$ en el primer miembro de la última igualdad, tendremos finalmente esta otra expresión diádica:

$$\left(c\,\frac{U_1}{U_2} \right) = \frac{\left(a\,U_1 \right)}{\left(b\,U_2 \right)} = \left(\frac{a}{b}\,\frac{U_1}{U_2} \right)$$

Y ya se observa entre los términos segundo y tercero de esta ecuación la **definición analítica de la división diádica** entre los concretos diádicos escalares ($a\,U_1$) y ($b\,U_2$), deducido mediante el razonamiento precedente. Conque se puede concluir que el cociente de esas dos díadas es igual a un elemento diádico cuyo primario es el cociente de los primarios de los factores y cuyo secundario es la división diádica de las unidades del dividendo y del divisor. Expresado analíticamente:

$$\frac{\left(a\,U_1 \right)}{\left(b\,U_2 \right)} = \left(\frac{a}{b}\,\frac{U_1}{U_2} \right)$$

Comprobamos de este modo, como para el resto de las operaciones anteriormente analizadas, que los símbolos de las unidades se comportan idealmente como los demás elementos de R, pero esta consecuencia no es debida a la lógica simbólica tradicional, y se insiste, sería un error craso e inadmisible considerarlo así, porque se ha justificado irrefutablemente que ese comportamiento formal se debe al concepto de multiplicación diádica mediante rectángulos abstractos, no es fruto de las propiedades de las operaciones en R. Se trata simplemente de un isomorfismo, no de una identidad en absoluto.

Por otra parte, se advierte que con la **doble barra** se ha simbolizado la división de concretos escalares analizada en este artículo, operación distinta del cociente de díadas homogéneas, que se ha convenido en representar con ese mismo signo. Y es que la diversidad de leyes algebraicas es tal que, aunque se busque la exhaustividad simbólica por claridad pedagógica, resulta inevitable y hasta a veces conveniente recurrir en cierto grado al principio de economía simbólica, pero ello sin que sea permisible confundir las operaciones diferentes señaladas con los mismos signos.

Article 11

MULTIPLICATION BETWEEN SCALAR
AND VECTOR DYADS

Newton's second law relates quantities of the vector magnitude called force to the product of the scalar magnitude called mass and the vector magnitude known as acceleration. However, a multiplication like this has never been defined, but an isomorphic behavior with the algebra of R has been tacitly assumed, which is an unacceptable misadventure that forces it to be established epistemically, as has been done with the others that the they precede in this work, because it appears constantly in the physical equations and, without such definitions, the meanings of the scientific laws are denatured and gloomy, apart from being expropriated of all logical foundation and scientific consistency.

So the multiplication of scalar dyads by other vectors has to compose scalar measurements $(a\ U_1)$ of $\{R, U_1\}$ with vector elements $(\overline{b}\ U_2)$ of $\{R^3, U_2\}$. You can distinguish this composition law using any sign, for example, «⊚». To establish the appropriate definition of this product form, the external law of the vector space R^3 over R and dyadic multiplication can be counted on, so there is no better formulation than these two equations:

$$(a\ U_1) \circledcirc (\overline{b}\ U_2) = [(a \bullet \overline{b})\,(U_1 * U_2)]$$

$$(\overline{b}\ U_2) \circledcirc (a\ U_1) = [(\overline{b} \bullet a)\,(U_2 * U_1)]$$

The sign «⊚» of the first members of these expressions symbolizes the composition law that is being defined in this article, the product of a scalar physical pair by another vector; the sign «•» placed in the factors $(a \bullet \overline{b})$ y $(\overline{b} \bullet a)$ of the second members indicates the external law of R^3 on R or product of a scalar by a vector; and the multiplications of the terms $(U_1 * U_2)$ and $(U_2 * U_1)$ mark the dyadic product of two scalar dyadic elements, defined in article 9.

However, as we are observing with insistent repetition, the stubborn vice of using the same «×» sign for all these multiplicative laws still prevails, in application of the easy tacit principle of symbolic economy, whose fatal illusory and equivocal effects are noted throughout of this work and in more detail in article 15.

Although it may be idle to the attentive reader, it is noted here that the product defined in this article allows, like any other, to establish division as a derivative operation. It is enough to consider the second member of the previous definition equations as a dividend and any of the factors of the first member as a divisor, resulting in the other factor of this being the quotient. Thus we find that the quotient between the vector dyad of the second member and the scalar dyad of the first member gives a vector quotient; as well as the quotient between the same vector dyad of the second member and the other vector of the first one results in a scalar quotient. Although yes, by the very definition of this product, the vectors of the first and second member must be collinear.

This division could be symbolized by any arbitrary sign; but, in order not to increase the operational symbology to infinity, it is enough to distinguish it as one more dyadic quotient, for example, with the double bar «//». However, the algebraist must know how to distinguish the various operations not by their symbols but by the elements they relate. Detailed symbology is more of a didactic element than necessary and can be cumbersome, hence the need for a certain symbolic economy, provided that some operations are not confused with others.

Artículo 11

MULTIPLICACIÓN ENTRE DÍADAS ESCALARES Y VECTORIALES

La *segunda ley de Newton* relaciona cantidades de la magnitud vectorial llamada fuerza con el producto entre la magnitud escalar denominada masa y la magnitud vectorial conocida por aceleración. Sin embargo, nunca se ha definido una multiplicación como esta, sino que se ha presumido tácitamente un comportamiento isomorfo con el álgebra de R, lo cual es un desaguisado inaceptable que obliga a asentarla epistémicamente, al igual que se ha hecho con las demás que la preceden en este trabajo, porque aparece constantemente en las ecuaciones físicas y, sin tales definiciones, los significados de las leyes científicas quedan desnaturalizados y sombríos, aparte de expropiados de todo fundamento lógico y consistencia científica.

Así que la multiplicación de díadas escalares por otras vectoriales ha de componer mediciones escalares $(a\ U_1)$ de $\{R, U_1\}$ con elementos vectoriales $(\overline{b}\ U_2)$ de $\{R^3, U_2\}$. Se puede distinguir esta ley de composición utilizando cualquier signo, por ejemplo, «⊚». Para establecer la definición conveniente de esta forma de producto se puede contar con la ley externa del espacio vectorial R^3 sobre R y la multiplicación diádica, por lo que no cabe mejor formulación que estas dos ecuaciones:

$$(a\ U_1)⊚(\overline{b}\ U_2) = [(a\bullet\overline{b})\ (U_1 * U_2)]$$

$$(\overline{b}\ U_2)⊚(a\ U_1) = [(\overline{b}\bullet a)\ (U_2 * U_1)]$$

El signo «⊚» de los primeros miembros de estas expresiones simboliza la ley de composición que se está definiendo en este artículo, el producto de un par físico escalar por otro vectorial; el signo «•» puesto en los factores $(a\bullet\overline{b})$ y $(\overline{b}\bullet a)$ de los segundos miembros señala la ley externa de R^3 sobre R o producto de un escalar por un vector; y las multiplicaciones de los términos $(U_1 * U_2)$ y $(U_2 * U_1)$ marcan el producto diádico de dos elementos diádicos escalares, definido en el artículo 9.

Sin embargo, como estamos observando con repetición insistente, aún prevalece el contumaz vicio de usar el mismo signo «×» para todas estas leyes multiplicativas, en aplicación del facilitón principio tácito de economía simbólica, cuyos fatales efectos ilusorios y equívocos se advierten a lo largo de este trabajo y con más detalle en el artículo 15.

Aunque pueda resultar ocioso al lector atento, se advierte aquí que el producto definido en este artículo permite, como cualquier otro, establecer como operación derivada la división. Basta para ello contemplar el segundo miembro de las ecuaciones de definición anteriores como un dividendo y cualquiera de los factores del primer miembro como un divisor, resultando el otro factor de este ser el cociente. Así encontramos que el cociente entre la díada vectorial del segundo miembro y la díada escalar del primero da un cociente vectorial; así como el cociente entre la misma díada vectorial del segundo miembro y la otra vectorial del primero da como resultado un cociente escalar. Aunque eso sí, por la propia definición de este producto, los vectores del primer y del segundo miembro han de ser colineales.

Esta división se podría simbolizar con cualquier signo arbitrario; pero, para no incrementar hasta el infinito la simbología operacional, basta distinguirla como un cociente diádico más, por ejemplo, con la doble barra «//». No obstante, el algebrista debe saber distinguir las diversas operaciones no por sus símbolos sino por los elementos que relacionan. La simbología pormenorizada es más un elemento didáctico que necesario y puede resultar farragosa, de donde surge la necesidad de cierta economía simbólica, siempre que no se confundan unas operaciones con otras.

Article 12

DEFINITION OF SCALAR PRODUCTS
AND VECTOR OF VECTOR DYADS

In Physics, vector magnitudes use mathematical vectors and the products between vectors called scalar product and vector product. It is usual to indicate the scalar with a mathematical point «·» and to distinguish with the angle «∧» the vector product of vectors. In turn, for the dyadic homonyms of quantities of vector magnitudes, the circle with a point «⊙» will be reserved for the scalar dyadic product and the circle with an asterisk «⊛» for the vector dyadic product.

There is no better way to define both the scalar product and the vector product of vector magnitudes than in terms of their counterparts of mathematical vectors. Starting with the scalar product of two vector concretes $(\overline{a}\ U_1)$ and $(\overline{b}\ U_2)$ of the dyadic or concrete sets $\{R^3,U_1\}$ and $\{R^3,U_2\}$, the first elements $\overline{a}$ and $\overline{b}$ the pairs are distinguished here as vectors of R^3. And with this, the **dot product of two vector dyads** must be defined with the epistemic equation:

$$(\overline{a}\ U_1)\odot(\overline{b}\ U_2)=[(\overline{a}\cdot\overline{b})\,(U_1*U_2)]$$

That is, by definition, the scalar product of two vector dyads measures a scalar magnitude with the scalar pair of the set $\{R,U_1*U_2\}$ such that its primary is the scalar product of the vector primaries of the factors and the unit is the product dyadic or concrete of the factor units.

Regarding the **vector product of two vector dyads**, the following definition equation will be had in an analogous way:

$$(\overline{a}\ U_1)\circledast(\overline{b}\ U_2)=[(\overline{a}\wedge\overline{b})\,(U_1*U_2)]$$

In this case, by definition, the vector product of two vector concretes measures a vector magnitude with the dyad of the set $\{R^3,U_1*U_2\}$ such that its primary is another vector equal to the vector product of the vectors that make up the primaries of the given concrete or dyads, and whose unit is the dyadic product of the factor units.

The two definition formulas above will facilitate the correct interpretation of the physical equations in which these composition laws intervene, such as the magnitudes of work and moment of a force, the work for the scalar product and the moment for the vector product.

On the other hand, these composition laws will not make sense for the case of scalar dyads.

Artículo 12

DEFINICIÓN DE LOS PRODUCTOS ESCALAR Y VECTORIAL DE DÍADAS VECTORIALES

En Física las magnitudes vectoriales se sirven de los vectores matemáticos y de los productos entre vectores denominados producto escalar y producto vectorial. Es usual indicar el escalar con un punto matemático «·» y distinguir con el ángulo «∧» el producto vectorial de vectores. A su vez, para los homónimos diádicos de cantidades de magnitudes vectoriales se reservarán el círculo con un punto «⊙» para el producto diádico escalar y el círculo con asterisco «⊛» para el producto diádico vectorial.

Tanto el producto escalar como el vectorial de magnitudes vectoriales no hay mejor manera de definirlos que en función de sus homónimos de los vectores matemáticos. Comenzando por el producto escalar de dos concretos vectoriales $(\overline{a}\ U_1)$ y $(\overline{b}\ U_2)$ de los conjuntos diádicos o concretos $\{R^3, U_1\}$ y $\{R^3, U_2\}$, se distinguen aquí los primeros elementos $\overline{a}$ y $\overline{b}$ de los pares como vectores de R^3. Y con ello debe definirse el **producto escalar de dos díadas vectoriales** con la ecuación epistémica:

$$(\overline{a}\ U_1)\odot(\overline{b}\ U_2)=[(\overline{a}\cdot\overline{b})\,(U_1*U_2)]$$

Es decir, por definición, el producto escalar de dos díadas vectoriales mide una magnitud escalar con el par escalar del conjunto $\{R, U_1*U_2\}$ tal que su primario sea el producto escalar de los primarios vectoriales de los factores y la unidad el producto diádico o concreto de las unidades de los factores.

En cuanto al **producto vectorial de dos díadas vectoriales** se tendrá de manera análoga la siguiente ecuación de definición:

$$(\overline{a}\ U_1)\circledast(\overline{b}\ U_2)=[(\overline{a}\wedge\overline{b})\,(U_1*U_2)]$$

En este caso, por definición, el producto vectorial de dos concretos vectoriales mide una magnitud vectorial con la díada del conjunto $\{R^3, U_1*U_2\}$ tal que su primario sea otro vector igual al producto vectorial de los vectores que integran los primarios de los concretos o díadas dados, y cuya unidad sea el producto diádico de las unidades de los factores.

Las dos fórmulas de definición anteriores facilitarán la interpretación correcta de las ecuaciones físicas en que intervengan estas leyes de composición, tales como las magnitudes trabajo y momento de una fuerza, el trabajo para el producto escalar y el momento para el producto vectorial.

En cambio, estas leyes de composición no tendrán sentido para el caso de díadas escalares.

Artículo 13

ALGEBRAIC STRUCTURE OF
THE DYADIC SETS

The scalar and vector dyadic sets associated with a given unit U, respectively symbolized $\{R,U\}$ and $\{R^3,U\}$, endowed with the corresponding additive internal law, scalar or vector, and indicated in both cases with the sign «$\oplus$» , form the **algebraic structures $R,U,\oplus\}$ and $\{R^3,U,\oplus\}$ with the abelian group properties**, because it is relatively easy to prove that they are defined everywhere with uniqueness and the commutative and associative properties are verified in both cases , existence of a neutral element and existence of a symmetrical element, as we do in the first part of this work.

The same does not happen with the dyadic multiplication «$*$», because in the case of a law of external composition, even if it is commutative and associative; however, unitary and inverse elements cannot exist in the same set. So **the structures $\{R,U,*\}$ and $\{R^3,U,*\}$ do not satisfy the group conditions**. In turn, with the two laws of composition indicated, **the structures $\{R,U,\oplus,*\}$ and $\{R^3,U,\oplus,*\}$ do not satisfy the properties of the field**.

Associating the abelian groups $\{R,U,\oplus\}$ and $\{R^3,U,\oplus\}$ with the field R of the real numbers and considering the respective external laws, both of which have been identified with the sign «$\circ$» in article 6, For both scalar and vector magnitudes, it is easily verified that these external composition laws are defined everywhere and they verify the following properties: they are associative with respect to the multiplication of the field R; they are modular, which means that the unit element of the body R leaves every element of R and R^3 invariant; they are distributive with respect to the additive laws of R and R^3; and they are distributive with respect to the additive law in the field R.

Consequently, **the abelian groups $\{R,U,\oplus\}$ and $\{R^3,U,\oplus\}$ present respective vector space structures on the field R of the real numbers**. This is why the resulting isomorphisms allow the dyadic elements to be operated as if their various components were elements of R, even though they are not really so and the stubborn tradition presumes it subliminally or arbitrarily. In this way, the question of why it is possible to operate with the magnitudes as is usually done is resolved and this insidious gap is overcome, which suggested diverse and rather esoteric explanations, lacking foundation and alien to the scientific method.

In short, it all comes down to solving the latent incongruous omission by establishing the necessary composition laws and in harmony with the usual algebraic structures. This leads to the isomorphism between the structure of the geometric segments $\{S, \oplus, \circ, *\}$ with the field of real numbers R, which with its two internal laws, additive and multiplicative, is also a vector space with its own body R as domain of operators and the same multiplication.

For its part, the set of geometric segments $\{S\}$ with its internal additive law «$\oplus$», its external multiplicative law by a scalar «$\circ$», together with the external generating multiplicative operation «$*$», constitutes a homologous structure of R.

In this way, the bijective map f that makes each segment $S \in \{S\}$ correspond to the real number $x \in R$ that indicates its measure with a certain unit such that $x = f(S)$ represents the isomorphism between $\{S\}$ and R. With the laws of composition defined in $\{S\}$ the map f is such that, given any segments S, S_1 and S_2 of $\{S\}$, it makes them correspond their measures x, x_1 and x_2 of R and the following properties are obtained for all α of R:

$$f(S_1 \oplus S_2) = x_1 + x_2 = f(S_1) + f(S_2)$$

$$f(\alpha \circ S) = \alpha \times x = \alpha \times f(S)$$

$$f(S_1 * S_2) = x_1 \times x_2 = f(S_1) \times f(S_2)$$

Thus the arithmetic of R and the non-arithmetic algebra of $\{S\}$ are connected, becoming isomorphic structures.

Note that the inverse or reciprocal application f^{-1} of R in $\{S\}$ is also an isomorphism, characterized by the relation $f^{-1}(x) = S$ for all $x \in R$ and all $S \in \{S\}$.

Artículo 13

ESTRUCTURA ALGEBRAICA DE
LOS CONJUNTOS DIÁDICOS

Los conjuntos diádicos escalares y vectoriales asociados a una unidad determinada U, simbolizados respectivamente $\{R, U\}$ y $\{R^3, U\}$, dotados con la ley interna aditiva correspondiente, escalar o vectorial, y señalada en ambos casos con el signo «⊕», forman las **estructuras algebraicas $\{R, U, \oplus\}$ y $\{R^3, U, \oplus\}$ con las propiedades de grupo abeliano**, porque resulta relativamente sencillo probar que están definidas por doquier con unicidad y se verifican en ambos casos las propiedades conmutativa, asociativa, existencia de elemento neutro y existencia de elemento simétrico, tal como hacemos en la primera parte de este trabajo.

No ocurre lo mismo con la multiplicación diádica «∗», porque tratándose de una ley de composición externa, aunque sea conmutativa y asociativa; sin embargo, no pueden existir los elementos unitario ni inversos en el mismo conjunto. Conque **las estructuras $\{R, U, *\}$ y $\{R^3, U, *\}$ no satisfacen las condiciones de grupo**. A su vez, con las dos leyes de composición indicadas, **las estructuras $\{R, U, \oplus, *\}$ y $\{R^3, U, \oplus, *\}$ no satisfacen las propiedades de cuerpo**.

Asociando a los grupos abelianos $\{R, U, \oplus\}$ y $\{R^3, U, \oplus\}$ el cuerpo R de los números reales y considerando la leyes externas respectivas, que se han identificado ambas con el signo «∘» en el artículo 6, tanto para magnitudes escalares como vectoriales, se comprueba con facilidad que estas leyes de composición externas están definidas por doquier y verifican las propiedades siguientes: son asociativas respecto de la multiplicación del cuerpo R; son modulares, lo que significa que el elemento unidad del cuerpo R deja invariante a todo elemento de R y de R^3; son distributivas respecto de las leyes aditivas de R y R^3; y son distributivas respecto de la ley aditiva en el cuerpo R.

En consecuencia, **los grupos abelianos $\{R, U, \oplus\}$ y $\{R^3, U, \oplus\}$ presentan sendas estructuras de espacio vectorial sobre el cuerpo R de los números reales**. Razón por la que los isomorfismos resultantes permiten operar con los elementos diádicos como si sus diversos componentes fuesen elementos de R, aunque no lo sean realmente y la pertinaz tradición lo presuma subliminal o arbitrariamente. De este modo queda resuelta la incógnita de por qué se puede operar con las magnitudes como se hace usualmente y se salva esta insidiosa laguna, que sugería explicaciones diversas y más bien esotéricas, carentes de fundamento y ajenas al método científico.

En resumen, todo se reduce a resolver la omisión incongruente latente estableciendo las leyes de composición necesarias y en armonía con las estructuras algebraicas usuales. Con ello se llega al isomorfismo entre la estructura de los segmentos geométricos $\{S, \oplus, \circ, *\}$ con el cuerpo de los números reales R, que con sus dos leyes internas, aditiva y multiplicativa, es también un espacio vectorial con el propio cuerpo R como dominio de operadores y la misma multiplicación.

Por su parte, el conjunto de los segmentos geométricos $\{S\}$ con su ley interna aditiva «$\oplus$», su ley externa multiplicativa por un escalar «$\circ$», junto con la operación multiplicativa externa generatriz «$*$», constituye una estructura homóloga de R.

De este modo, la aplicación biyectiva f que hace corresponder a cada segmento $S \in \{S\}$ el número real $x \in R$ que indique su medida con cierta unidad tal que $x = f(S)$ representa el isomorfismo entre $\{S\}$ y R. Con las leyes de composición definidas en $\{S\}$ la aplicación f es tal que, dados los segmentos cualesquiera S, S_1 y S_2 de $\{S\}$, les hace corresponder sus medidas x, x_1 y x_2 de R y se tienen las siguientes propiedades para todo α de R:

$$f(S_1 \oplus S_2) = x_1 + x_2 = f(S_1) + f(S_2)$$

$$f(\alpha \circ S) = \alpha \times x = \alpha \times f(S)$$

$$f(S_1 * S_2) = x_1 \times x_2 = f(S_1) \times f(S_2)$$

Así resultan conectadas la aritmética de R y el álgebra no aritmética de $\{S\}$, convirtiéndose en estructuras isomorfas.

Nótese que la aplicación inversa o recíproca f^{-1} de R en $\{S\}$ es también un isomorfismo, caracterizado por la relación $f^{-1}(x) = S$ para todo $x \in R$ y todo $S \in \{S\}$.

Article 14

Thales' Theorem[35], in its classical formulation, is based on the theory of ratios and proportions, with its specific notions of equality and addition of segments. In sum, **the traditional theory assumes replacing the segments by their measure in a certain unit of length**, which can be any or even abstract, but always the same for all the segments considered. That is, the segments or lengths are assumed to be uniform. With this artifice the ratios and proportions between segments, which are geometric, are reduced to abstract numerical ratios and proportions.

However, it is clear that a segment is not just a number, but a quantity of length. It is a dyad with the form $(a \, U_L)$, where the primary a is R and indicates the measure of the segment in the unit of length U_L. Well, given two segments expressed with uniform dyads $(a \, U_L)$ and $(b \, U_L)$, classical Mathematics defines the ratio between them, identifying it with the arithmetic of its primaries a/b. With this, geometric algebra is replaced by symbolic algebra, described by the measures of the segments in a certain unit of uniform length.

In the same way, the geometric proportionality of segments is replaced by the proportionality of real numbers, so that two given segments will be said to be proportional to two others when the numbers that express their measures in the same unit of length form a proportion in the set R That is, the segments $(a \, U_L)$ and $(b \, U_L)$ are said to be proportional to the $(p \, U_L)$ and $(q \, U_L)$, if and only if the arithmetic proportion $a/b = p/q$ is verified. And this is where this traditional method is childish, because it is not at all justified why the unit of length is outside the defined proportion. And this even despite the fact that, as is known, and it will be recalled here, the conclusion is correct, even

---

if it is unfounded and is not epistemic, which leads to overlooking certain intuitive and latent presuppositions, which become evident when a dyadic analysis of the geometric proportionality of segments is performed.

The first question to elucidate is which dyadic operations affect it. Well, taking into account that it relates only lengths, it should be pointed out to the addition of segments of article 4, the symbol «⊕», to its derivative operation or multiplication by a scalar of article 6, with the symbol «○», and to the dyadic division associated with this one of the symbol «//» of article 7, where it has been proven that the **dyadic quotient of uniform scalar elements**, that is, referring to the same unit, must result in a real number, which will be precisely the quotient in R of the primaries between dividend and divisor. So that the proportionality of segments will be deduced in $\mathscr{D}$ by the following analytical argumentation:

$$\frac{\left(a\,U_L\right)}{\left(b\,U_L\right)} = \frac{\left(p\,U_L\right)}{\left(q\,U_L\right)} = \frac{a}{b} = \frac{p}{q}$$

So the geometric or dyadic proportionality is, in effect, reduced to the arithmetic proportionality of the measures of the segments in the same unit of length. But with the nuance, nothing banal from an epistemological point of view, that, just as the mathematical tradition presupposes or postulates it, the dyadic algebra deduces it unequivocally, without giving the option to subjective intuition or arbitrariness.

So the geometric or dyadic proportionality is, in effect, reduced to the arithmetic proportionality of the measures of the segments in the same unit of length. But with the nuance, nothing banal from an epistemological point of view, that, just as the mathematical tradition presupposes or postulates it, the dyadic algebra deduces it unequivocally, without giving the option to subjective intuition or arbitrariness.

A question that arises immediately is whether in $\mathscr{D}$ the proportions also verify as in R the property that the product of the means is equal to the product of the extremes. To verify this, it must be remembered that the product of segments is the geometric one of article 8, generalized to any magnitude in article 9, so that, given the proportion of segments in the last equation, the geometric products $(a\,U_L)*(q\,U_L)$ and $(b\,U_L)*(p\,U_L)$ to check if they are the same. Well, operating according to the dyadic laws, we have the equations:

$$(a\,U_L)*(q\,U_L)=(a\times q)\,(U_L*U_L)$$

$$(b\,U_L)*(p\,U_L)=(b\times p)\,(U_L*U_L)$$

Both products are uniform, because they refer to the compound unit $U_L*U_L=U_L^2$, so to be equal, according to the dyadic criterion of equality, they must have the same primaries, and, in effect, they do, because in R the property under study $a\times q=b\times p$ is satisfied, which allows to conclude that the proportions of segments also satisfy the condition that the product of the means is equal to that of the extremes:

$$\text{If } \frac{(a\,U_L)}{(b\,U_L)}=\frac{(p\,U_L)}{(q\,U_L)}\text{, then, }(a\,U_L)*(q\,U_L)=(b\,U_L)*(p\,U_L)$$

It can be easily verified that the converse statement also holds, so that if four segments satisfy the equality $(a\,U_L)*(q\,U_L)=(b\,U_L)*(p\,U_L)$, they must be in the corresponding dyadic proportion.

Segment proportionality reveals an important fact, and it is that it establishes a subtle relationship between the uniform dyadic division of article 7 and the dyadic multiplication between scalar magnitudes of articles 8 and 9.

On the other hand, it should not be forgotten that geometric proportionality is the basis of metric geometry and, therefore, also of trigonometry. Today this fundamental fact has been

The trigonometric ratios
Paradigm of dyadic algebra

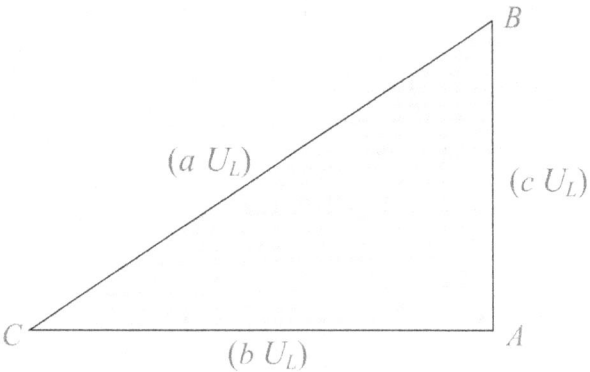

Trigonometric ratios such as the sine, the cosine or the tangent are not arithmetic divisions in themselves, because they do not relate pure numbers, but segments or quantities of lengths, so they strictly belong to geometric algebra and indicate dyadic divisions between magnitudes, in accordance with the operation analyzed in section 7, which serves as fundamental support for the proportionality of segments.

Figure 12

forgotten, with the somewhat tricky simplification of substituting the segments that are made up in the different operations by their measurements in the same unit of length. And this, although operationally correct by chance, is intellectually and pedagogically toxic, because it loses sight of the true significance of geometric algebra which, although isomorphic in many respects with that of R, is a different algebra due to its specific nature. So, it is convenient not to forget things as simple as that the trigonometric ratios are not in themselves arithmetic ratios, but geometric or dyadic divisions of article 7, which analytically, in the right triangle of figure 12 would be expressed as follows:

$$\operatorname{sen} C = \frac{AB}{BC} = \frac{(c\,U_L)}{(a\,U_L)} = \frac{c}{a}$$

$$\cos C = \frac{AC}{BC} = \frac{(b\,U_L)}{(a\,U_L)} = \frac{b}{a}$$

$$\operatorname{tg} C = \frac{AB}{AC} = \frac{(c\,U_L)}{(b\,U_L)} = \frac{c}{b}$$

With capital letters A, B and C the angles and vertices of the triangle have been designated, with the pairs of letters AB, AC and BC the segments that form the sides of the triangle, which in turn are expressed by the dyads $AB=(c\ U_L)$, $AC=(b\ U_L)$ and $BC=(a\ U_L)$.

Artículo 14

ANÁLISIS DIÁDICO DEL *TEOREMA DE TALES*

El *Teorema de Tales*[36], en su formulación clásica, se basa en la teoría de las razones y proporciones, con sus específicas nociones de igualdad y adición de segmentos. En suma, **la teoría tradicional asume sustituir los segmentos por su medida en una determinada unidad de longitud**, que puede ser cualquiera e incluso abstracta, pero siempre la misma para todos los segmentos considerados. Es decir, se presupone que los segmentos o longitudes son uniformes. Con este artificio se reducen las razones y proporciones entre segmentos, que son geométricas, a razones y proporciones numéricas abstractas.

Sin embargo, es evidente que un segmento no es un número sin más, sino una cantidad de longitud. Es una díada con la forma $(a\ U_L)$, donde el primario a es de R e indica la medida del segmento en la unidad de longitud U_L. Pues bien, dados dos segmentos expresados con díadas uniformes $(a\ U_L)$ y $(b\ U_L)$, la Matemática clásica define la razón entre ellos identificándola con la aritmética de sus primarios a/b. Con ello, el álgebra geométrica es sustituida por el álgebra simbólica, descrita por las medidas de los segmentos en una cierta unidad de longitud uniforme.

De la misma manera, la proporcionalidad geométrica de segmentos queda sustituida por la proporcionalidad de números reales, de modo que dos segmentos dados se dirán proporcionales a otros dos cuando los números que expresan sus medidas en la misma unidad de longitud formen proporción en el conjunto R. Es decir, los segmentos $(a\ U_L)$ y $(b\ U_L)$ se dicen proporcionales a los $(p\ U_L)$ y $(q\ U_L)$, si y solo si se verifica la proporción aritmética $a/b=p/q$. Y aquí es donde este método tradicional peca de pueril, porque no queda en absoluto justificado por qué la unidad de longitud sea ajena a la proporción definida. Y ello incluso a pesar de que, como es sabido, y se va a recordar aquí, la conclusión sea

---

[36] La deducción clásica del *Teorema de Tales* se puede encontrar en la «Lección 26» de *Matematizar 1*.

correcta, aunque esté infundada y no sea epistémica, lo que lleva a pasar por alto determinadas presuposiciones intuitivas y latentes, que quedan en evidencia cuando se realiza un análisis diádico de la proporcionalidad geométrica de segmentos.

La primera cuestión a dilucidar es qué operaciones diádicas la afectan. Pues bien, teniendo en cuenta que relaciona solo longitudes, debe señalarse a la adición de segmentos del artículo 4, de símbolo «⊕», a su operación derivada o multiplicación por un escalar del artículo 6, con el símbolo «∘», y a la división diádica asociada a esta de símbolo «⫽» del artículo 7, donde se ha acreditado que el **cociente diádico de elementos escalares uniformes**, es decir, referidos a la misma unidad, ha de dar como resultado un número real, que será precisamente el cociente en R de los primarios entre dividendo y divisor. De modo que la proporcionalidad de segmentos quedará deducida en $\mathscr{D}$ por la argumentación analítica siguiente:

$$\frac{\left(a\,U_L\right)}{\left(b\,U_L\right)} = \frac{\left(p\,U_L\right)}{\left(q\,U_L\right)} = \frac{a}{b} = \frac{p}{q}$$

Así que la proporcionalidad geométrica o diádica se reduce, en efecto, a la proporcionalidad aritmética de las medidas de los segmentos en la misma unidad de longitud. Pero con el matiz, nada banal desde un punto de vista epistemológico, de que, así como la tradición matemática lo presupone o postula, el álgebra diádica lo deduce inequívocamente, sin dar opción a la intuición ni a la arbitrariedad subjetivas.

Una interrogante que se suscita inmediatamente es si en $\mathscr{D}$ las proporciones también verifican como en R la propiedad de que el producto de los medios sea igual al producto de los extremos. Para comprobarlo hay que recordar que el producto de segmentos es el geométrico del artículo 8, generalizado a cualquier magnitud en el artículo 9, por lo que, dada la proporción de segmentos de la última ecuación, se pueden calcular por separado los productos geométricos $(a\ U_L)*(q\ U_L)$ y $(b\ U_L)*(p\ U_L)$ para comprobar si resultan iguales. Pues bien, operando según las leyes diádicas, se tienen las ecuaciones:

$$(a \; U_L) * (q \; U_L) = (a \times q)(U_L * U_L)$$

$$(b \; U_L) * (p \; U_L) = (b \times p)(U_L * U_L)$$

Ambos productos son uniformes, porque se refieren a la unidad compuesta $U_L * U_L = U_L^2$, por lo que para ser iguales, de acuerdo con el criterio diádico de igualdad, deben tener los mismos primarios, Y, en efecto, los tienen, porque en R se cumple la propiedad en estudio $a \times q = b \times p$, lo que permite concluir que las proporciones de segmentos también satisfacen la condición de que el producto de los medios es igual al de los extremos:

Si $\dfrac{(a\,U_L)}{(b\,U_L)} = \dfrac{(p\,U_L)}{(q\,U_L)}$, entonces, $(a \; U_L)*(q \; U_L)=(b \; U_L)*(p \; U_L)$

Se puede comprobar fácilmente que el enunciado recíproco también se verifica, de modo que, si cuatro segmentos satisfacen la igualdad $(a \; U_L)*(q \; U_L)=(b \; U_L)*(p \; U_L)$, deben estar en la proporción diádica correspondiente.

La proporcionalidad de segmentos revela un hecho importante, y es que establece una relación sutil entre la división diádica uniforme del artículo 7 y la multiplicación diádica entre magnitudes escalares de los artículos 8 y 9.

Por otra parte, no debe olvidarse que la proporcionalidad geométrica es la base de la geometría métrica y, por tanto, también de la trigonometría. En la actualidad se ha olvidado este hecho fundamental, con la simplificación algo tramposa de sustituir los segmentos que se componen en las distintas operaciones por sus medidas en la misma unidad de longitud. Y ello, si bien operativamente resulta casualmente correcto, intelectual y pedagógicamente es tóxico, porque se pierde de vista la auténtica significación del álgebra geométrica que, aunque isomorfa en muchos aspectos con la de R, es un álgebra diferente por su naturaleza específica. De modo que, conviene no olvidar cosas tan simples como que las razones trigonométricas no son en sí mismas razones aritméticas, sino divisiones geométricas o diádicas del artículo 7, lo que analíticamente, en el triángulo rectángulo de la figura 12 quedaría expresado así:

$$sen\ C = \frac{AB}{BC} = \frac{\left(c\,U_L\right)}{\left(a\,U_L\right)} = \frac{c}{a}$$

$$cos\ C = \frac{AC}{BC} = \frac{\left(b\,U_L\right)}{\left(a\,U_L\right)} = \frac{b}{a}$$

$$tg\ C = \frac{AB}{AC} = \frac{\left(c\,U_L\right)}{\left(b\,U_L\right)} = \frac{c}{b}$$

Las razones trigonométricas
Paradigma del álgebra diádica

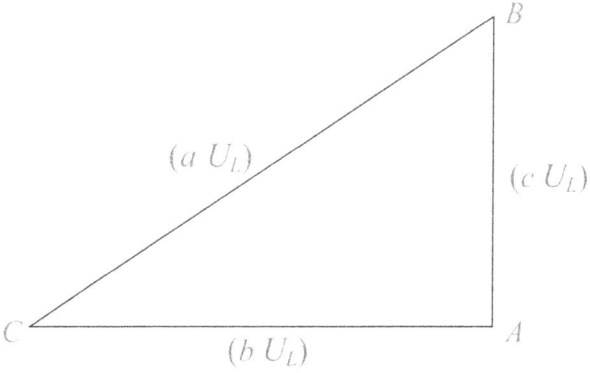

Las razones trigonométricas como el seno, el coseno o la tangente, no son divisiones aritméticas en sí mismas, porque no relacionan números puros, sino segmentos o cantidades de longitudes, por lo que en rigor pertenecen al álgebra geométrica e indican divisiones diádicas entre magnitudes, de acuerdo con la operación analizada en el apartado 7, que sirve de soporte fundamental a la proporcionalidad de segmentos.

Figura 12

Con letras mayúsculas A, B y C se han designado los ángulos y vértices del triángulo, con las parejas de letras AB, AC y BC los segmentos que forman los lados del triángulo, que a su vez quedan expresados por las díadas $AB=(c\ U_L)$, $AC=(b\ U_L)$ y $BC=(a\ U_L)$.

Article 15

THE PYTHAGOREAM THEOREM:
THE FIRST DYADIC FORM OF MATHEMATICS

Egyptian builders and surveyors used a very simple and ingenious instrument, a knotted string that marked equal lengths or segments. With it they formed a triangle with sides 3, 4 and 5 of those segments, which turned out to be right, and thus they were able to draw perpendicular alignments. The Greek mathematician Pythagoras, born in Samos in 580 BC, investigated this property, known to the Egyptians for that singular triangle with sides proportional to 3, 4 and 5, and generalized it to all triangles such that one of its angles be straight, formulating his famous Pythagorean Theorem[37].

Pythagoras used geometric algebra to compose segments and thus concluded that **the area of the square built on the hypotenuse of a right triangle is equal to the sum of the areas of the squares built on the legs**. Such a way of operating with segments, consisting of constructing squares with them, is the same way that the geometric multiplication of lengths is defined in article 8, which is generalized to the dyadic multiplication of any scalar magnitudes in article 9. Therefore The Pythagorean Theorem seems to be the first dyadic formulation of Mathematics, as explained in figure 13.

However, although the classics differentiated in this way the geometric operations with segments from the arithmetic ones, at present the property analyzed in article 8, on the geometric experiment with areas and volumes, has been used to substitute geometric algebra for the arithmetic, operating only with numbers that represent the measures of the segments, ignoring the segments themselves. With this, the Pythagorean Theorem is usually formulated referring to R operations with this

---

[37] Various deductions of the Pythagorean Theorem can be found in «Lesson 29» of Mathematize 1.

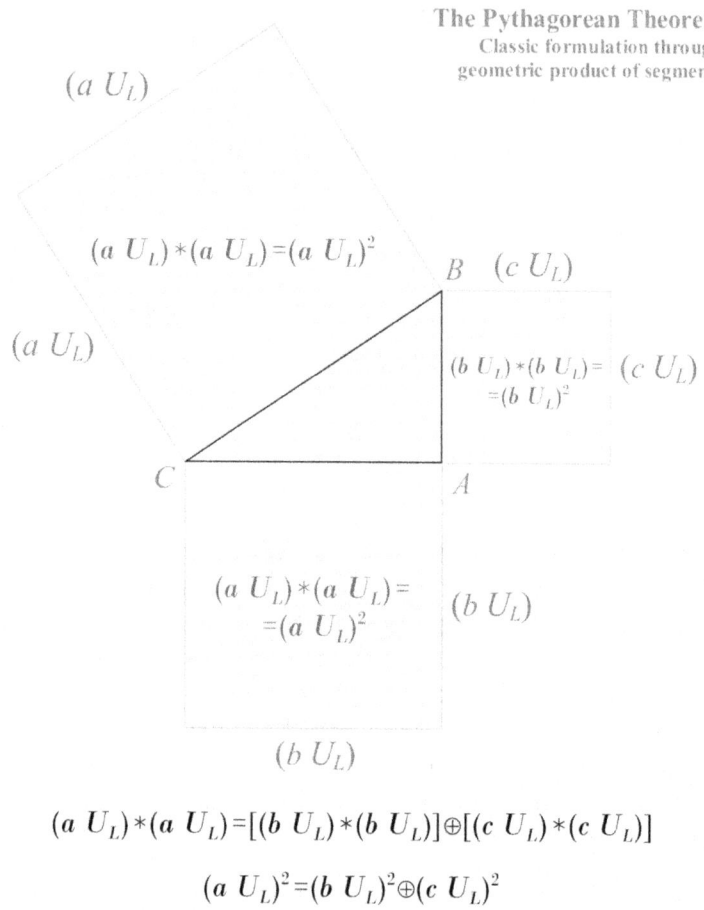

The Pythagorean Theorem
Classic formulation through
geometric product of segments

$(a\ U_L)*(a\ U_L)=[(b\ U_L)*(b\ U_L)]\oplus[(c\ U_L)*(c\ U_L)]$

$(a\ U_L)^2=(b\ U_L)^2\oplus(c\ U_L)^2$

In every right triangle, the area of the square built on the hypotenuse is equal to the sum of the areas of the squares built on the legs.

Figure 13

statement: «**In every right triangle the hypotenuse squared is equal to the sum of the squares of the legs, analytically $a^2=b^2+c^2$**». And this arithmetic simplification, which omits the geometric origin of such an important property, loses much of the meaning it encompasses.

Classical geometric algebra
Book II of Euclid's Elements
First proposition or distributive property of multiplication
regarding the addition of segments

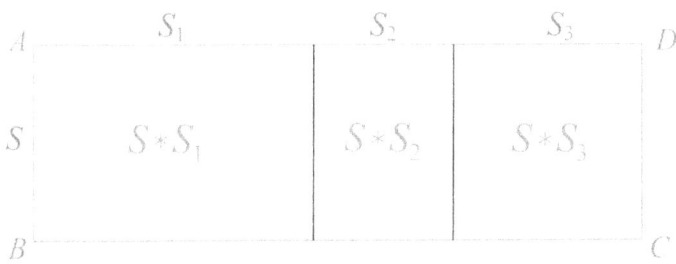

$$S*(S_1 \oplus S_2 \oplus S_3) = (S*S_1) \oplus (S*S_2) \oplus (S*S_3)$$

The geometric multiplication of two segments, which consists of forming rectangles with them, satisfies the distributive property with respect to the dyadic addition of lengths or areas.

Figure 14

The importance that the Greeks conferred on geometric algebra is evident in Euclid's Book II of the Elements. Although it is one of the shortest, with only 14 propositions, none of which is present in modern textbooks, for the classics its statements had great significance. Such disagreement between the ancient and modern criteria is due to the imposing hypnotic force of symbolic logic, which has led to the foundations or meanings of symbols being ruined, more than arbitrarily, spuriously giving them their own substantivity and undermining the mathematical quality.

The first proposition of Book II states: «**If we have two straight lines and we cut one of them into any number of segments, then the rectangle contained by the two straight lines is equal to the**

423

rectangles contained by the straight line that was not cut and each one of the previous segments». The meaning of this proposition can be better understood with the example in figure 14:

Let AB and AD be two segments; take AD and form the arbitrary sum with, for example, the three segments S_1, S_2 and S_3, so that we have $AD = S_1 \oplus S_2 \oplus S_3$, where «$\oplus$» indicates the geometric addition of lengths; the rectangle $ABCD$ will be the geometric product of the segments AB and AD, which is symbolized $AB * AD$ or $S * AD$, if $S = AB$ is identified, and where the symbol «$*$» indicates the geometric multiplication of segments; the rectangle $S * AD$ is the sum of the interior rectangles described analytically by $S * S_1 \oplus S * S_2 \oplus S * S_3$; so that we have the conclusion:

$$S * (S_1 \oplus S_2 \oplus S_3) = (S * S_1) \oplus (S * S_2) \oplus (S * S_3)$$

And this is nothing but the distributive **property of multiplication with respect to dyadic addition** for segments or areas, operations proper to geometric algebra.

Another didactic case of the effect of loss of geometric meaning caused by symbolic algebra is the **square of a binomial**. Modern algebra simply writes it abstractly in the form $(a+b)^2 = a^2 + b^2 + 2 \times a \times b$. Well, in ancient geometric algebra this property is described in the fourth proposition of the referred Book II of Euclid: **«If a straight line is cut in an arbitrary way, then, the square built on the total is equal to the squares on the two segments and twice the rectangle contained by both segments».** The meaning of this proposition is clarified with figure 15:

Let segment AB and decompose into the sum of two arbitrary segments S_1 y S_2; form the square $ABCD$, which is decomposed into the two squares and two rectangles indicated in the figure. The area of the square $ABCD$ represents the geometric product $((S_1 \oplus S_2) * (S_1 \oplus S_2) = (S_1 \oplus S_2)^2$ and is equal to the sum of the area indicated by $S_1 * S_1 = S_1^2$, of the area $S_2 * S_2 = S_2^2$ and twice the area of the rectangle identified with $S_1 * S_2$; this sum can be symbolized by the following equation:

$$(S_1 \oplus S_2)^2 = S_1^2 \oplus S_2^2 \oplus 2 \circ (S_1 * S_2)$$

Note that the operation «○» corresponds to the multiplication of a scalar by a dyad, defined in article 6. And thus we have the geometric or dyadic shape of the square of a binomial without losing an iota of its geometric meaning.

There is no doubt that the visual evidence of geometric reasoning sheds a lot of light on the different observed properties, which are hidden in modern abstract expressions, which is why the high school students of our time are deprived of

Geometric meaning of
square of a binomial

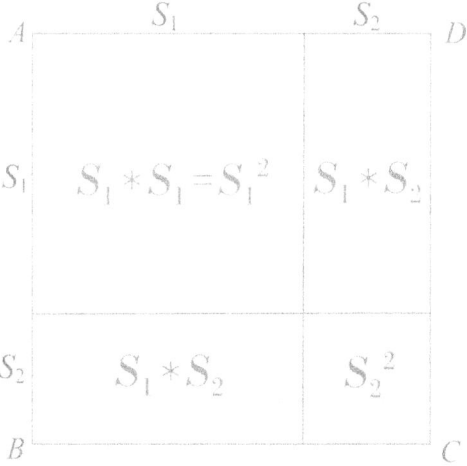

The area of the square $ABCD$ is the sum of the four interior areas into which it is decomposed. In this way the geometric or dyadic expression of the square of an abstract binomial $(a+b)^2 = a^2 + b^2 + 2 \times a \times b$ results:

$$(S_1 \oplus S_2)^2 = S_1^2 \oplus S_2^2 \oplus 2 \circ (S_1 * S_2)$$

Figure 15

the ability to understand the rules algebraic that apply with a mechanism almost unworthy of intelligent beings.

With these examples it is clear that **in Euclid's time the magnitudes were already represented as segments subject to the axioms and theorems of geometry**. It is true that at that time algebraic structures, as they are conceived today, did not exist, which has led some to affirm that the Greeks lacked algebra, which is not entirely true, because, although incomplete and primitive, algebra Geometric is there, but yes, **formulated in propositions of ordinary language** and, therefore, curtailed the power of abstraction inherent to the symbology of modern Mathematics. However, that same symbolic and logical power provoke a fascination that is very difficult to contain, inadvertently inducing almost invisible presuppositions, as would be the case of the arbitrary reduction of operations with magnitudes only to their numerical part, losing the geometric meaning along the way. of the symbolic abstractions thus formed, which can simplify the manipulation of these graphic elements, but clearly detrimental to the complete knowledge of mathematical and physical phenomena. There is no doubt that a Greek geometrist versed in the 14 theorems of **Euclid's geometric algebra** would be far more skilled in the practice of measurement than a modern expert geometrist.

Hence, the methodology of this First Algebra of Magnitudes tries to combine the virtues of both techniques, the classical, with its eloquent geometric visualization, and the symbolic, which provides structures and logical methods that were very valuable before unknown, in order to weave a modern physical algebra, abstract and ilative, but well grounded in basic geometric algebra, and thereby **substantiate, explain and define operations with magnitudes without omissions, inconsistencies or latent presuppositions.**

Artículo 15

EL *TEOREMA DE PITÁGORAS*
LA PRIMERA FORMA DIÁDICA DE LA MATEMÁTICA

Los constructores y agrimensores egipcios utilizaban un instrumento muy simple y de gran ingenio, un cordel anudado que marcaba longitudes o segmentos iguales. Con él formaban un triángulo de lados 3, 4 y 5 de esos segmentos, que resultaba ser rectángulo, y así eran capaces de trazar alineaciones perpendiculares. El matemático griego Pitágoras, nacido en Samos el año 580 antes de Cristo, investigó dicha propiedad, conocida por los egipcios para ese triángulo singular de lados proporcionales a 3, 4 y 5, y la generalizó a todos los triángulos tales que uno de sus ángulos sea recto, formulando su famoso *Teorema de Pitágoras*[38].

Pitágoras utilizó el álgebra geométrica para componer segmentos y así concluyó que **el área del cuadrado construido sobre la hipotenusa de un triángulo rectángulo es igual a la suma de las áreas de los cuadrados construidos sobre los catetos**. Tal forma de operar con segmentos, consistente en construir con ellos cuadrados, es la misma con que se define la multiplicación geométrica de longitudes en el artículo 8, que se generaliza a la multiplicación diádica de magnitudes escalares cualesquiera en el artículo 9. Por tanto, el *Teorema de Pitágoras* parece ser la primera formulación diádica de la Matemática, tal como se explica en la figura 13.

No obstante, si bien los clásicos diferenciaban de ese modo las operaciones geométricas con segmentos de las aritméticas, en la actualidad se ha aprovechado la propiedad analizada en el artículo 8, sobre el experimento geométrico con áreas y volúmenes, para sustituir el álgebra geométrica por la aritmética, operando únicamente con números que representan las medidas de los segmentos, obviando los propios segmentos. Con ello, el *Teorema*

---

[38] Diversas deducciones del *Teorema de Pitágoras* se pueden encontrar en la «Lección 29» de *Matematizar 1*.

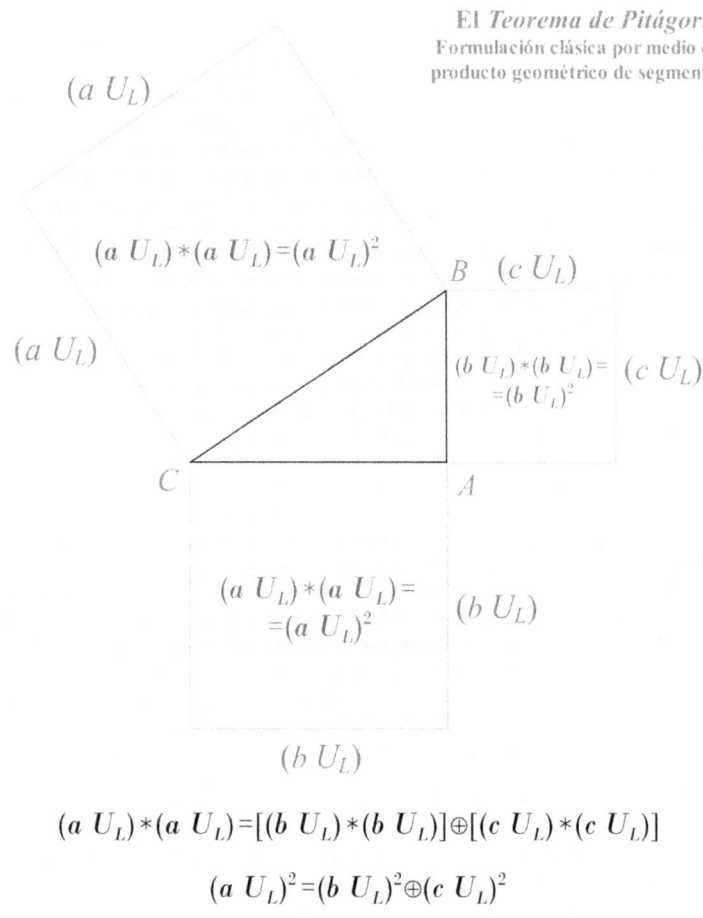

El *Teorema de Pitágoras*
Formulación clásica por medio del
producto geométrico de segmentos

$$(a\ U_L)*(a\ U_L)=[(b\ U_L)*(b\ U_L)]\oplus[(c\ U_L)*(c\ U_L)]$$

$$(a\ U_L)^2=(b\ U_L)^2\oplus(c\ U_L)^2$$

En todo triángulo rectángulo, el área del cuadrado construido sobre la hipotenusa es igual a la suma de las áreas de los cuadrados construidos sobre los catetos.

Figura 13

de Pitágoras suele formularse aludiendo a operaciones de R con este enunciado: «**En todo triángulo rectángulo la hipotenusa al cuadrado es igual a la suma de los cuadrados de los catetos, en forma analítica $a^2=b^2+c^2$**». Y esta simplificación aritmética, que omite el origen geométrico de una propiedad tan importante, hace perder gran parte del significado que engloba.

La importancia que los griegos conferían al álgebra geométrica queda patente en el Libro II de los *Elementos* de Euclides. Pese a que es uno de los más cortos, con solo 14 proposiciones, ninguna de las cuales está presente en los libros de texto modernos, para los clásicos sus enunciados tenían una gran trascendencia. Tal disconformidad entre los criterios antiguos y modernos es debida a la imponente fuerza hipnótica de la lógica simbólica, que ha llevado a arrumbar, más que arbitrariamente, los fundamentos o significados de los símbolos, otorgándoles espuriamente sustantividad propia y menoscabando la calidad matemática.

La proposición primera del Libro II enuncia: **«Si tenemos dos líneas rectas y cortamos una de ellas en un número cualquiera de segmentos, entonces, el rectángulo contenido por las dos líneas rectas es igual a los rectángulos contenidos por la línea recta que no fue cortada y cada uno de los segmentos anteriores».** El significado de esta proposición se puede comprender mejor con el ejemplo de la figura 14:

Sean dos segmentos AB y AD; tómese AD y fórmese la suma arbitraria con, por ejemplo, los tres segmentos S_1, S_2 y S_3, de modo que se tenga $AD = S_1 \oplus S_2 \oplus S_3$, donde «$\oplus$» indica la adición geométrica de longitudes; el rectángulo $ABCD$ será el producto geométrico de los segmentos AB y AD, que se simboliza $AB * AD$ o $S * AD$, si se identifica $S = AB$, y donde el símbolo «$*$» señala la multiplicación geométrica de segmentos; el rectángulo $S * AD$ es la suma de los rectángulos interiores descrita en forma analítica por $S * S_1 \oplus S * S_2 \oplus S * S_3$; de modo que se tiene la conclusión:

$$S * (S_1 \oplus S_2 \oplus S_3) = (S * S_1) \oplus (S * S_2) \oplus (S * S_3)$$

Y esta no es sino la **propiedad distributiva de la multiplicación respecto de la adición diádicas** para segmentos o áreas, operaciones propias del álgebra geométrica.

Otro caso didáctico del efecto de pérdida de significado geométrico provocado por el álgebra simbólica es el **cuadrado de**

Álgebra geométrica clásica
Libro II de los *Elementos* de Euclides
Proposición primera o propiedad distributiva de la
multiplicación respecto de la adición de segmentos

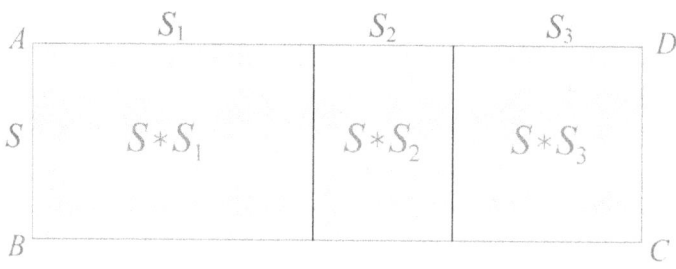

$$S*(S_1 \oplus S_2 \oplus S_3) = (S*S_1) \oplus (S*S_2) \oplus (S*S_3)$$

La multiplicación geométrica de dos segmentos, que consiste en la
formación con ellos de rectángulos, satisface la propiedad distributiva
respecto de la adición diádica de longitudes o áreas.

Figura 14

un binomio. El álgebra moderna se limita a escribirlo en abstracto
con la forma $(a+b)^2 = a^2 + b^2 + 2 \times a \times b$. Pues bien, en el álgebra
geométrica antigua se describe esta propiedad en la proposición
cuarta del referido Libro II de Euclides: **«Si una linea recta se
corta de una manera arbitraria, entonces, el cuadrado construido
sobre el total es igual a los cuadrados sobre los dos segmentos y dos
veces el rectángulo contenido por ambos segmentos».** El significado
de esta proposición queda aclarado con la figura 15:

Sea el segmento AB y descompóngase en la suma de dos
segmentos arbitrarios S_1 y S_2; fórmese el cuadrado $ABCD$, que se
descompone en los dos cuadrados y dos rectángulos que se indican
en la figura. El área del cuadrado $ABCD$ representa el producto

geométrico $(S_1 \oplus S_2) * (S_1 \oplus S_2) = (S_1 \oplus S_2)^2$ y es igual a la suma del área indicadas por $S_1 * S_1 = S_1^2$, del área $S_2 * S_2 = S_2^2$ y dos veces el área del rectángulo identificado con $S_1 * S_2$; esta suma se puede simbolizar con la ecuación siguiente:

$$(S_1 \oplus S_2)^2 = S_1^2 \oplus S_2^2 \oplus 2 \circ (S_1 * S_2)$$

Nótese que la operación «$\circ$» corresponde a la multiplicación de un escalar por una díada, definida en el artículo 6. Y así se tiene la forma geométrica o diádica del cuadrado de un binomio sin perder un ápice de su significado geométrico.

No cabe duda de que la evidencia visual de los razonamientos geométricos arrojan mucha luz sobre las diferentes propiedades

Significado geométrico del
cuadrado de un binomio

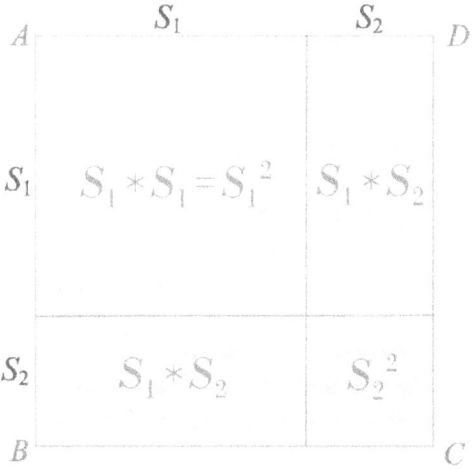

El área del cuadrado ABCD es la suma de las cuatro áreas interiores en que se descompone. De esta manera resulta la expresión geométrica o diádica del cuadrado de un binomio abstracto $(a+b)^2 = a^2 + b^2 + 2 \times a \times b$:

$$(S_1 \oplus S_2)^2 = S_1^2 \oplus S_2^2 \oplus 2 \circ (S_1 * S_2)$$

Figura 15

431

observadas, que se ocultan en las expresiones abstractas modernas, por lo que los estudiantes de enseñanza media de nuestro tiempo son privados de la capacidad de comprender las reglas algebraicas que aplican con un mecanicismo casi indigno de seres inteligentes.

Con estos ejemplos queda patente que **en tiempos de Euclides las magnitudes ya se representaban como segmentos sujetos a los axiomas y teoremas de la geometría.** Es cierto que en esa época no existían las estructuras algebraicas, tal como se conciben hoy en día, lo que ha llevado a algunos a afirmar que los griegos carecían de álgebra, cosa no del todo cierta, porque, aunque incompleta y primitiva, el álgebra geométrica está ahí, pero eso sí, **formulada en proposiciones del lenguaje ordinario** y, por tanto, cercenado el poder de abstracción propio de la simbología de la Matemática moderna. Sin embargo, ese mismo poder simbólico y lógico provocan una fascinación muy difícil de contener, induciendo sin querer presuposiciones casi invisibles, como sería el caso de la reducción arbitraria de las operaciones con magnitudes únicamente a su parte numérica, perdiéndose en el camino el significado geométrico de las abstracciones simbólicas así formadas, lo cual puede simplificar la manipulación de estos elementos gráficos, pero en claro detrimento del conocimiento completo de los fenómenos matemáticos y físicos. No hay duda de que un geómetra griego versado en los 14 teoremas del **álgebra geométrica de Euclides** sería mucho más habilidoso en la práctica de la medida que un geómetra experto de la actualidad.

De ahí que la metodología de esta *Primera álgebra de magnitudes* intente combinar las virtudes de ambas técnicas, la clásica, con su elocuente visualización geométrica, y la simbólica, que aporta estructuras y métodos lógicos muy valiosos antes desconocidos, para así tejer un álgebra física moderna, abstracta e ilativa, pero bien cimentada en el álgebra geométrica básica, y con ello **fundamentar, explicar y definir sin omisiones ni incongruencias ni presuposiciones latentes las operaciones con magnitudes.**

Article 16

The dyadic multiplication of scalar magnitudes of article 9 and its derived operation, the scalar dyadic division of article 10, are those that correspond to physical quantities such as, for example, density. Its analysis could be done like this: let (M kg) be the dyad that expresses the mass of a body indicated in kilograms, let (V m^3) be the volume of the body in cubic meters. It is not forbidden to define a compound magnitude, which is called a density and indicated, for example, with the dyad (d U_d), where U_d represent the unit in which density must be measured to meet the definition of this magnitude. Defining the density as the dyadic quotient between the mass of a body and the volume it occupies, analytically the following reasoning will be had:

$$\left(d\,U_{d}\right) = \frac{\left(M\,kg\right)}{\left(V\,m^{3}\right)} = \left(\frac{M}{V}\ \frac{kg}{m^{3}}\right)$$

The uniform equality criterion of the dyads of the first and last member requires that the respective primaries and secondaries be equal, which leads to two identities, the first of R and the second of dyadic algebra, which to understand each other we have been symbolizing with the letter $\mathscr{D}$, and thus, if the U_d unit of the first member is identified with that of the second $kg/\!/m^3$, it results:

$$\text{If } U_{d} = \frac{kg}{m^{3}} \text{ in } \mathscr{D}, \text{ then, } d = \frac{M}{V} \text{ in } R$$

The physical meaning of density, which is a compound magnitude, by definition, from mass and volume, is obtained without more than considering the unit volume of a cubic meter and finding out its mass, resulting in that, being $V=1$, the density will indicate precisely the mass that corresponds to each cubic meter, which is commonly expressed as «mass per unit volume».

Other forms of physical laws are those involving products of scalar and vector magnitudes, such as Newton's second law. This is the multiplication of article 11. The logical scheme in this case should be the following: let it be a mass given by the scalar dyad $(M\ kg)$ and let it be the vector dyad $(\overline{a}\ m/\!/s^2)$ that indicates the acceleration of its movement. Note that the unit of acceleration is a dyadic quotient. Let be the vector dyad $(\overline{F}\ U_F)$, where U_F indicates the unit of uniform force that corresponds to satisfy the equality represented by Newton's second law, an identity that should be written in dyadic algebra $\mathscr{D}$ with the form $(\overline{F}\ U_F)=(M\ kg)\odot(\overline{a}\ m/\!/s^2)$, where the multiplication indicated «$\odot$» corresponds to the composition law of article 11. The definition of this operation allows us to write the initial equation with the form $(\overline{F}\ U_F)=[(M\bullet\overline{a})\ (kg*m/\!/s^2)]$, in which three laws of composition appear: the multiplication «$\bullet$» of a real number by a vector, the scalar dyadic multiplication «$*$» and the scalar dyadic division «$/\!/$». The criterion of equality of uniform dyads, that is to say, referring to the same unit, allows the last formula to be divided into its two components, one from the algebra of R and the other from the dyadic algebra $\mathscr{D}$, resulting:

$$\text{If } U_F = \frac{kg*m}{s^2} \text{ in } \mathscr{D}, \text{ then, } \overline{F} = M\bullet\overline{a} \text{ in } \mathbb{R}^3$$

It is customary to name the unit v with the term newton, abbreviated N, whose physical meaning corresponds to the force that must be applied to the mass of one kilogram to induce an acceleration of one $m/\!/s^2$ in it, which will be read «meter per second squared», meaning scalar dyadic division.

Another capital concept of Physics is the **work of a force**, defined as the dyadic scalar product of the force magnitude by the length magnitude traveled by its point of application. What in $\mathscr{D}$ will be written explicitly, without the traditional symbolic presuppositions, $(T\ U_T)=(\overline{F}\ N)\odot(\overline{e}\ m)$, where $(T\ U_T)$ denotes the scalar dyad that measures the work done by a quantity of force indicated with the dyad vector $(\overline{F}\ N)$ when its point of application is displaced as determined by the vector dyad $(\overline{e}\ m)$, here m symbolizing the standard length unit called meter, not the mass magnitude. The corresponding multiplication is «⊙», that is, the scalar dyadic of article 12. The definition of this composition law allows us to write the work $(T\ U_T)=(\overline{F}\cdot\overline{e})\ (N*m)$. The multiplication of the mathematical point «·» is the scalar product of vectors in R^3, while the asterisk «*» symbolizes the scalar dyadic product. As in the previous cases, the equality criterion in $\mathscr{D}$ determines the formulation of the compound unit U_T and the resulting primary:

$$\text{If } U_T = N*m \text{ in } \mathscr{D}, \text{ then, } T=\overline{F}\cdot\overline{e}$$

The compound unit $U_T = N*m$ in Physics is called the joule, and it means the work or energy produced by a force of one newton when its point of application is displaced by an amount of length equal to one meter.

The last significant example of a physical equation to be analyzed is that of the **moment magnitude of a force**. In dyadic algebra $\mathscr{D}$ the moment $(\overline{\mu}U_\mu)$ of a force with respect to a point must be defined as the vector dyadic product between the vector concrete that indicates the position vector $(\overline{r}\ m)$ of the point of application of the force, with respect to that other to which refers to the moment, where m is the standard meter, and the vector dyad that represents the force $(\overline{F}\ N)$. In $\mathscr{D}$ analytics it will be written with the explicit form $(\overline{\mu}U_\mu)=(\overline{r}\ m)\circledast(\overline{F}\ N)$, where «⊛» is the vector dyadic multiplication of article 12. According to the definition of this composition law, it is permissible to put $(\overline{\mu}\ U_\mu)=[(\overline{r}\wedge\overline{F})\ (m*N)]$. The multiplication «∧» is the vector product of vectors in R^3, while the asterisk «*» symbolizes the scalar dyadic product. The equality criterion in $\mathscr{D}$ sets the compound unit U_μ and its dyadic pair:

$$\text{If } U_\mu = m*N \text{ in } \mathscr{D}, \text{ then, } \overline{\mu}=\overline{r}\wedge\overline{F} \text{ en } R^3$$

Artículo 16

FORMA DIÁDICA DE LAS ECUACIONES FÍSICAS

La multiplicación diádica de magnitudes escalares del artículo 9 y su operación derivada, la división diádica escalar del artículo 10, son las que corresponden a magnitudes físicas como, por ejemplo, la **densidad**. Su análisis podría hacerse así: sea $(M \, kg)$ la díada que expresa la masa de un cuerpo indicada en kilogramos, sea $(V \, m^3)$ el volumen del cuerpo en metros cúbicos. No está prohibido definir una magnitud compuesta, que se denomina densidad e indicada, por ejemplo, con la díada $(d \, U_d)$, donde U_d represente la unidad en que deba medirse la densidad para cumplir con la definición de esta magnitud. Definiendo la densidad como el cociente diádico entre la masa de un cuerpo y el volumen que ocupa, analíticamente se tendrá el siguiente razonamiento:

$$\left(d \, U_d\right) = \frac{\left(M \, kg\right)}{\left(V \, m^3\right)} = \left(\frac{M}{V} \, \frac{kg}{m^3}\right)$$

El criterio de igualdad uniforme de las díadas del primer miembro y del último exige que sean iguales los primarios y los secundarios respectivos, lo que conduce a dos identidades, la primera de R y la segunda del álgebra diádica, que para entendernos venimos simbolizando con la letra $\mathscr{D}$, y con ello, si la unidad U_d del primer miembro se identifica con la del segundo $kg/\!/m^3$, resulta:

$$\text{Si } U_d = \frac{kg}{m^3} \text{ en } \mathscr{D}, \text{ entonces, } d = \frac{M}{V} \text{ en } R$$

El significado físico de la densidad, que es una magnitud compuesta, por definición, a partir de la masa y del volumen, se obtiene sin más que considerar el volumen unitario de un metro cúbico y averiguar su masa, resultando que, siendo $V=1$, la densidad indicará precisamente la masa que corresponda a cada metro cúbico, lo que se expresa comúnmente como «masa por unidad de volumen».

Otras formas de leyes físicas son la que implican productos de magnitudes escalares y vectoriales, tales como la *segunda ley de Newton*. Se trata de la multiplicación del artículo 11. El esquema lógico en este caso habría de ser el siguiente: sea una masa dada por la díada escalar ($M\ kg$) y sea la díada vectorial ($\overline{a}\ m/\!/s^2$) que indique la aceleración de su movimiento. Obsérvese que la unidad de aceleración es un cociente diádico. Sea la díada vectorial ($\overline{F}\ U_F$), donde U_F indique la unidad de fuerza uniforme que corresponda para que se satisfaga la igualdad que representa la *segunda ley de Newton*, identidad que debería escribirse en el álgebra $\mathscr{D}$ con la forma ($\overline{F}\ U_F$)=($M\ kg$)⊙($\overline{a}\ m/\!/s^2$), donde la multiplicación «⊙» se corresponde con la ley de composición del artículo 11. La definición de esta operación permite escribir la ecuación inicial con la forma ($\overline{F}\ U_F$)=[(M•$\overline{a}$) ($kg*m/\!/s^2$)], en la que aparecen tres leyes de composición: la multiplicación «•» de un número real por un vector, la multiplicación diádica escalar «*» y la división diádica escalar «/\!/». El criterio de igualdad de díadas uniformes, es decir, referidas a la misma unidad, permite desdoblar la última fórmula en sus dos componentes, una del álgebra de R y la otra del álgebra diádica $\mathscr{D}$, resultando:

$$\text{Si } U_F = \frac{kg*m}{s^2} \text{ en } \mathscr{D}, \text{ entonces, } \overline{F} = M • \overline{a} \text{ en } R^3$$

Se acostumbra a nombrar la unidad U_F con el término newton, abreviadamente N, cuyo significado físico corresponde a la fuerza que debe aplicarse a la masa de un kilogramo para inducir en ella una aceleración de un $m/\!/s^2$, que se leerá «metro por segundo al cuadrado», con el significado de división diádica escalar.

Otro concepto capital de la Física es el **trabajo de una fuerza**, definido como el producto escalar diádico de la magnitud fuerza por la magnitud longitud recorrida por su punto de aplicación. Lo que en $\mathscr{D}$ se escribirá explícitamente, sin las presuposiciones simbólicas tradicionales, ($T\ U_T$)=($\overline{F}\ N$)⊙($\overline{e}\ m$), donde ($T\ U_T$) denota la díada escalar que mide el trabajo realizado por una cantidad de fuerza señalada con la díada vectorial ($\overline{F}\ N$) cuando su punto de aplicación se desplaza según determina la díada vectorial $(\overline{e}\ m)$, simbolizando aquí m la unidad de longitud patrón llamada metro, no la magnitud masa. La multiplicación que corresponde es «⊙», es decir, la diádica escalar del artículo 12. La definición de esta ley de composición permite escribir el trabajo $(T\ U_T)=(\overline{F}\cdot\overline{e})\ (N*m)$. La multiplicación del punto matemático «·» es el producto escalar de vectores en R^3, mientras que el asterisco «*» simboliza el producto diádico escalar. Como en los casos anteriores, el criterio de igualdad en $\mathscr{D}$ determina la formulación de la unidad compuesta U_T y el primario resultante:

$$\text{Si } U_T = N*m \text{ en } \mathscr{D}, \text{ entonces, } T=\overline{F}\cdot\overline{e}$$

La unidad compuesta $U_T = N*m$ en Física se llama julio, y significa el trabajo o energía producido por una fuerza de un newton cuando su punto de aplicación se desplaza una cantidad de longitud igual a un metro.

El último ejemplo significativo de ecuación física que se analizará, es el de la magnitud **momento de una fuerza**. En el álgebra diádica $\mathscr{D}$ el momento $(\overline{\mu}\ U_\mu)$ de una fuerza respecto de un punto debe definirse como el producto diádico vectorial entre el concreto vectorial que indique el vector posición $(\overline{r}\ m)$ del punto de aplicación de la fuerza, respecto de aquel otro a que se refiera el momento, siendo m el metro patrón, y la díada vectorial que represente la fuerza $(\overline{F}\ N)$. En analítica de $\mathscr{D}$ se escribirá con la forma explícita $(\overline{\mu}\ U_\mu)=(\overline{r}\ m)\circledast(\overline{F}\ N)$, donde «⊛» es la multiplicación diádica vectorial del artículo 12. De acuerdo con la definición de esta ley de composición, es permisible poner $(\overline{\mu}\ U_\mu)=[(\overline{r}\wedge\overline{F})\ (m*N)]$. La multiplicación «∧» es el producto vectorial de vectores en R^3, mientras que el asterisco «*» simboliza el producto diádico escalar. El criterio de igualdad en $\mathscr{D}$ fija la unidad compuesta U_μ y su par diádico:

$$\text{Si } U_\mu=m*N \text{ en } \mathscr{D}, \text{ entonces, } \overline{\mu}=\overline{r}\wedge\overline{F} \text{ en } R^3$$

Article 17

EFFECTS OF THE PRINCIPLE OF SYMBOLIC ECONOMY

Throughout the preceding compendium it has been observed how the geometric algebra of segments or lengths is generalized in the abstract, giving rise to the generic algebra of concretes or dyads, as mathematical representatives of the quantities of physical magnitudes. It has also been warned about the hypnotic effect that can be produced by availing itself of the symbolic economy, understood as the simplification of signs for the different operations of the same species, such as the additive ones, all denoted with the typical cross «+», the multiplicative ones indicated generically, for example, with the sign «×», subtraction with the dash «−», or divisions with the slash «/». To break this spell and warn for pedagogical purposes about how easy it is to be fascinated by it and believe that what really remains in the dark is understood, an effort has been made of symbolic detail, to make explicit the maximum number of distinguishable laws of composition each other, as well as the relationships that arise between them; although, given their large number, as the symbology is limited and it would not be useful to take such differentiation to the absolute extreme, it is inevitable and even convenient that some share common signs, which is not an obstacle for the phenomenon to be explained with sufficient didactic clarity. Take the expression in $\mathscr{D}$ as an example:

$$[p \circ (\overline{a}\ U_1)] \circledast [q \circ (\overline{b}\ U_2)] = (p \times q) \circ [(\overline{a}\ U_1) \circledast (\overline{b}\ U_2)]$$

The principle of symbolic economy allows to symbolize all the laws of composition of the same multiplicative species with the same character «×», and with this the traditional notation results:

$$(p \times \overline{a}\ U_1) \times (q \times \overline{b}\ U_2) = (p \times q) \times [(\overline{a}\ U_1) \times (\overline{b}\ U_2)]$$

Observing this last expression, unless one has algebraic expertise, it is difficult to escape the illusion caused by the constant sign of the sign «×» and it is easy to believe that the property that describes equality is evident by the laws of R^3.

However, this is not the case, because what are related are physical dyads, and the complete meaning of equality is given by the different composition laws that comprise the equation itself and specifically defined between the sets R, $\{R^3,U_1\}$, $\{R^3,U_2\}$ y $\{R^3,U_1*U_2\}$, so the formula must be interpreted based on them.

For a better overview of the **spell of symbolic reduction, so toxic for the learning of Physics, the scientific accuracy and the precise meaning of the compound magnitudes**, the symbols of the operations that intervene in the dyadic algebra can be detailed, represented with the sign $\mathscr{D}$, unlike the traditional structures of R, C and R³ or any other. In this way the following synoptic scheme results:

Type of dyadic composition law / Article of section XXVIII		Ordinary number algebra (see note)		Dyadic algebra or physical	With the principle of symbolic economy
		In R y C	In R³	In $\mathscr{D}$	
Scalar magnitudes and vector	Addition (4)	+	+	⊕	+
	Subtraction (5)	−	−	⊖	−
	Multiplication by a number (6)	×	•	○	×
	Homogeneous division (7)	/ ÷		// ≑	/ ÷
Scalar magnitudes	Heterogeneous multiplication (9)	×		∗	×
	Heterogeneous division (10)			// ≑	/ ÷
Vector magnitudes	Product of mixed magnitudes (11)			◎	×
	Scalar product (12)		•	⊙	•
	Vector product (12)		∧	⊛	×

(Note) The symbols of the operations in R, C and R³ obviously refer to the addition, subtraction, multiplication and division of these algebraic structures, not to the dyadic or concrete ones that are defined in the articles of the first column.

Artículo 17

EFECTOS DEL PRINCIPIO DE ECONOMÍA SIMBÓLICA

A lo largo del compendio precedente se ha observado cómo se generaliza en abstracto el álgebra geométrica de segmentos o longitudes, dando lugar al álgebra genérica de los concretos o díadas, como representantes matemáticos de las cantidades de las magnitudes físicas. También se ha advertido sobre el efecto hipnótico que puede producir acogerse a la economía simbólica, entendida como la simplificación de signos para las distintas operaciones de la misma especie, tales como las aditivas, denotadas todas con la típica cruz «+», las multiplicativas indicadas genéricamente, por ejemplo, con el aspa «×», las restas con el guion «−», o las divisiones con la barra «/». Para romper ese hechizo y advertir a efectos pedagógicos sobre lo fácil que resulta dejarse fascinar por él y creer que se comprenda lo que realmente permanezca en la oscuridad, se ha hecho un esfuerzo de detalle simbólico, para explicitar el máximo número de leyes de composición distinguibles entre sí, así como las relaciones que surgen entre ellas; aunque, dado su gran número, como la simbología es limitada y tampoco tendría utilidad llevar al extremo absoluto tal diferenciación, es inevitable y hasta conveniente que algunas compartan signos comunes, lo cual no es obstáculo para que el fenómeno pueda explicarse con suficiente claridad didáctica. Sea como ejemplo la expresión en $\mathscr{D}$:

$$[p\circ(\overline{a}\ U_1)]\circledast[q\circ(\overline{b}\ U_2)]=(p\times q)\circ[(\overline{a}\ U_1)\circledast(\overline{b}\ U_2)]$$

El principio de economía simbólica permite simbolizar todas las leyes de composición de la misma especie multiplicativa con el mismo carácter «×», y con ello resulta la notación tradicional:

$$(p\times\overline{a}\ U_1)\times(q\times\overline{b}\ U_2)=(p\times q)\times[(\overline{a}\ U_1)\times(\overline{b}\ U_2)]$$

Observando esta última expresión, salvo que se tenga pericia algebraica, resulta difícil sustraerse a la ilusión que provoca el signo constante del aspa «×» y se tiende a creer con facilidad que la propiedad que describe la igualdad sea evidente por las leyes propias de R^3. Sin embargo, no es así, porque lo que se relacionan son díadas físicas, y el significado completo de la igualdad viene dado por las diferentes leyes de composición que comprende la propia ecuación y específicamente definidas entre los conjuntos R, $\{R^3, U_1\}$, $\{R^3, U_2\}$ y $\{R^3, U_1 * U_2\}$, por lo que la fórmula debe interpretarse en función de ellas.

Para una mejor visión de conjunto del **hechizo de la reducción simbólica, tan tóxico para el aprendizaje de la Física, la exactitud científica y la significación precisa de las magnitudes compuestas**, se pueden detallar los símbolos de las operaciones que intervienen en el álgebra diádica, representada con el signo $\mathscr{D}$, a diferencia de las estructuras tradicionales de R, C y R^3 o cualquier otra. De este modo resulta el esquema sinóptico siguiente:

Tipo de ley de composición diádica Artículo del apartado XXVIII		Álgebra numérica ordinaria (ver nota)		Álgebra diádica o física	Con el principio de economía simbólica
		En R y C	En R^3	En $\mathscr{D}$	
Magnitudes escalares y vectoriales	Adición (4)	+	+	$\oplus$	+
	Sustracción (5)	−	−	$\ominus$	−
	Multiplicación por un número (6)	×	•	○	×
	División homogénea (7)	/ ÷		// ÷	/ ÷
Magnitudes escalares	Multiplicación heterogénea (9)	×		$*$	×
	División heterogénea (10)			// ÷	/ ÷
Magnitudes vectoriales	Producto de magnitudes mixtas (11)			◎	×
	Producto escalar (12)		•	⊙	•
	Producto vectorial (12)		∧	⊛	×

(Nota) Los símbolos de las operaciones en R, C y R^3 obviamente se refieren a la adición, sustracción, multiplicación y división propias de estas estructuras algebraicas, no a las diádicas o concretas que se definen en los artículos de la primera columna.

Article 18

Physics teachers take for granted that unit abbreviations operate with the same algebra of abstract numbers, and on this tacit assumption, without justifying it in any way, they teach their classes and completely omit all specific algebra for magnitudes, ignoring , as if they did not exist, the philosophical problems related to the magnitudes and their composition laws, teaching the physical operations in an intuitive, subjective and arbitrary way, sowing in the students, even without knowing it, seeds of ignorance and confusion that vitiate all knowledge acquired with this lagoon pending clarification. Thus the teaching quality is debased, because the key to a thorough understanding is not to advance at all without first having precisely defined all of the foregoing, and even more so, if possible, in the case of something so fundamental to understand and develop natural laws such as magnitudes, your measurements and your operations.

And even more serious is the stubborn negligence of the International System of Units and of all the great scientists who propitiate and tolerate the already repeated false hypothesis that physical quantities behave with the Abelian multiplicative group structure, hypothesis that we have put in evidence with the unappealable configuration of multiplicative operations as external composition laws, and therefore lacking in unitary and inverse elements. All this makes up a poisoned panorama that traps the foundations of Physics and prevents the development of coherent and precise models.

Let's look at the following experiment: imagine a mass of one kilogram arranged materially by means of a weight, a length of one meter measured with a ruler, and a quantity of time of one second marked by a clock; it is essential to have prepared teaching answers to explain the multiplication of a kilogram by a second, for example, or even to solve if it is possible to divide a kilogram by a second or a meter by a kilogram, or if such divisions can only be conceived in certain cases or if, on the contrary, they will always be possible.

And this is necessary because, if these omnipresent operations are not rigorously justified, then what foundation and meaning could be attributed to compound maagnitudes and units, scientific laws and physical equations? Is not all knowledge empty? imparted without having saved this gap? Physical algebra responds to these kinds of questions, which gives full meaning to the various compound magnitudes and formulations of natural laws, saving that pernicious omission of the essential **generating external composition laws**. We will not get tired of repeating it, because the need for these algebraic elements is very evident.

Hence, it is **convenient for the quality of education** to solve this **capital pedagogical defect**, whose permanence denatures scientific language and its real meaning, insidiously depriving students of their rights not to be intoxicated with **latent presuppositions** and to receive complete information on any subject of curricular interest, since it is a core content of the sciences, which stimulates creative and free talent, saving the prevailing ignorance and confusion.

It is not an accessory or superfluous subject, physical algebra is essential, it is an necessary principle. Its omission disqualifies and precarious all the scientific knowledge built without that nuclear pillar. Science should not allow such a serious incongruity, as Mathematics would not accept that arithmetic be thrown away, because, just as the laws of numerical composition solidly base all mathematical structures, from the most basic to the most abstract, the algebra of magnitudes is the origin of all physical formulations, without exception. So they lose all their meaning when their composition laws are disregarded.

So why has the physical tradition indulged in this insidious and elemental **epistemological heresy**? It's a mystery. But the unequivocal thing is that, once the incongruity and its resolution have been discovered, it must be saved, first, in the interests of the logical coherence of scientific theories, and second, to chart the course of new investigations oriented to non-Euclidean algebras, who knows in this case? first moment of dyadic innovation to what new domains, models or discoveries it can lead.

Artículo 18

DISCUSIÓN Y CONCLUSIONES

Los profesores de Física dan por sentado que las abreviaturas de unidades operen con la misma álgebra de los números abstractos, y sobre esta presunción tácita, sin justificarla en modo alguno, imparten sus clases y omiten absolutamente toda álgebra específica para las magnitudes, pasando por alto, como si no existiesen, los problemas filosóficos atinentes a las magnitudes y sus leyes de composición, enseñando las operaciones concretas de manera intuitiva, subjetiva y arbitraria, sembrando en los alumnos, aun sin saberlo, semillas de ignorancia y confusión que vician todo el conocimiento adquirido con esta laguna pendiente de ser clarificada. Así **se envilece la calidad docente**, porque la clave del entendimiento cabal es no avanzar en absoluto sin antes haber definido con precisión todo lo precedente, y más, si cabe, tratándose de algo tan fundamental para comprender y desarrollar las leyes naturales como las magnitudes, sus mediciones y sus operaciones.

Y más grave aún es la contumaz negligencia del Sistema Internacional de Unidades y de todos los grandes científicos que propician y toleran la hipótesis falsa ya repetida sobre que las magnitudes físicas se comporten con la estructura de grupo multiplicativo abeliano, hipótesis que hemos puesto en evidencia con la inapelable configuración de las operaciones multiplicativas como leyes de composición externas, y por tanto, carentes de elementos unitarios e inversos. Todo ello compone un panorama envenenado que entrampa los fundamentos de la Física e impide el desarrollo de modelos coherentes y precisos.

Veamos el siguiente experimento: imagínense dispuestos materialmente una masa de un kilogramo mediante una pesa, una longitud de un metro medida con una regla y una cantidad de tiempo de un segundo marcada por un reloj; es imprescindible tener preparadas respuestas docentes para explicar la multiplicación de un kilogramo por un segundo, por ejemplo, o incluso resolver si es posible dividir un kilogramo entre un

446

segundo o un metro entre un kilogramo, o si tales divisiones solo se pueden concebir en determinados casos o si, por el contrario, siempre serán posibles.

Y ello es necesario porque, si estas operaciones omnipresentes no se justifican con todo rigor, entonces, ¿qué fundamento y significado se podría atribuir a las magnitudes y unidades compuestas, las leyes científicas y las ecuaciones físicas?, ¿no resulta vacuo todo el conocimiento impartido sin haber salvado esta laguna? A esta clase de cuestiones responde el álgebra física, que da pleno sentido a las diversas magnitudes compuestas y formulaciones de las leyes naturales, salvando esa perniciosa omisión de las imprescindibles **leyes de composición externas generatrices**. No nos cansaremos de repetirlo, porque es muy evidente la necesidad de estos elementos algebraicos.

De ahí que **conviene a la calidad de la educación** solventar este **defecto pedagógico capital**, cuya permanencia desnaturaliza de raíz el lenguaje científico y su significado real, privando insidiosamente a los estudiantes de sus derechos a no ser intoxicados con **presuposiciones latentes** y a recibir información cabal sobre cualquier materia de interés curricular, pues se trata de un contenido troncal de las ciencias, que estimula el talento creativo y libre, salvando la ignorancia y la confusión imperantes.

No se trata de una materia accesoria ni superflua, el álgebra física es esencial, es un principio imprescindible. Su omisión descalifica y precariza todo el saber científico construido sin ese pilar nuclear. La ciencia no debería permitir una incongruencia tan grave, como la Matemática no aceptaría que se arrumbase la aritmética, pues, así como las leyes de composición numéricas fundamentan con solidez todas las estructuras matemáticas, desde las más básicas hasta las más abstractas, el álgebra de magnitudes es el origen de todas las formulaciones físicas, sin excepción. Por lo que estas pierden todo su sentido cuando se desprecian sus leyes de composición.

Entonces, ¿por qué la tradición física ha incurrido en esta **herejía epistemológica** tan insidiosa como elemental? Es un misterio. Pero lo inequívoco es que, descubierta la incongruencia y su resolución, debe salvarse, primero, en interés de la coherencia lógica de las teorías científicas, y segundo, para trazar el rumbo de nuevas investigaciones orientadas a álgebras no euclidianas, que quién sabe en este primer momento de la innovación diádica a qué nuevos dominios, modelos o descubrimientos pueden conducir.

Section XXIX

THE BLACK LAGOON OF MATH
ORIGIN OF THE «ARITHMETIZATION» OF PHYSICS

Here we reveal the vice of «arithmetization» of modern mathematics, which has been forgotten by supine ignorance of the geometric algebra handed down by the Greeks and analytically updated in the present work, giving it the form of a dyadic algebraic structure. This alarming and intolerable vice has poisoned and curtailed Physics with the same gap for the rest of the measurable magnitudes and those affine to length. Obviously, it is not sustainable to deprive Physics of such a fundamental tool as its own algebra, as all awake and honest intelligences will easily appreciate, examining the development of the following articles:

Number		Page
19	Epitome of the black lagoon	450
20	Equality and addition of segments	455
21	The proporcionality of segments bases the mathematical metric	459
22	The blot the classic texts:Math has bypassing segment multiplications	471
23	Saving the lagoon with the omitted geometric multiplication	475
24	The trascendental experiment with areas and volumes	481
25	Relationship betwen proporcionality of segments and geometric multiplication	485
26	The non-arithmetic composite rule of three Foundation of physical equations	491

EPITOME OF THE BLACK LAGOON

In this section, in which the reader is supposed to have sufficient knowledge of mathematical foundations, a transcendental gap in modern mathematics is exposed, which debases metric operations from its base. The contamination process is as follows: first, the theory of geometric equality and addition of segments leads to prove the theorem of the mean parallel, and this rationally justifies the proportionality of segments of Thales' Theorem. And this done, the blur is produced, because Mathematics correctly accredits that the proportionality of segments implies the numerical proportionality of their measurements in a determined arbitrary unit of length; but then he concludes without more than that, being such proportions of a numerical nature, as well as in arithmetic the product of the extremes of all proportions must be equal to that of the means, in the same way, the proportionality of segments should meet the same condition , «because if». And this is how one falls into the error of confusing arithmetic multiplication with segment multiplication, which are very different operations, and arises like a charm the undesirable «arithmetization» of the most fundamental magnitude of all: the length. The first is a law of internal composition, since every product of numbers is another number; but not so for the geometric multiplication of segments that, as we will not tire of repeating, is a non-arithmetic operation, it is geometric, and such that it generates new magnitudes, the area with two factors or the volume with three. None of these geometric factors, which are quantities of length, can assume the multiplier function on which arithmetic multiplication is based. Which means that this is a law of external composition totally different from the ordinary product. Following this erroneous scheme, modern mathematics ignores and forgets the indispensable geometric multiplication of segments inherited from the classical Greeks, the basis of the later analytical formulation of the product of lengths and of the entire mathematical metric, which in turn is the basis of multiplicative operations with physical quantities, as we have detailed in detail in what precedes. Here we epistemically expose this substantial error, which debases current Mathematics and spreads insidiously to Physics, recovering the lost values of classical geometry.

Apartado XXIX

LA LAGUNA NEGRA DE LA MATEMÁTICA
ORIGEN DE LA «ARITMETIZACIÓN» DE LA FÍSICA

Aquí ponemos de manifiesto el vicio de «aritmetización» de la Matemática moderna, que se ha olvidado por ignorancia supina del álgebra geométrica legada por los griegos y actualizada analíticamente en el presente trabajo, dándole forma de estructura algebraica diádica. Dicho vicio alarmante e intolerable ha intoxicado y cercenado la Física con la misma laguna para el resto de las magnitudes medibles y afines a la longitud. Obviamente, no es sostenible privar a la Física de una herramienta tan fundamental como es su propia álgebra, como fácilmente apreciarán todas las inteligencias despiertas y honestas, que examinen el desarrollo de los siguientes artículos:

Número		Página
19	Epítome de la laguna negra	453
20	Igualdad y adición de segmentos	457
21	La proporcionalidad de segmentos fundamenta la métrica matemática	465
22	El borrón de los textos clásicos: La Matemática ha pasado por alto la multiplicación de segmentos	473
23	Salvando la laguna con la soslayada multiplicación geométrica	478
24	El trascendental experimento con áreas y volúmenes	483
25	Relación entre proporcionalidad de segmentos y multiplicación geométrica	488
26	La regla de tres compuesta no aritmética Fundamento de las ecuaciones físicas	495

Artículo 19

EPÍTOME DE LA LAGUNA NEGRA

En este apartado, en que se suponen al lector conocimientos suficientes de los fundamentos matemáticos, se expone un vacío trascendental de la Matemática moderna, que envilece las operaciones métricas desde su base. El proceso de contaminación es el siguiente: en primer lugar, la teoría de igualdad y adición geométrica de segmentos conduce a probar el teorema de la paralela media, y este justifica racionalmente la proporcionalidad de segmentos del *Teorema de Tales*. Y hecho esto, se produce el borrón, porque la Matemática acredita correctamente que la proporcionalidad de segmentos implica la proporcionalidad numérica de sus medidas en determinada unidad de longitud arbitraria; pero a continuación concluye sin más que, siendo tales proporciones de índole numérica, así como en aritmética el producto de los extremos de toda proporción ha de ser igual al de los medios, de la misma forma, la proporcionalidad de segmentos debería cumplir esa misma condición, «porque sí». Y así es como se cae en el error de confundir la multiplicación aritmética con la multiplicación de segmentos, que son operaciones muy diferentes, y surge como por ensalmo la indeseable «aritmetización» de la magnitud más fundamental de todas: la longitud. La primera es una ley de composición interna, pues todo producto de números es otro número; pero no así para la multiplicación geométrica de segmentos que, como no nos cansaremos de repetir, es una operación no aritmética, es geométrica, y tal que engendra nuevas magnitudes, el área con dos factores o el volumen con tres. Ninguno de esos factores geométricos, que son cantidades de longitud, puede asumir la función de multiplicador en que se basa la multiplicación aritmética. Lo que conlleva que esta sea una ley de composición externa totalmente distinta del producto ordinario. Siguiendo este esquema erróneo, la Matemática moderna pasa por alto y olvida la indispensable multiplicación geométrica de segmentos heredada de los griegos clásicos, base de la posterior formulación analítica del producto de longitudes y de toda la métrica matemática, que a su vez es la base de las operaciones multiplicativas con magnitudes físicas, tal como hemos detallado minuciosamente en lo que precede. A continuación exponemos epistémicamente este sustancial error, que envilece la Matemática actual y se propaga insidiosamente a la Física, recuperando los valores perdidos de la geometría clásica.

Article 20

EQUALITY AND ADDITION OF SEGMENTS

Segment equality is defined in geometry with the condition that the compared segments can be made to coincide by means of a movement, which can be represented by a transformation of the space itself[39].

In turn, the addition of segments is conceived geometrically as the graphical operation that provides the segment sum by

Display of the sum of segments

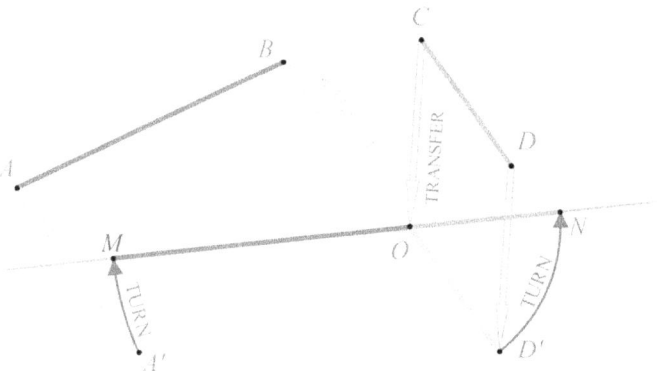

Given two segments *AB* and *CD*, representatives of the classes of the congruent with them, and given any line and a point *O* of it, the sum of the given segments is defined as the segment *MN* that results from the juxtaposition of those on the given line and starting from *O*. Each segment is transported on the line as follows: a translation that carries *B* over *O* and *A* over *A'*, a turn that carries *A'* over *M*, a translation that carries *C* over *O* and *D* over *D'*, and a spin that takes *D'* over *N*. So we say that *MN* is the sum of *AB* and *CD*, we write it *MN=AB⊕CD*. The representative chosen from each class to compose the sum is indifferent, because the segments of the same class are equal. Segments *AB* and *MO* belong to the same class, because they are congruent. Like the *CD* and *ON* segments. Juxtaposition consists, then, in arranging on the line considered two congruent segments with the given addends, one after the other and on a different side of a point *O*, so that the union segment of these two is called the sum.

Figure 16

---

39 In Mathematize 1, «Lesson 16, Movements and congruence or geometric equality», this topic can be consulted at length.

juxtaposing the addends[40], as described in figure 16. Symbolizing {S} the set of all segments, the analytic form this operation can be symbolized with a specific sign, for example «⊕»; although the usual thing in Mathematics is to apply the principle of symbolic economy and represent all additive operations with the same cross «+». However, the expert mind must clearly distinguish that, depending on the elements that are composed, the corresponding law of composition is not arithmetic, but its own, which must have been expressly defined previously. Under these conditions, keeping the symbolic difference, to avoid confusion, the addition of segments would be analytically defined in the following way: given two segments S_1 and S_2 of the set {S}, the addition is an internal composition law that applies the Cartesian product {S}×{S} in {S} and such that the segment sum S of {S} is obtained by graphical juxtaposition of the addends, the function indicated by the equation $S_1 \oplus S_2 = S$ being analytically described.

---

[40]   In Mathematize 1, «Lesson 22, Metric Geometry, Sum of segments and angles», the composition law of the geometric addition of segments is developed.

Artículo 20

IGUALDAD Y ADICIÓN DE SEGMENTOS

La igualdad de segmentos se define en geometría con la condición de que los segmentos comparados se puedan hacer coincidir mediante un movimiento[41], que podrá ser representado mediante una transformación del espacio en sí mismo.

A su vez, la adición de segmentos se concibe geométricamente como la operación gráfica que proporciona el segmento suma

Visualización de la suma de
segmentos

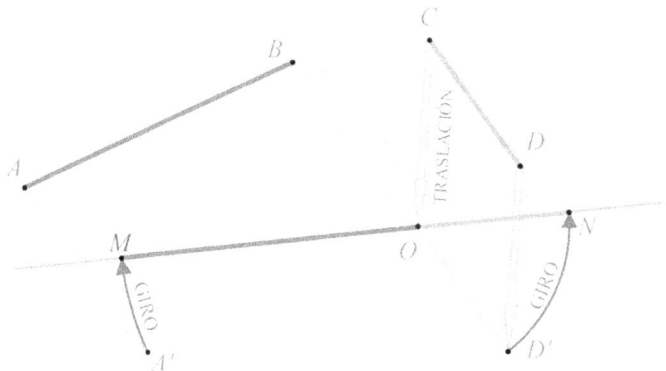

Dados dos segmentos *AB* y *CD*, representantes de las clases de los congruentes con ellos, y dada un recta cualquiera y un punto *O* de ella, la suma de los segmentos dados se define como el segmento *MN* que resulta de la **yuxtaposición** de aquéllos sobre la recta dada y a partir de *O*. Cada segmento se transporta sobre la recta del siguiente modo: una traslación que lleva *B* sobre *O* y *A* sobre *A'*, un giro que lleva *A'* sobre *M*, una traslación que lleva *C* sobre *O* y *D* sobre *D'*, y un giro que lleva *D'* sobre *N*. Decimos así que *MN* es la suma de *AB* y *CD*, lo escribimos *MN=AB⊕CD*. Es indiferente el representante que se elija de cada clase para componer la suma, porque los segmentos de la misma clase son iguales. Los segmentos *AB* y *MO* pertenecen a la misma clase, porque son congruentes. Al igual que los segmentos *CD* y *ON*. La yuxtaposición consiste, pues, en disponer sobre la recta considerada dos segmentos congruentes con los sumandos dados, uno a continuación del otro y a distinto lado de un punto *O*, de modo que el segmento unión de éstos dos se denomina suma.

Figura 16

---

[41]   En *Matematizar 1*, «Lección 16, Movimientos y congruencia o igualdad geométrica», se puede consultar con extensión este tema.

mediante la yuxtaposición de los sumandos[42], tal como se describe en la figura 16. Simbolizando $\{S\}$ el conjunto de todos los segmentos, la forma analítica de esta operación se puede simbolizar con un signo específico, por ejemplo «⊕»; aunque lo usual en Matemáticas sea aplicar el principio de economía simbólica y representar todas las operaciones aditivas con la misma cruz «+». No obstante, la mente experta debe distinguir con claridad que, en función de los elementos que sean compuestos, la ley de composición que les corresponde no es la aritmética, sino la suya propia, que debe haber sido expresamente definida previamente. En estas condiciones, manteniendo la diferencia simbólica, para evitar confusiones, la adición de segmentos quedaría definida analíticamente en la forma siguiente: dados dos segmentos S_1 y S_2 del conjunto $\{S\}$, la adición es una **ley de composición interna** que aplica el producto cartesiano $\{S\} \times \{S\}$ en $\{S\}$ y tal que el segmento suma S de $\{S\}$ se obtiene por yuxtaposición gráfica de los sumandos, quedando descrita analíticamente la función indicada con la ecuación $S_1 \oplus S_2 = S$.

---

[42] En *Matematizar 1*, «Lección 22, Geometría métrica, Suma de segmentos y ángulos», se encuentra desarrollada la ley de composición de la adición geométrica de segmentos.

Article 21

Once the geometric equality and addition of segments are defined, these being understood as figures formed by certain sets of points, following the dictates of elementary geometry, the mean parallel theorem, described in figure 17, makes it possible to divide a segment into equal parts, as indicated in figure 18, and this operation allows us to conclude Thales's famous Theorem on the proportionality of segments formed on certain lines sectioned by others parallel to each other, in accordance with what is determined by geometry and shown in figure 19[43].

Segment proportionality is the daughter of addition, because it arises from a sum in which an addend is repeated. Thus, by reiterating the sum of the same segment S a certain number of times λ, the resulting segment can be indicated analytically with the form $\lambda \circ S$, where the sign «$\circ$» indicates the multiplication of the real number λ by the segment S, making the numerical factor times multiplier and segment multiply. If λ were integer, the interpretation of the product $\lambda \circ S$ does not offer difficulty. If λ were rational, such as $\lambda = a/b$, with a and b integers, the product $(a/b) \circ S$ must indicate the operation of dividing the segment S into b equal segments, taking one of them and adding it a itself sometimes. In this way, the product $\lambda \circ S$ can always be found for any real λ. Note that the operation «$\circ$» is not the arithmetic «$\times$», because it only composes numbers, whose product is another number, and the other one composes numbers and segments to result in a new segment. So arithmetic multiplication is an internal composition law, while the operation $\lambda \circ S$ is an external law.

---

[43] In Mathematize 1, «Lesson 26, Thales' Theorem», the complete reasoning that concludes the proportionality of segments is developed.

Let P be the segment resulting from the product $\lambda \circ S$, which will be written with the traditional equal sign with the form $\lambda \circ S = P$. Nothing prevents observing this expression and considering that P is a dividend, that S is a divisor and that λ is a quotient, with which the product could be written with another equivalent symbolic form such as $P /\!/ S = \lambda$, and thus it would be defined the division of segments «$/\!/$», whose quotient will always be a real number λ. This division has been indicated with the double bar «$/\!/$» to distinguish it with regard to arithmetic «$/$», because this confusion is the cause of the mathematical blur that is being described in this investigation.

Therefore, the ratio of two segments always results in a real number, so that if two ratios are equal, what could be called a proportion is formed with them, but it is not a numerical proportion, but of segments . So the proportionality of Thales, although it is conventionally represented by the arithmetic sign of the dividing bar «$/$», in reality it is necessary to appreciate that it is the geometric operation «$/\!/$», derived from the product of a scalar by a segment. Well, this very elementary confusion is the poison that poisons Mathematics in a tragic way, as will be seen immediately.

The differences between addition and scalar multiplication are also evident by the different algebraic structures that they generate. The addition of segments, which is a mapping of the Cartesian product $\{S\} \times \{S\}$ in $\{S\}$, is a law of internal composition and it can be easily verified that it satisfies the properties necessary to endow the set $\{S\}$ with the Abelian group structure. Instead, multiplication by a scalar is an external law, indicated by an application of the Cartesian product $R \times \{S\}$ en $\{S\}$, where R is the field of real numbers. It can be easily verified that this operation is such that it endows the set $\{S\}$ with the algebraic structure of a vector space over R. However, for this, it is necessary to define the concept of an **opposite or negative segment**, and nothing prevents establishing it as that which it is added by juxtaposition in the opposite sense to that considered positive, since in the line there are two directions to carry out an addition; if the first addend is greater than the second, the sum will result in a positive segment; if the first addend is less than the second, the sum will be in a negative segment; and all this by definition of this law of internal composition. This is equivalent to recognizing in the segment the attribute of **sense or linear order**, so that the equality of segments requires not only congruence but also the same order between their points or, in other words, that the segments have the same sense, although they may differ in direction, an attribute that is characteristic of vectors.

Similarly, multiplication by a negative scalar must consider that the sign of the product is opposite to that of the segment that appears as multiplying. This conceptualization is the origin that justifies the Cartesian reference systems that give way to analytical geometry.

Middle parallel of a trapezoid and triangle

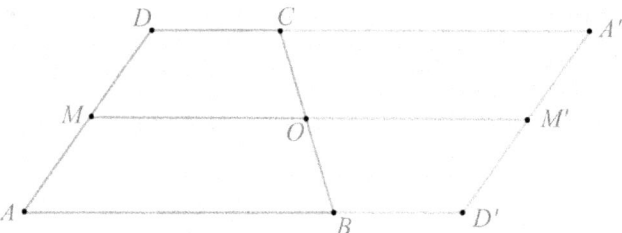

Let be the trapezoid *ABCD*, let *MO* be the segment joining the midpoints of the non-parallel sides *BC* and *AD*, the center *O* symmetric figure of the trapezoid *ABCD* is *A'CBD'*, the central symmetry is such that homologous lines are parallel, so that the segments *DC* and *BD'* are parallel, and the *AB* and *CA'* also, as well as the segments *AD* and *AD'*, the indicated central symmetries give rise to the parallelogram *AD'A'D* with center of symmetry *O*. The homologous points *A*, *D'* and *M'* will result and, being by hypothesis *MM'=MO*, we have that the central symmetry establishes that their homologues *A'M'=M'D'* are equal; thus it turns out to be *MM'* the mean parallel of the parallelogram, so this mean parallel must be parallel to the sides *DA'* and *AD'*, reaching the conclusion that the segment *MO* that joins the midpoints of the non-parallel sides of the trapezoid *ABCD* must be parallel to *AB* and *DC*, it must also be *MO=OM'*, due to the central symmetry, and resulting in *MM'=AB⊕BD'* or *MM'=DC⊕CA'* indistinctly, since *DA'=MM'=AD'* and the segments *AB* with *CA'* being homologous and *DC* with *BD'*, *MM'=AB⊕DC* and *MM'=2*MO*; so, finally we have that, *MO=(AB⊕DC)/2*; and thus we can state that **the segment that joins the midpoints of the non-parallel sides of a trapezoid is the average parallel to its parallel sides, called bases, and measures half their sum.**

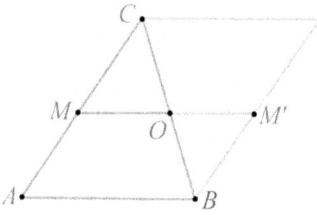

Let us be the triangle *ABC*, let us form the segment *MO* that joins the midpoints of the sides *BC* and *AC*, we are facing a particular case of the previous one, so we could extend to it the deduced property for the trapezoid without further ado. The center *O* symmetry transforms triangle *ABC* into triangle *A'CB*.

The homologous lines of the central symmetry are parallel, therefore, the segments *AB* with *CA'* and *AC* with *A'B* are parallel; it thus turns out that the figure *ABA'C* is a parallelogram, with all its properties; as the segments *A'M'* and *M'B* are equal, because they are symmetric homologous of the equal segments *CM* and *MA*, it turns out that the segment *MM'* joins the midpoints of the sides *AC* and *BA'* of the parallelogram *ABA'C*, then, *MM'* must be parallel to *AB* and *CA'* and equal to them; since *MM'=2*MO*, it follows that *MO=AB//2*; with which the following statement is true: **the segment that joins the midpoints of two sides of a triangle is parallel to the third side and equal to half of it.**

Figure 17

Division of a segment
in equal parts

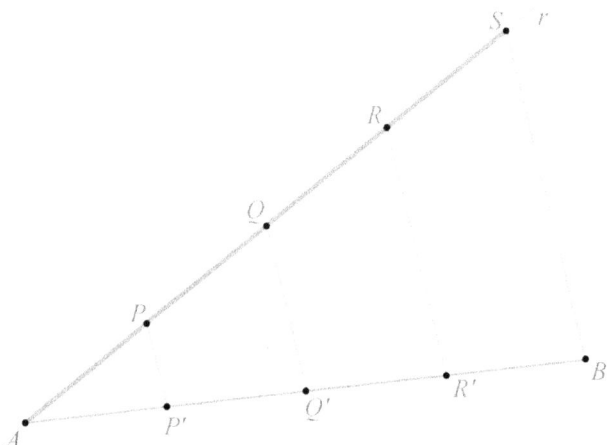

Given a segment *AB*, we try to divide it, for example, into four equal parts. To do this, we will repeatedly use the theorem of the mean parallel in the various triangles that can be conceived. First, we draw any ray *Ar*, which does not coincide with *AB*; second, we take any segment and carry it from *A* and over *r* four times, one after the other, thus we obtain the points *P*, *Q*, *R* and *S*; third, we join the last of these points, which in this case is *S*, with *B*; fourth, we draw by *P*, *Q* and *R* parallel to *SB*, and we obtain the points *P'*, *Q'* and *R'*. It thus turns out that in triangle *AQQ'* the segment *PP'* is mean parallel, with *P* being the midpoint of *AQ*, because that is how we have constructed it, then *P'* is the midpoint of *AQ'*, and it turns out that *AP'=P'Q'*; we observe something similar with the segment *QQ'*, which is the mean parallel of the trapezoid *P'Q'RP*, since *Q* is the midpoint of the side *PR*, because we have established this by carrying equal segments on *r*, then, it also turns out that *P'Q'=Q'R'*; and, finally, in an analogous way it is also obtained that it is also *Q'R'=R'B*, by identical reasoning scheme with the trapezoid *Q'BSQ*; concluding that segment *AB* has thus been divided into four equal segments, *AP'*, *P'Q'*, *Q'R'* and *R'B*. The determination of the segments on the line *AB* by drawing parallels from the line *r* is called the parallel projection of the segments of one line onto the other in the direction of the drawn parallels. This allows us to affirm the following statement: **given two lines, the parallel projection of equal segments of one line determines equal segments on the other.** And from here to Thales' Theorem there is almost nothing left.

Figue 18

Thales Theorem

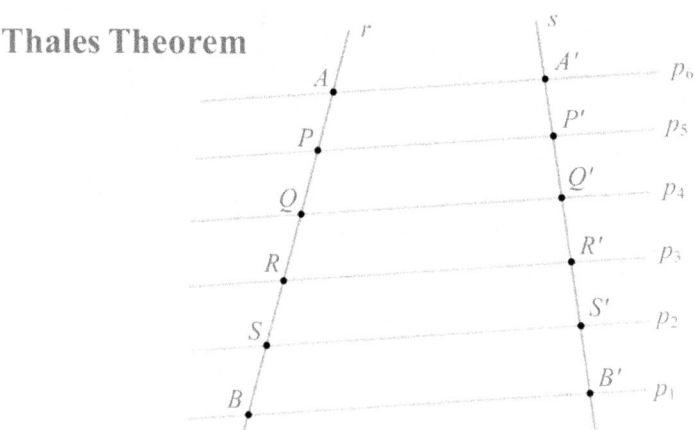

Let the lines r and s be the segment AB of the line r; it is possible to divide it into equal parts, in the figure it has been divided into five; suppose that the points that result from dividing the segment AB in equal parts are the P, Q, R and S; let the points A', B', P', Q', R' and S' be projected in parallel from their homologues A, B, P, Q, R and S, by means of the given system of parallels p_1, p_2, p_3, p_4, p_5 y p_6; under these conditions, segment AB is the sum of segments AP, PQ, QR, RS and SB, by definition of the sum of segments; as all these addend segments are equal by hypothesis, by definition of the product of a segment by a number, we have that sum can be expressed as the product $AB=5\circ AP$, with AP being the segment that we take as representative of all equal addends to express this operation, which would be identical if any other of the equal addends were taken; any other segment of the same line r, such as PS, for example, is the sum of the segments PQ, QR and RS, equal to each other and also equal to AP, then, we can write $PS=3\circ AP$. The quotient of segments is defined with the same appearance as the quotient of numbers, although with its specific meaning, in this case, the quotient between segments $PS//AB$, with a double bar to differentiate it from the arithmetic quotient, is the rational number 3/5, because, by definition of segment division, $PS=3/5\circ AB$. So $3/5\circ AB$ means dividing AB into 5 equal segments and multiplying the resulting segment by 3, or adding it 3 times. Since segment AB divided by 5 is precisely AP, it turns out that $3/5\circ AB=3\circ AP$, and $3\circ AP$ turns out to be PS, then, the expression $PS//AB=3/5$ is correct; that is, the PS and AB segments are in ratio 3/5. Let us now see in what ratio are the segments projected parallel on s by the corresponding parallels. The projected segments are $P'S'$ and $A'B'$, the segments $A'P'$, $P'Q'$, $Q'R'$, $R'S'$ and $S'B$ are all the same, so we can take one of them to represent them all in the sums that we are going to establish, let $A'P'$ this representative; we can write that $P'S'=3\circ A'P'$ y $A'B'=5\circ A'P'$; so the segments projected in parallel from r to s are in the ratio $P'S'//A'B'=3/5$, the same rational number as the ratio $PS//AB$. It turns out, then, that these two ratios form a proportion and we can safely write that $PS//AB=P'S'//A'B'$. As this same result would be had whatever the segments were taken on the line r and their projected in s, we arrive at the following statement, which is the well-known Thales Theorem: **if two lines are cut by other lines parallel to each other, the segments that they determine on one line are proportional to the segments they project onto the other line.** Analytically, multiple proportions will result like these: $AP//AQ=A'P'//A'Q'$, $PR//AQ=P'R'//A'Q'$, $QB//PQ=Q'B'//P'Q'$, etc.

Figure 19

Artículo 21

LA PROPORCIONALIDAD DE SEGMENTOS
FUNDAMENTA LA MÉTRICA MATEMÁTICA

Definidas la igualdad y la adición geométricas de segmentos, entendidos estos como figuras formadas por ciertos conjuntos de puntos, siguiendo los dictados de la geometría elemental, el teorema de la paralela media, descrito en la figura 17, posibilita dividir un segmento en partes iguales, tal como se indica en la figura 18, y esta operación permite concluir el famoso *Teorema de Tales* sobre la proporcionalidad de segmentos formados sobre ciertas rectas seccionadas por otras paralelas entre sí, de conformidad con lo determinado por la geometría y expuesto en la figura 19[44].

La proporcionalidad de segmentos es hija de la adición, porque nace de una suma en que se repita un sumando. Así, reiterando la suma del mismo segmento S un cierto número de veces λ, el segmento resultante se puede indicar analíticamente con la forma $\lambda \circ S$, donde el signo «$\circ$» señale la multiplicación del número real λ por el segmento S, haciendo el factor numérico las veces de multiplicador y el segmento de multiplicando. Si λ fuese entero, la interpretación del producto $\lambda \circ S$ no ofrece dificultad. Si λ fuese racional, tal como $\lambda = a/b$, con a y b enteros, el producto $(a/b) \circ S$ debe indicar la operación de dividir el segmento S en b segmentos iguales, tomar uno de ellos y sumarlo consigo mismo a veces. De este modo, siempre se podrá encontrar el producto $\lambda \circ S$ para cualquier real λ. Nótese que la operación «$\circ$» no es la aritmética «$\times$», porque esta solo compone números, cuyo producto es otro número, y la otra compone números y segmentos para dar como resultado un nuevo segmento. Así que la multiplicación aritmética es una ley de composición interna, mientras que la operación $\lambda \circ S$ es una ley externa.

---

[44] En *Matematizar 1*, «Lección 26. *Teorema de Tales*», se encuentra desarrollado el razonamiento completo que concluye la proporcionalidad de segmentos.

Sea P el segmento resultante del producto $\lambda \circ S$, lo cual se escribirá con el signo igual tradicional con la forma $\lambda \circ S = P$. Nada impide observar esta expresión y considerar que P sea un dividendo, que S sea un divisor y que λ sea un cociente, con lo cual se podría escribir el producto con otra forma simbólica equivalente tal como $P /\!/ S = \lambda$, y así quedaría definida la división de segmentos «$/\!/$», cuyo cociente será siempre un número real λ. Se ha indicado esta división con la doble barra «$/\!/$» para distinguirla a propósito de la aritmética «$/$», porque esa confusión es la causante del borrón matemático que se está describiendo en esta investigación.

Por tanto, la razón de dos segmentos siempre da como resultado un número real, de modo que, si dos razones son iguales, se forma con ellas lo que se podría llamar una proporción, pero no se trata de una proporción numérica, sino de segmentos. Así que la proporcionalidad de Tales, aunque convencionalmente se represente con el signo aritmético de la barra divisoria «$/$», en realidad hay que saber apreciar que se trata de la operación geométrica «$/\!/$», derivada del producto de un escalar por un segmento. Pues bien, esta confusión tan elemental es el tóxico que envenena la Matemática de manera trágica, como enseguida se verá.

Las diferencias entre la adición y la multiplicación escalar quedan también patentes por las distintas estructuras algebraicas que engendran. La adición de segmentos, que es una aplicación del producto cartesiano $\{S\} \times \{S\}$ en $\{S\}$, es una ley de composición interna y se puede comprobar con facilidad que satisface las propiedades necesarias para dotar al conjunto $\{S\}$ de la estructura de grupo abeliano. En cambio, la multiplicación por un escalar es una ley externa, indicada por una aplicación del producto cartesiano $R \times \{S\}$ en $\{S\}$, siendo R el cuerpo de los números reales. Puede comprobarse con facilidad que esta operación es tal que dota al conjunto $\{S\}$ de la estructura algebraica de espacio vectorial sobre R. No obstante, para ello, es preciso definir el concepto de **segmento opuesto o negativo**, y nada impide establecerlo como aquel que se sume por yuxtaposición en sentido contario al considerado positivo, dado que en la recta existen dos sentidos para efectuar una adición; si el primer sumando es mayor que el segundo, la suma resultará en un segmento positivo; si el primer sumando es menor que el segundo, la suma quedará en un segmento negativo; y todo ello por definición de esta ley de composición interna. Ello equivale a reconocer en el segmento el atributo del **sentido u orden lineal**, de manera que la igualdad de segmentos exija no solo la congruencia sino, además, el mismo orden entre sus puntos o, dicho de otro modo, que los segmentos tengan el mismo sentido, aunque puedan diferir en dirección, atributo este propio de los vectores.

De manera análoga, la multiplicación por un escalar negativo habrá de considerar que el signo del producto sea opuesto al del segmento que figure como multiplicando. Esta conceptuación es el origen que justifica los sistemas cartesianos de referencia que dan paso a la geometría analítica.

Paralela media de un trapecio y de un triángulo

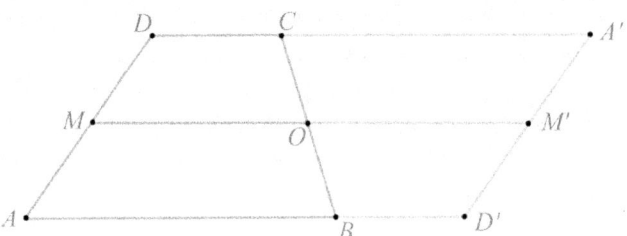

Sea el trapecio *ABCD*, sea *MO* el segmento que une los puntos medios de los lados no paralelos *BC* y *AD*; la figura simétrica de centro *O* del trapecio *ABCD* es la *A'CBD'*, la simetría central es tal que rectas homólogas son paralelas, por lo que los segmentos *DC* y *BD'* son paralelos, y los *AB* y *CA'* también, así como los segmentos *AD* y *A'D'*, las indicadas simetrías centrales, dan lugar al paralelogramo *AD'A'D* con centro de simetría *O*, resultarán los puntos homólogos *A'*, *D'* y *M'* y, siendo por hipótesis *AM=MD*, tenemos que la simetría central establece que son iguales sus homólogos *A'M'=M'D'*, así resulta ser *MM'* la paralela media del paralelogramo, conque esta paralela media ha de ser paralela a los lados *DA'* y *AD'*, llegando a la conclusión de que el segmento *MO* que une los puntos medios de los lados no paralelos del trapecio *ABCD* ha de ser paralelo a *AB* y *DC*, también ha de ser *MO=OM'*, por la simetría central, y resultando *MM'=AB ⊕ BD'* o *MM'=DC ⊕ CA'* indistintamente, por ser *DA'=MM'=AD'*, y siendo homólogos los segmentos *AB* con *CA'* y *DC* con *BD'*, resulta que *MM'=AB⊕DC* y *MM'=2∘MO*; conque, finalmente se tiene que, *MO=(AB ⊕ DC)/2*, y así podemos enunciar que el segmento que une los puntos medios de los lados no paralelos de un trapecio es la paralela media a sus lados paralelos, llamados bases, y mide la mitad que su suma.

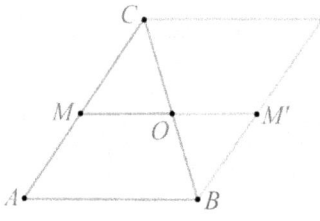

Sea el triángulo *ABC*, formemos el segmento *MO* que une los puntos medios de los lados *BC* y *AC*, estamos ante un caso particular del anterior, por lo que podríamos extender a él sin más la propiedad deducida para el trapecio. La simetría de centro *O* transforma el triángulo *ABC* en el triángulo *A'CB*.

Las rectas homólogas de la simetría central son paralelas, por tanto, son paralelos los segmentos *AB* con *CA'* y *AC* con *A'B*; resulta así que la figura *ABA'C* es un paralelogramo, con todas sus propiedades; como son iguales los segmentos *A'M'* y *M'B*, porque son simétricos homólogos de los segmentos iguales *CM* y *MA*, resulta que el segmento *MM'* une los puntos medios de los lados *AC* y *BA'* del paralelogramo *ABA'C*, luego, *MM'* ha de ser paralelo a *AB* y a *CA'* e iguales a ellos; como *MM'=2∘MO*, resulta que *MO=AB//2*; con lo cual es cierto el enunciado siguiente: **el segmento que une los puntos medios de dos lados de un triángulo es paralelo al tercer lado e igual a la mitad de éste.**

Figura 17

468

División de un segmento
en partes iguales

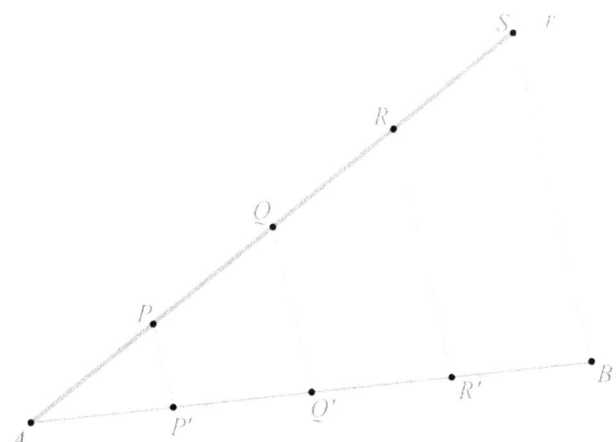

Dado un segmento *AB*, se trata de dividirlo, por ejemplo, en cuatro partes iguales. Para ello, utilizaremos reiteradamente el teorema de la paralela media en los diversos triángulos que se pueden concebir. Primero, trazamos una semirrecta *Ar* cualquiera, que no coincida con *AB*; segundo, tomamos un segmento cualquiera y lo llevamos a partir de *A* y sobre *r* cuatro veces, uno a continuación del otro, así obtenemos los puntos *P*, *Q*, *R* y *S*; tercero, unimos el último de estos puntos, que en este caso es *S*, con *B*; cuarto, trazamos por *P*, *Q* y *R* paralelas a *SB*, y obtenemos los puntos *P'*, *Q'* y *R'*; resulta así que en el triángulo *AQQ'* el segmento *PP'* es paralela media, siendo *P* el punto medio de *AQ*, porque así lo hemos construido, luego, *P'* es el punto medio de *AQ'*, y resulta que *AP'=P'Q'*; algo parecido observamos con el segmento *QQ'*, que es paralela media del trapecio *P'Q'RP*, puesto que *Q* es el punto medio del lado *PR*, porque así lo hemos establecido llevando sobre *r* segmentos iguales, luego, también resulta que *P'Q'=Q'R'*; y, finalmente, de manera análoga se tiene que también es *Q'R'=R'B*, por idéntico esquema de razonamiento con el trapecio *Q'BSQ*; concluyendo que el segmento *AB* ha quedado así dividido en cuatro segmentos iguales, los *AP'*, *P'Q'*, *Q'R'* y *R'B*. La determinación de los segmentos en la recta *AB* mediante el trazado de paralelas desde la recta *r* se denomina **proyección paralela** de los segmentos de una recta sobre la otra en la dirección de las paralelas trazadas. Esto nos permite afirmar el siguiente enunciado: **dadas dos rectas, la proyección paralela de segmentos iguales de una recta determina sobre la otra segmentos iguales**. Y de aquí al *Teorema de Tales* ya no queda casi nada.

Figura 18

Teorema de Tales

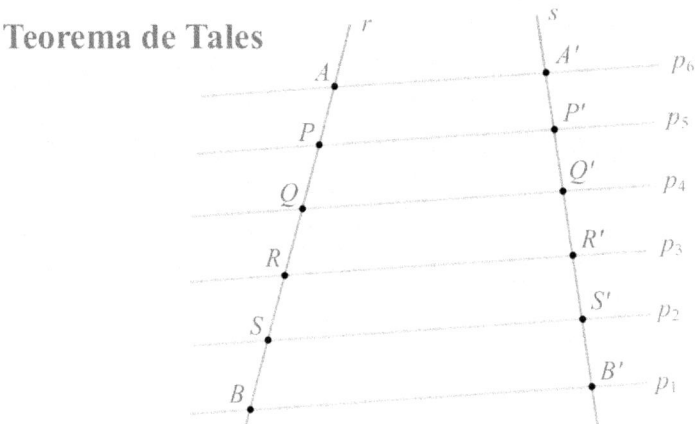

Sean las rectas *r* y *s*, sea el segmento *AB* de la recta *r*; es posible dividirlo en partes iguales, en la figura se ha dividido en cinco; supongamos que los puntos que resultan de dividir el segmento *AB* en partes iguales sean los *P*, *Q*, *R* y *S*; sean los puntos *A'*, *B'*, *P'*, *Q'*, *R'* y *S'* los proyectados paralelamente desde sus homólogos *A*, *B*, *P*, *Q*, *R* y *S*, por medio del sistema dado de paralelas p_1, p_2, p_3, p_4, p_5 y p_6; en estas condiciones, el segmento *AB* es la suma de los segmentos *AP*, *PQ*, *QR*, *RS* y *SB*, por definición de suma de segmentos; como todos estos segmentos sumandos son iguales por hipótesis, por definición de producto de un segmento por un número, tenemos que esa suma se pueda expresar como el producto *AB*=5°*AP*, siendo *AP* el segmento que tomamos como representante de todos los sumandos iguales para expresar esta operación, que sería idéntica si se tomase cualquier otro de los sumandos iguales; otro segmento cualquiera de la misma recta *r*, tal como *PS*, por ejemplo, es la suma de los segmentos *PQ*, *QR* y *RS*, iguales entre sí y también iguales a *AP*, luego, se podrá escribir *PS*=3°*AP*. El cociente de segmentos se define con la misma apariencia que el cociente de números, aunque con su significado específico, en este caso, el cociente entre segmentos *PS*//*AB*, con doble barra para diferenciarlo del cociente aritmético, es el número racional 3/5, porque, por definición de división de segmentos, *PS*=3/5°*AB*. Así que 3/5°*AB* significa dividir *AB* en 5 segmentos iguales y el segmento resultante multiplicarlo por 3, o sumarlo 3 veces. Como el segmento *AB* dividido entre 5 es precisamente *AP*, resulta que 3/5°*AB*=3°*AP*, y 3°*AP* resulta ser *PS*, luego, es correcta la expresión *PS*//*AB*=3/5; es decir, los segmentos *PS* y *AB* están en la razón 3/5. Veamos ahora en qué razón están los segmentos proyectados paralelamente sobre *s* por las paralelas correspondientes. Los segmentos proyectados son *P'S'* y *A'B'*, los segmentos *A'P'*, *P'Q'*, *Q'R'*, *R'S'* y *S'B* son todos iguales, por lo que podemos tomar a uno de ellos para representarlos a todos en las sumas que vamos a establecer, sea *A'P'* este representante; podremos escribir que *P'S'*=3°*A'P'* y *A'B'*=5°*A'P'*, así que los segmentos proyectados paralelamente de *r* en *s* están en la razón *P'S'*//*A'B'*=3/5, el mismo número racional que la razón *PS*//*AB*. Resulta, entonces, que esas dos razones forman proporción y podremos escribir tranquilamente que *PS*//*AB*=*P'S'*//*A'B'*. Como este mismo resultado se tendría cualesquiera que fueran los segmentos que se tomasen en la recta *r* y sus proyectados en *s*, llegamos al siguiente enunciado, que es el conocidísimo *Teorema de Tales*: **si dos rectas se cortan por otras rectas paralelas entre sí, los segmentos que éstas determinan en una recta son proporcionales con los segmentos que proyectan sobre la otra recta.** Analíticamente, resultarán múltiples proporciones como éstas: *AP*//*AQ*=*A'P'*//*A'Q'*, *PR*//*AQ*=*P'R'*//*A'Q'*, *QB*//*PQ*=*Q'B'*//*P'Q'*, etc.

Figura 19

Article 22

Let us see the common plot scheme of geometry texts with the case of the prestigious Course in Metric Geometry, Volume I, by Professor Pedro Puig Adam. In «Lesson 22» (p. 129) on the Pythagorean Theorem he formulates an enigmatic warning: «Preliminary warning about the **product of segments**».

Once the proportion of segments is established, he declares **without demonstrating** that «any proportion between segments can be interpreted as a proportion between their measures», and continues:

> In this way we will interpret the segmental proportions in this lesson and in the following ones, as soon as **we equalize the products of means and extremes in them**. Thus, from now on, wherever the reader sees a product of segments written or enunciated $\overline{A\,B} \cdot \overline{A\,C}$ must understand as such the number product of the measures of AB and AC with the same unit.

In this way, the proportionality of segments, which is a singular geometric operation, is confused in the exemplary texts in a disturbing way with the product of segments that generates new magnitudes, which is a **very different composition law, since it relates geometric figures to each other. with different magnitudes: lengths, areas and volumes**.

The algebraic error of the previous approach is obvious: the proportionality of segments only implies that the reasons that make up the relationship are such that they are equivalent to the same real number, according to the homogeneous dyadic division of section XI; but to infer from this that the product of the means is equal to that of the extremes is to fall into an inadmissible assumption, because the product of segments, as elementary geometric figures, cannot be assimilated «just for the sake of it» to the numerical product, but must be expressly defined separately, since the segments are not numbers in themselves, they are geometric figures that include indeterminable quantities of a fundamental magnitude: length.

Puig Adam is not the only one who fell into this trap. None of us have overcome the temptation to «arithmetize» everything without thinking about what we were doing. Even the most illustrious mathematicians like David Hilbert, who sought to give analytical form to the multiplication of segments, was not able to free himself from the spell of arithmetic. His text *Fundamentals of Geometry*, published in 1899, is considered his most important contribution to modern mathematics, incorporating the formal axiomatic method.

In this investigation, Hilbert proposes the product of segments by means of an «arithmetized» multiplication based on the following geometric figure: let us take two secant lines at point O, on one of them we carry the segment $\overline{OA}$ and on the other we carry the segments $\overline{OB}$ and $\overline{OU}$, the latter taken as unit; let us draw through B the parallel to AU, which will intersect the line OA at the point P. Hilber defines the product of the segments $\overline{OA}$ and $\overline{OB}$ as the segment $\overline{OP}$. It is evident that this construction incurs the same error as Puig Adam, since it identifies the segments with their measurements and with this the geometric multiplication of segments remains undefined, which we already know is strictly a law of external generating composition that produces a new geometric magnitude called area, so the product of segments can never give rise to another segment.

This is how Puig Adam and David Hilbert and with them all of us confuse the additive operations of sections V to XI without distinguishing them from the so-called generating multiplicative operations of sections XII to XVII, forgetting the latter, which are essential to compose geometric and physical magnitudes.

In short, given a proportion of segments $S_1/\!/S_2 = S_3/\!/S_4$, it is not correct to infer that the product of the extremes is equal to the product of the means $S_1 \times S_4 = S_2 \times S_3$, as in the proportions arithmetic, because the products $S_1 \times S_4$ and $S_2 \times S_3$ lack meaning for the segments, if the generating multiplicative laws have not previously been defined to compose magnitudes.

Artículo 22

EL BORRÓN DE LOS TEXTOS CLÁSICOS: LA MATEMÁTICA HA PASADO POR ALTO LAS MULTIPLICACIONES DE SEGMENTOS

Veamos el esquema argumental común a los textos de geometría con el caso del prestigioso *Curso de geometría métrica*, Tomo I, del profesor Pedro Puig Adam. En la «Lección 22» (p. 129) sobre el *Teorema de Pitágoras* formula una enigmática prevención: «Advertencia preliminar sobre el **producto de segmentos**».

Establecida la proporción de segmentos, declara **sin demostrarlo** que «toda proporción entre segmentos puede interpretarse como proporción entre sus medidas», y continúa:

> En esta forma interpretaremos las proporciones segmentarias en esta lección y en las sucesivas, en cuanto **igualemos en ellas los productos de medios y extremos**. Así, pues, de aquí en adelante donde vea el lector escrito o enunciado un producto de segmentos $\overline{AB} \cdot \overline{AC}$ deberá entender como tal el número producto de las medidas de AB y AC con una misma unidad.

De este modo, la proporcionalidad de segmentos, que es una operación geométrica singular, se confunde en los textos ejemplares de manera perturbadora con el producto de segmentos que genera nuevas magnitudes, que es una **ley de composición muy distinta, pues relaciona entre sí figuras geométricas con magnitudes diferentes: longitudes, áreas y volúmenes**.

El error algebraico del planteamiento anterior es obvio: la proporcionalidad de segmentos solo implica que las razones que conformen la relación son tales que equivalen al mismo número real, de acuerdo con la división diádica homogénea del apartado XI; pero inferir de ello que el producto de los medios sea igual al de los extremos es caer en una suposición inadmisible, porque el producto de segmentos, en tanto que figuras geométricas elementales, no puede asimilarse «porque sí» al producto numérico, sino que debe definirse expresamente aparte, puesto que los segmentos no son números en sí mismos, son figuras geométricas que incluyen cantidades indeterminables de una magnitud fundamental: la longitud.

Puig Adam no es el único que cayó en esa trampa. Ninguno hemos vencido la tentación de «aritmetizarlo» todo sin pensar en lo que hacíamos. Incluso los más ilustres matemáticos como David Hilbert, que buscó dar forma analítica a la multiplicación de segmentos, no fue capaz de librarse del embrujo de la aritmética. Su texto *Fundamentos de la geometría*, publicado en 1899, es considerado su contribución más importante a la Matemática moderna, incorporando el método axiomático formal.

En esa investigación Hilbert propone el producto de segmentos mediante una multiplicación «aritmetizada» basada en la siguiente figura geométrica: tomemos dos rectas secantes en el punto O, sobre una de ellas llevemos el segmento $\overline{OA}$ y sobre la otra llevemos los segmentos $\overline{OB}$ y $\overline{OU}$, este último tomado como unidad; tracemos por B la paralela a AU, que cortará a la recta OA en el punto P. Hilber define el producto de los segmentos $\overline{OA}$ y $\overline{OB}$ como el segmento $\overline{OP}$. Es evidente que esta construcción incurre en el mismo error que Puig Adam, pues identifica los segmentos con sus medidas y con ello queda indefinida la multiplicación geométrica de segmentos, que ya sabemos es en rigor una ley de composición externa generatriz que produce una nueva magnitud geométrica denominada área, por lo que el producto de segmentos nunca puede dar lugar a otro segmento.

Así es como Puig Adam y David Hilbert y con ellos todos nosotros confundimos las operaciones aditivas de los apartados V a XI sin distinguirlas de las llamadas multiplicativas generatrices de los apartados XII a XVII, olvidando estas últimas, que son esenciales para componer magnitudes geométricas y físicas.

En suma, dada una proporción de segmentos $S_1 /\!/ S_2 = S_3 /\!/ S_4$, no es correcto inferir que se cumpla sin más que el producto de los extremos sea igual al producto de los medios $S_1 \times S_4 = S_2 \times S_3$, como en las proporciones aritméticas, porque los productos $S_1 \times S_4$ y $S_2 \times S_3$ carecen de significado para los segmentos, si previamente no se han definido las leyes multiplicativas generatrices para componer magnitudes.

Article 23

It has been previously concluded that the proportions of segments do not allow at all to establish the multiplication of these elementary figures, which represent quantities of lengths, so the assumption that in such proportions the product of the means is equal to that of the extremes, does not it only has no foundation, but must be excluded a priori, without first having defined the multiplicative composition laws that compose segments with mathematical rigor. To do this, without forgetting that what is being handled are geometric figures, the multiplication of two segments can be conceived as the geometric operation consisting of forming with them another rectangular figure whose dimensions are precisely the multiplied segments. By definition of this composition law, the result of such multiplication or product will be precisely that rectangle, which will host a certain quantity of a new magnitude composed or derived from length, called **area or surface**, as graphically described in figure 20.

In turn, if instead of two segments three are multiplied, the result or product can be identified, by definition, with a geometric figure called a straight parallelepiped, whose edges are precisely the multiplied segments, giving rise to a body that accommodates a certain quantity of a new magnitude, composed with the length or derived from it, which is called **volume**, a composition law that is also described in the same figure 20. To represent these operations, so different from the multiplication of numbers, the mathematical asterisk can be used « * », Although the principle of symbolic economy, which uses the same sign «×» to represent all multiplicative laws, tends to indicate them all with this same spelling; but this should not prevent the intellects from distinguishing them and knowing how to differentiate them, because they are independent operations.

**Graphic definition of
segment multiplication**

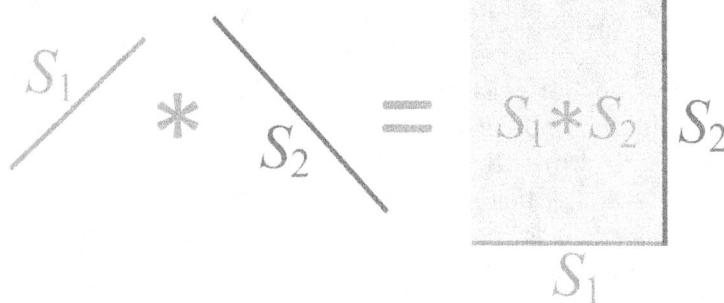

The graphic or geometric product of two segments is not another segment, but, by definition, it is a new geometric figure called a rectangle, giving rise to a compound or derived quantity of length called **surface or area**.

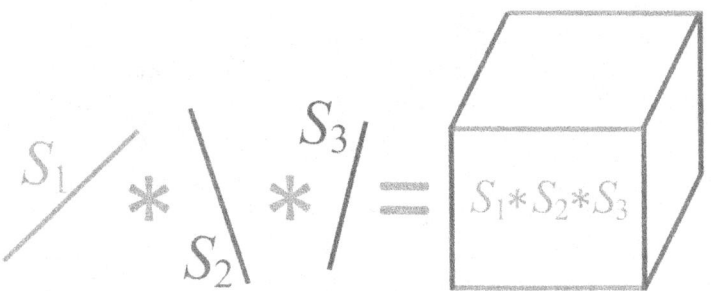

The graphic or geometric product of three segments is not another segment, but, by definition, is a new geometric figure called a right parallelepiped, giving rise to a compound or derived quantity of length called **volume**.

Figure 20

An important observation about these notions of segment multiplication is that, considering that the products do not refer to lengths, but rather indicate areas or volumes, that is, different magnitudes of length, **they are composition laws of an external nature, not internal,** which reveals a notable conceptual difference with numerical operations. This is clear when specifying the applications that define them: with two factors it is an application of $\{S\} \times \{S\}$ in $\{S * S\}$, where simply $\{S * S\}$ indicates the set of all possible rectangles formed with segments, which can also be denoted by the exponential form $\{S^2\}$; in turn, with three factors the geometric product is an application of the Cartesian product $\{S\} \times \{S\} \times \{S\}$ in $\{S * S * S\}$, with this symbol representing all the possible right parallelepipeds formed with segments, in exponential notation $\{S^3\}$.

Artículo 23

SALVANDO LA LAGUNA CON LA SOSLAYADA
MULTIPLICACIÓN GEOMÉTRICA

Se ha concluido anteriormente que las proporciones de segmentos no permiten en absoluto establecer la multiplicación de estas figuras elementales, que representan cantidades de longitudes, por lo que la suposición de que en tales proporciones el producto de los medios sea igual al de los extremos, no solo no tiene fundamento, sino que debe excluirse a priori, sin antes haber definido las leyes de composición multiplicativas que compongan segmentos con rigor matemático. Para ello, sin olvidar que lo que se está manejando son figuras geométricas, la multiplicación de dos segmentos puede concebirse como la operación geométrica consistente en formar con ellos otra figura rectangular cuyas dimensiones sean precisamente los segmentos multiplicados. Por definición de esta ley de composición, el resultado de tal multiplicación o producto será precisamente ese rectángulo, que acogerá una determinada cantidad de una nueva magnitud compuesta o derivada de la longitud, denominada **área o superficie**, tal como se describe gráficamente en la figura 20.

A su vez, si en vez de dos segmentos se multiplicasen tres, el resultado o producto se puede identificar, por definición, con una figura geométrica denominada paralelepípedo recto, cuyas aristas sean precisamente los segmentos multiplicados, dando lugar a un cuerpo que acoja una determinada cantidad de una nueva magnitud, compuesta con la longitud o derivada de ella, que se denomina **volumen**, ley de composición que se describe también en la misma figura 20. Para representar estas operaciones tan distintas de la multiplicación de números se puede utilizar el asterisco matemático «*», aunque el principio de economía simbólica, que se sirve del mismo signo «×» para representar todas las leyes multiplicativas, tienda a indicarlas todas con esta misma grafía; pero ello no habría de impedir que los intelectos las distingan y sepan diferenciarlas, porque son operaciones independientes.

**Definición gráfica de
multiplicación de segmentos**

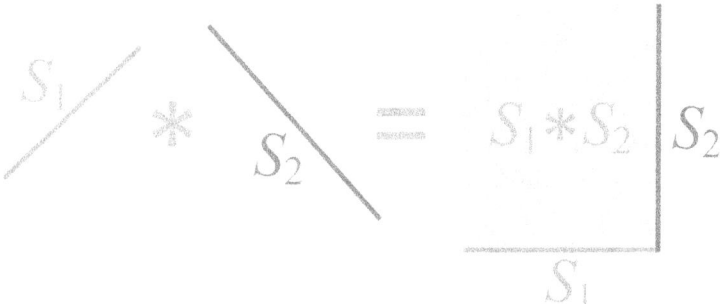

El producto gráfico o geométrico de dos segmentos no es otro segmento, sino que, por definición, es una nueva figura geométrica llamada rectángulo, dando lugar a una magnitud compuesta o derivada de la longitud denominada **superficie o área**.

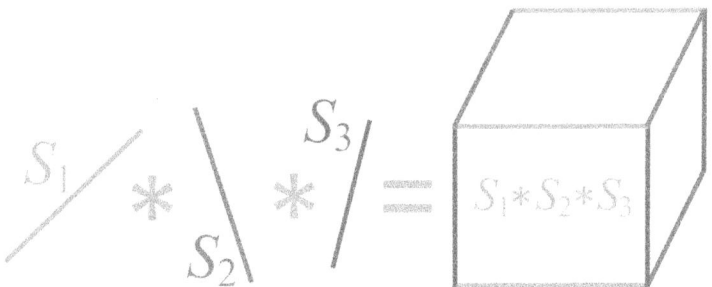

El producto gráfico o geométrico de tres segmentos no es otro segmento, sino que, por definición, es una nueva figura geométrica llamada paralelepípedo recto, dando lugar a una magnitud compuesta o derivada de la longitud denominada **volumen**.

Figura 20

Una observación importante sobre estas nociones de multiplicación de segmentos es que, considerando que los productos no se refieren a longitudes, sino que indican áreas o volúmenes, es decir, magnitudes diferentes de la longitud, **se trata de leyes de composición de naturaleza externa, no interna**, lo cual revela una notable diferencia conceptual con las operaciones numéricas. Ello queda patente al precisar las aplicaciones que las definen: con dos factores se trata de una aplicación de $\{S\} \times \{S\}$ en $\{S*S\}$, donde simplemente $\{S*S\}$ indique el conjunto de todos los rectángulos posibles formados con segmentos, que también puede denotarse con la forma exponencial $\{S^2\}$; a su vez, con tres factores el producto geométrico es una aplicación del producto cartesiano $\{S\} \times \{S\} \times \{S\}$ en $\{S*S*S\}$, con este símbolo representando a todos los posibles paralelepípedos rectos formados con segmentos, en notación exponencial $\{S^3\}$.

Article 24

Once the mathematical gap of geometric multiplication has been saved, it is observed that this operation, given its graphic nature, does not by itself provide information regarding the quantities of lengths of the segments or regarding the quantities of surface or volume. Therefore, it is necessary to undertake **measurement** procedures that allow quantifying and operating analytically with such magnitudes.

To establish a criterion for measuring the lengths of the segments, any one of them is used that is taken as a unit, which will be such that it will comprise a quantity of uncountable length, indeterminacy that will be saved by assigning it a symbol that represents it, for example , U_L. With this, the measurement of any segment S can be indicated with the pair or dyad (λ, U_L), where λ is a real number that indicates the times that the unit of length U_L is included in S or, expressed with the symbology that has already been developed, the meaning of (λ, U_L) is to verify the symbolic equation $S = \lambda \circ U_L$, where the operation «$\circ$» indicates the multiplication of a scalar by a segment, defined in article 6. So we can admit the equality of meaning between the dyad (λ, U_L), which represents the amount of length of the segment S, with the multiplication $\lambda \circ U_L$, or what is equivalent, with the sense of equivalence of meanings we can write the equality $(\lambda, U_L) = \lambda \circ U_L$.

Under these conditions, the geometric experiment with the areas is described in figure 10 of article 8 and the one corresponding to the volumes in figure 11. The result of them is evidence that, by analytically describing the length quantities of the segments that are multiplied by dyads referring to certain units of length, it is found that with two factors, the measure of the area that integrates the product of two segments is equal to a dyad in which the first of its elements is the real number obtained by multiplying the numerical elements of the factor dyads, and the second of the product dyad elements is the amount of area, not numerically expressible, of the unit rectangle that results from geometrically multiplying the units of length in which the factors are expressed. In terms of mathematical analytics, it would have been, if the multiplied segments S_1 and S_2 had been expressed in terms of the respective units of length U_{L1} and U_{L2} with the respective dyads (a, U_{L1}) and (b, U_{L2}), where a and b are the numbers real values indicated by the measurements of the segments S_1 and S_2 using the units U_{L1} and U_{L2}, it will be found that the amount of area generated by the geometric product of those segments would be given by the dyad $[(a \times b), (U_{L1} * U_{L2})]$, whose meaning is that said product area is $a \times b$ times the amount of surface area of the unit rectangle generated by the unit segments U_{L1} and U_{L2}, that is, the rectangle symbolized by $U_{L1} * U_{L2}$.

With three multiplied segments, the procedure is totally analogous, although in this case the magnitude generated is a volume instead of a surface, a volume that will be measured with the compound unit with the geometric product of the length units of the factors $U_{L1} * U_{L2} * U_{L3}$. And from all this we can conclude the analytical forms of these two geometric multiplications, which remain in this way for two and three factors:

$$(a, U_{L1}) * (b, U_{L2}) = [(a \times b), (U_{L1} * U_{L2})]$$

$$(a, U_{L1}) * (b, U_{L2}) * (c, U_{L3}) = [(a \times b \times c), (U_{L1} * U_{L2} * U_{L3})]$$

At this point, nothing prevents generalizing the multiplication of segments to the case of n factors, which analytically does not offer the least difficulty, since it is enough to establish the following equality:

$$(a_1, U_{L1}) * (a_2, U_{L2}) * \ldots * (a_n, U_{Ln}) =$$
$$= [(a_1 \times a_2 \times \ldots \times a_n), (U_{L1} * U_{L2} * \ldots * U_{ln})]$$

Artículo 24

EL TRASCENDENTAL EXPERIMENTO
CON ÁREAS Y VOLÚMENES
(Artículo 8, figuras 10 y 11)

Salvada la laguna matemática de la multiplicación geométrica, se observa que esta operación, dado su carácter gráfico, no aporta por sí misma información en cuanto a las cantidades de longitudes de los segmentos ni en cuanto a las cantidades de superficie ni de volumen. Por tanto, es preciso acometer procedimientos de **medida** que permitan cuantificar y operar analíticamente con tales magnitudes.

Para establecer un criterio de medida de las longitudes de los segmentos se recurre a uno cualquiera de ellos que se tome como unidad, que será tal que comprenderá una cantidad de longitud no numerable, indeterminación que se salvará asignándole un símbolo que la represente, por ejemplo, U_L. Con ello, la medida de cualquier segmento S se podrá indicar con la pareja o díada (λ, U_L), donde λ sea un número real que indica las veces que en S esté comprendida la unidad de longitud U_L o, expresado con la simbología que ya se ha desarrollado, el significado de (λ, U_L) es que se verifique la ecuación simbólica $S = \lambda \circ U_L$, donde la operación «$\circ$» indica la multiplicación de un escalar por un segmento, definida en el artículo 6. Así que podemos admitir la igualdad de significado entre la díada (λ, U_L), que representa la cantidad de longitud del segmento S, con la multiplicación $\lambda \circ U_L$, o lo que es equivalente, con el sentido de equivalencia de significados se puede escribir la igualdad $(\lambda, U_L) = \lambda \circ U_L$.

En estas condiciones, el experimento geométrico con las áreas queda descrito en la figura 10 del artículo 8 y el correspondiente a los volúmenes en la figura 11. El resultado que de ellos se obtiene evidencia que, describiendo analíticamente las cantidades de longitud de los segmentos que se multipliquen mediante díadas referidas a ciertas unidades de longitud, se encuentra que con dos factores, la medida del área que integra el producto de dos segmentos es igual a una díada en que el primero de sus elementos sea el número real que se obtiene al multiplicar los elementos numéricos de las díadas de los factores, y el segundo de los elementos de la díada del producto es la cantidad de área, no expresable numéricamente, del rectángulo unitario que resulte de multiplicar geométricamente las unidades de longitud en que se expresen los factores. En términos de analítica matemática se tendría que, si los segmentos multiplicados S_1 y S_2 se hubieran expresado en función de sendas unidades de longitud U_{L1} y U_{L2} con las respectivas díadas (a, U_{L1}) y (b, U_{L2}), donde a y b son los números reales que indican las medidas de los segmentos S_1 y S_2 mediante las unidades U_{L1} y U_{L2}, resultará que la cantidad de área generada por el producto geométrico de esos segmentos vendría dada por la díada $[(a \times b), (U_{L1} * U_{L2})]$, cuyo significado es que dicha área producto es $a \times b$ veces la cantidad de superficie del rectángulo unitario engendrado por los segmentos unitarios U_{L1} y U_{L2}, es decir, el rectángulo simbolizado por $U_{L1} * U_{L2}$.

Con tres segmentos multiplicados el procedimiento es totalmente análogo, aunque en este caso la magnitud generada sea un volumen en vez de una superficie, volumen que se medirá con la unidad compuesta con el producto geométrico de las unidades de longitud de los factores $U_{L1} * U_{L2} * U_{L3}$. Y de todo ello se pueden concluir las formas analíticas de estas dos multiplicaciones geométricas, que quedan de este modo para dos y tres factores:

$$(a, U_{L1}) * (b, U_{L2}) = [(a \times b), (U_{L1} * U_{L2})]$$

$$(a, U_{L1}) * (b, U_{L2}) * (c, U_{L3}) = [(a \times b \times c), (U_{L1} * U_{L2} * U_{L3})]$$

Llegados a este punto, nada impide generalizar la multiplicación de segmentos al caso de n factores, lo que analíticamente no ofrece la menor dificultad, pues basta establecer la igualdad siguiente:

$$(a_1, U_{L1}) * (a_2, U_{L2}) * \ldots * (a_n, U_{Ln}) =$$
$$= [(a_1 \times a_2 \times \ldots \times a_n), (U_{L1} * U_{L2} * \ldots * U_{ln})]$$

RELATIONSHIP BETWEEN PROPORTIONALITY OF
SEGMENTS AND GEOMETRIC MULTIPLICATION

It has been observed that the proportionality of segments arises from multiplication by a scalar. This operation has been symbolized with the notation $\lambda \circ S = P$, and it has been established that for this reason it is said that the segments S and P are in the ratio $P /\!/ S$ of the scalar λ. Thus, when two ratios correspond to the same scalar λ, it will be said that they form a proportion, and $S_1 /\!/ S_2 = S_3 /\!/ S_4 = \lambda$ will be written.

Expressing the four segments of a proportion in the same unit of length U_L, being a_1, a_2, a_3 and a_4 the respective measurements of the segments with said unit U_L, we have the identities $S_1 = a_1 \circ U_L$, $S_2 = a_2 \circ U_L$, $S_3 = a_3 \circ U_L$ and $S_4 = a_4 \circ U_L$. In turn, the initial proportionality determines that $a_1 \circ U_L = \lambda \circ (a_2 \circ U_L)$. The meaning of the operation «$\circ$» is that of an abbreviated sum of segments, so we can write $\lambda \circ (a_2 \circ U_L) = (\lambda \times a_2) \circ U_L$, because the measure of the segment $\lambda \circ (a_2 \circ U_L)$ with the unit U_L must agree axiomatically with the arithmetic product $\lambda \times a_2$. All of this leads us to the equality $a_1 \circ U_L = (\lambda \times a_2) \circ U_L$. And, the units of length U_L of both members being equal, it can only be concluded that $a_1 = \lambda \times a_2$, based on the equality of segments. So it can be stated that, if two segments S_1 and S_2 are in the geometric ratio λ, their measures a_1 and a_2 in the same unit of length U_L are in the same arithmetic ratio λ.

In this way, the first happy assumption that makes up the blur of the model texts, represented by Puig Adam and David Hilbert, is evidenced, and it is proven that every geometric ratio of segments measured with the same unit can be said to generate an equal arithmetic ratio between their measurements. And in the same way, if two segment ratios form a geometric proportion, it is certain that their measurements with the same unit of length form an arithmetic proportion.

Next, given a geometric proportion of segments $S_1 /\!/ S_2 = S_3 /\!/ S_4 = \lambda$, let's analyze what happens to the geometric product of the extremes $S_1 * S_4$ and that of the means $S_2 * S_3$. Geometric multiplication allows writing $S_1 * S_4 = (a_1 \times a_4) \circ U_L$ and $S_2 * S_3 = (a_2 \times a_3) \circ U_L$. Considering that, by hypothesis, the given segments form a geometric proportion, as has just been proven, their measurements with the same unit will form an arithmetic proportion, and it is verified that $a_1 \times a_4 = a_2 \times a_3$, and from here we have that $S_1 * S_4 = S_2 * S_3$. Then, given the geometric proportion $S_1 /\!/ S_2 = S_3 /\!/ S_4$, it is verified that the geometric product of the ends is equal to that of the means, without forgetting that these two products are not ordinary, but rather represent areas that turn out to be equal.

With this approach, the defects of the assumptions of Puig Adam and David Hilbert and of the classical texts emerge. Their error is evident when considering that the proportion between the measurements of the segments with the same unit of length can be inferred that of the segments, because in no case do they define geometric multiplication as they should and, in the absence of this, such assumption is an inadmissible excess. Hence, they must resort to the rather tricky device of substituting the segments for their measurements. However, with the argument of these articles it is proven that such an agreement, apart from being incorrect, is not necessary, because it is enough to fill in the gap in force until now with the operations that had been unduly omitted, that is, those inherent to geometric multiplication of segments, and with this the range of operations and properties is completed with full logical coherence, the current classic deficiency being resolved.

The very texts of Mathematics, led in this investigation by Puig Adam's *Course in Metric Geometry* and David Hilbert's *Fundamentals of Geometry*, make full proof that Mathematics has bypassed geometric multiplication, ignoring this law of generatrix composition and essential to operate properly with the quantities of a fundamental magnitude: the length.

From this point on, this negligent oversight has blurred all the mathematical developments that rely on metrics: vector spaces, which are based on the concept of the modulus of a vector; Euclidean spaces, which are based on the interior connection or scalar product; the tensor spaces and, in short, as has been indicated, all those innumerable algebraic structures that use the notion of metric or distance between their points.

As Physics makes use of Mathematics almost blindly, hiding itself in that comfort, it has not noticed the malformation that affects its matrix science and, simply conforming to what Mathematics offered it, has left pending justification and development, not only operations with quantities of lengths, which are the most fundamental physical magnitude of all, but also those corresponding to any other magnitude. Due to this neglect, Physics drags from the beginning a fundamental pending subject, which is the epistemic algebra of magnitudes. This is how the mathematical virus has infected Physics and both sciences must heal and create antibodies, which is by no means impossible, because here we demonstrate how to do it. And, as long as the treatment for a disease is known, it does not seem doubtful that it should be prescribed.

With this it will be observed that the algebras of magnitudes will revolutionize Physics, because the current pending subject relegates the compound magnitudes to a merely symbolic plane, giving priority to the search for strictly arithmetic and rather childish numerical proportionalities. However, what Physics needs is that the magnitudes be the object of investigation in themselves, to discover their true nature and properties, and thus be able to adapt the relevant composition laws to reality for each physical experience, structuring them appropriately.

A first notable result of this research is the «dysmetric» observation, which leads Physics to the innovative and unexplored field of «dysmetric» spaces, which are called upon to improve physical models and which constitute a much more powerful tool than the current one tacit isometry to more accurately represent natural phenomena, as outlined in the second volume of this work.

RELACIÓN ENTRE PROPORCIONALIDAD DE
SEGMENTOS Y MULTIPLICACIÓN GEOMÉTRICA

Se ha observado que la proporcionalidad de segmentos nace de la multiplicación por un escalar. Se ha simbolizado esta operación con la notación $\lambda \circ S = P$, y se ha establecido que por ello se diga que los segmentos S y P están en la razón $P /\!/ S$ del escalar λ. Así, cuando a dos razones les corresponda el mismo escalar λ se dirá que forman proporción, y se escribirá $S_1 /\!/ S_2 = S_3 /\!/ S_4 = \lambda$.

Expresando los cuatro segmentos de una proporción en la misma unidad de longitud U_L, siendo a_1, a_2, a_3 y a_4 las medidas respectivas de los segmentos con dicha unidad U_L, se tienen las identidades $S_1 = a_1 \circ U_L$, $S_2 = a_2 \circ U_L$, $S_3 = a_3 \circ U_L$ y $S_4 = a_4 \circ U_L$. A su vez, la proporcionalidad inicial determina que $a_1 \circ U_L = \lambda \circ (a_2 \circ U_L)$. El significado de la operación «$\circ$» es el de una suma abreviada de segmentos, por lo que se podrá escribir $\lambda \circ (a_2 \circ U_L) = (\lambda \times a_2) \circ U_L$, porque la medida del segmento $\lambda \circ (a_2 \circ U_L)$ con la unidad U_L debe coincidir axiomáticamente con el producto aritmético $\lambda \times a_2$. Todo ello nos conduce a la igualdad $a_1 \circ U_L = (\lambda \times a_2) \circ U_L$. Y, siendo iguales las unidades de longitud U_L de ambos miembros, no puede sino concluirse que $a_1 = \lambda \times a_2$, con fundamento en la igualdad de segmentos. Conque se puede afirmar que, si dos segmentos S_1 y S_2 están en la razón geométrica λ, sus medidas a_1 y a_2 en la misma unidad de longitud U_L están en esa misma razón aritmética λ.

De este modo queda en evidencia la primera alegre suposición que conforma el borrón de los textos modélicos, representados por Puig Adam y David Hilbert, y se acredita que toda razón geométrica de segmentos medidos con la misma unidad puede afirmarse que engendra una razón aritmética igual entre sus medidas. Y de la misma forma, si dos razones de segmentos forman proporción geométrica, es seguro que sus medidas con una misma unidad de longitud forman proporción aritmética.

A continuación, dada una proporción geométrica de segmentos $S_1 /\!/ S_2 = S_3 /\!/ S_4 = \lambda$, analicemos qué sucede con el producto geométrico de los extremos $S_1 * S_4$ y el de los medios $S_2 * S_3$. La multiplicación geométrica permite escribir $S_1*S_4=(a_1\times a_4)\circ U_L$ y $S_2*S_3=(a_2\times a_3)\circ U_L$. Considerando que, por hipótesis, los segmentos dados forman proporción geométrica, como se acaba de acreditar, sus medidas con la misma unidad formarán proporción aritmética, y se verifica que $a_1\times a_4=a_2\times a_3$, y de aquí se tiene que $S_1*S_4=S_2*S_3$. Luego, dada la proporción geométrica $S_1/\!/S_2=S_3/\!/S_4$, se verifica que el producto geométrico de los extremos es igual al de los medios, sin olvidar que estos dos productos no son ordinarios, sino que representan sendas áreas que resultan ser iguales.

Con este planteamiento afloran los defectos de las suposiciones de Puig Adam y David Hilbert y de los textos clásicos. Queda patente el error de estos al considerar que de la proporción entre las medidas de los segmentos con la misma unidad de longitud se pueda inferir la de los segmentos, porque en ningún caso definen como debieran la multiplicación geométrica y, en ausencia de esta, tal suposición es un exceso inadmisible. De ahí que deban recurrir al artificio más bien tramposo de sustituir los segmentos por sus medidas. Sin embargo, con el argumento de estos artículos se acredita que tal convenio, aparte de incorrecto, no es necesario, porque basta con completar la laguna hasta ahora vigente con las operaciones que se habían omitido indebidamente, esto es, las inherentes a la multiplicación geométrica de segmentos, y con ello se completa el abanico de operaciones y propiedades con plena coherencia lógica, quedando resuelta la carencia clásica actual.

Los textos de Matemáticas, liderados por el *Curso de geometría métrica* de Puig Adam y *Fundamentos de la geometría* de David Hilbert, hacen prueba plena de que la Matemática ha pasado de largo por la multiplicación geométrica, ignorando esta ley de composición generatriz e indispensable para operar debidamente con las cantidades de una magnitud fundamental: la longitud.

Este descuido negligente ha emborronado desde este punto todos los desarrollos matemáticos que se apoyan en la métrica: los espacios vectoriales, que se basan en el concepto de modulo de un vector; los espacios euclidianos, que tienen por fundamento la conexión interior o producto escalar; los espacios tensoriales y, en suma, como se ha indicado, todas esas innumerables estructuras algebraicas que se sirven de la noción de métrica o distancia entre sus puntos.

Como la Física se sirve casi a ciegas de la Matemática, escudándose en esa comodidad, no ha reparado en la malformación que afecta a su ciencia matriz y, conformándose sin más con lo que la Matemática le ofrecía, ha dejado pendiente de justificación y desarrollo, no solo las operaciones con cantidades de longitudes, que son la magnitud física más fundamental de todas, sino además las correspondientes a cualquier otra magnitud. Por este descuido la Física arrastra desde el principio una asignatura pendiente fundamental, que es el álgebra epistémica de magnitudes. Así es como el virus matemático ha infectado la Física y ambas ciencias deben sanarse y crear anticuerpos, lo cual no es ni mucho menos imposible, porque aquí se demuestra cómo hacerlo. Y, siempre que se conozca el tratamiento para una enfermedad, no parece dudoso que sea obligado prescribirlo.

Con ello se observará que las álgebras de magnitudes revolucionarán la Física, porque la asignatura pendiente vigente relega las magnitudes compuestas a un plano meramente simbólico, dando prioridad a la búsqueda de proporcionalidades numéricas estrictamente aritméticas y más bien pueriles. Sin embargo, lo que la Física necesita es que las magnitudes sean objeto de investigación en sí mismas, para descubrir su verdadera naturaleza y propiedades, y así poder adaptar a la realidad las leyes de composición pertinentes para cada experiencia física, estructurándolas convenientemente.

Un primer resultado notable de esa investigación es la observación «dismétrica», que conduce a la Física al innovador e inexplorado campo de los espacios «dismétricos», que están llamados a mejorar los modelos físicos y que constituyen una herramienta mucho más potente que la actual isometría tácita para representar con mayor precisión los fenómenos naturales, tal como se esboza en el segundo volumen de esta obra.

Article 26

THE NON-ARITHMETIC COMPOSITE RULE OF THREE
FOUNDATION OF PHYSICAL EQUATIONS

We are going to examine the basic problems known by the name «compound rule of three», which are taught and faulty solved by arithmetic proportions, resulting in misunderstanding errors due to the «arithmetization» error. However, we will observe that, applying the algebra of magnitudes, these cases are solved with great ease and full mathematical rigor, without leaving room for logical gaps, as the reader can verify for himself, solving the problems by the classical method, which is It is known, and comparing it with what is explained in this article. We will also see that these elementary cases reflect with singular clarity the formation of physical equations in general. To do this, we will operate on specific examples and apply the operations that we have called «homogeneous multiplication and division», sections X and XI, or articles 6 and 7 of section XXVIII, as well as «heterogeneous multiplication and division», sections XII and XVI, or articles 9 and 10 of the aforementioned section XXVIII. We will not enter into repetitive demonstrations, but we will refer to the properties already exposed in these sections.

Consider the following «direct compound rule of three» problem: 5 *bottles* of 2 *liters* each filled with a liquid weigh 15 *kilograms*. How much do 2 *bottles* of 3 *liters* each weigh? It is understood that the weight is that of the liquid without the container.

The generating magnitudes are the *number of bottles* and the *volume*. The magnitude generated is the *weight*. So the problem equations must have the form *bottles*∗*volume*=*weight*. The statement «5 *bottles* of 2 *liters* each filled with a liquid weigh 15 *kilograms*» can be written as a product of dyadic algebra as follows:

$$5 \; bottle \; * \; 2 \; liter \; = \; 15 \; kilogram$$

Similarly, the ordinary language statement «How much do 2 *bottles* of 3 *liters* each weigh?», can be written algebraically with the following dyadic equation:

2 *bottle* * 3 *liter* = x *kilogram* (generated unknown)

Where *x* is the unknown measure of the weight of the 2 3-*liter* *bottles*. Thus, if these two equations are raised, operating with them, they become these two:

5×2 *bottle* *liter = 15 *kilogram*

2×3 *bottle* *liter = x *kilogram*

We can dyadically divide these two equations member by member and we get:

$$\frac{5 \times 2 \ bottle*liter}{2 \times 3 \ bottle*liter} = \frac{15 \ kilogram}{x \ kilogram}$$

Thus we have formed two equal dyadic ratios, so they form a non-arithmetic proportion. The two ratios are such that their numerators and denominators consist of dyads with the same secondary, so they are homogeneous, and their quotient must be equal to the arithmetic ratio of their primaries. Thus it turns out that the previous dyadic proportion justifies entering the following arithmetic proportion:

$$\frac{5 \times 2}{2 \times 3} = \frac{15}{x}$$

Solving, we have x=9, that is, 2 *bottles* of 3 *liters* each weigh 9 *kilograms*, which is the solution to the problem.

We observe that dyadic algebra allows us to relate units as different as the *bottle*, the *liter* and the *kilogram*. And this phenomenon is the one that occurs permanently in Physics when it relates fundamental units of length, mass or time, such as the *meter*, the *kilogram* and the *second*, or any others.

In such a way that in this example we could conceive with relative value limited to the assumption the dimensional formulas of section XXVIII, such as [WEIGHT]=BOTTLE∗VOLUME. We will not escape the incongruity that it may seem for Physics to consider *weight*, associated with *mass*, a fundamental magnitude, as a unit generated by the generating product of *bottles* and *volume*. However, its algebraic validity is unquestionable. The freedom in the choice of generating magnitudes is evident with a first reverse variant case: 5 *bottles* of 2 *liters* each filled with a liquid weigh 15 *kilograms*. How many liters are the *bottles* if 2 weigh 9 *kilograms*? The generating equations are here: 5 *bottlle* ∗ 2 *liter* = 15 *kilogram* and 2 *botella* ∗ x *litro* = 9 *kilogram*. The arithmetic ratio is $(5×2)/(2×x)=15/9$, so $x=3$. Each bottle will have a volume of 3 *liters*, like the first statement.

The second inverse variant would have the *number of bottles* as unknown. We would ask how many 3 *liter* bottles weigh 9 *kilograms*? The generating equations in this case are: 5 *bottle* ∗ 2 *liter* = 15 *kilogram* and x *bottle* ∗ 3 *liter* = 9 *kilogram*. It turns out $(5×2)/(x×3)=15/9$, so $x=2$. The solution is 2 *bottles*, logically, coinciding with the first statement.

Let's look at another example, in this case of the «inverse compound rule of three»: A passenger transport system is capable of moving 400 *passengers* a distance of 500 *kilometers* in 8 *hours*. How many *passengers* can be transported a distance of 300 *kilometers* in 12 *hours*? Simplifying the notation by using the abbreviations of units *v* for *passengers*, *km* for *kilometers* and *h* for *hours*, the two previous statements can be translated into the mathematical language of the algebra of magnitudes by means of the following expressions:

$$400\ v * 500\ km = 8\ h$$

$$x\ v * 300\ km = 12\ h \text{ (generating unknown)}$$

Operating with the dyadic product, these two equations are transformed into these others:

$$400 \times 500 \, v*km = 8 \, h$$

$$x \times 300 \, v*km = 12 \, h$$

Both expressions have in their primaries the same compound unit $v*km$ in the first member and the time h in the second. Therefore, dividing them member by member, two equal dyadic ratios result, that is, a non-arithmetic dyadic proportion, which is reduced to the arithmetic proportion of their primaries, that is:

$$\frac{400 \times 500 \, v*km}{x \times 300 \, v*km} = \frac{8 \, h}{12 \, h} \Rightarrow \frac{400 \times 500}{x \times 300} = \frac{8}{12} \Rightarrow x = 1,000$$

Therefore, the answer to the problem is that the number of transportable *passengers* is 1,000. In this case we also find the same observation about the relative validity of the dimensional equation that operates in the problem with the related magnitudes, whose expression would be:

$$[TIME] = PASSENGER * LENGTH$$

The physical incongruity of an expression like this is manifest, since time is an independent fundamental magnitude and in the previous form it appears as a magnitude composed of two others. Again we come across the necessary delicacy to interpret the relationships between the magnitudes that appear in the mathematical formulations of natural phenomena.

The examples analyzed in this article belong to the «direct and inverse compound rule of three» assumptions, according to current terminology. The classical rules for solving these problems are all abstruse and incomplete, due to the vice of «arithmetization». However, with the algebra of magnitudes, all logical steps are fully justified by mathematical laws and it is not necessary to resort to mysterious old-fashioned rules or intuitive reasoning, solving problems of this type in a very direct and easy way. And this serves very well as a didactic prelude in the handling and interpretation of the most complex physical laws and equations.

Artículo 26

LA REGLA DE TRES COMPUESTA NO ARITMÉTICA FUNDAMENTO DE LAS ECUACIONES FÍSICAS

Vamos a examinar los problemas básicos conocidos por la denominación «regla de tres compuesta», que se enseñan y resuelven defectuosamente mediante proporciones aritméticas, lo cual produce problemas de incomprensión por el error de la «aritmetización». Sin embargo, observaremos que, aplicando el álgebra de magnitudes, estos casos se solucionan con suma facilidad y pleno rigor matemático, sin dejar espacio a lagunas lógicas, como puede comprobar el lector por sí mismo, resolviendo los problemas por el método clásico, que se supone conocido, y comparándolo con lo explicado en este artículo. Veremos también que estos casos elementales reflejan con singular claridad la formación de las ecuaciones físicas en general. Para ello operaremos sobre ejemplos concretos y aplicaremos las operaciones que hemos denominado «multiplicación y división homogéneas», apartados X y XI, o artículos 6 y 7 del apartado XXVIII, así como la «multiplicación y división heterogéneas», apartados XII y XVI, o artículos 9 y 10 del citado apartado XXVIII. No entraremos en demostraciones repetitivas, sino que nos remitiremos a las propiedades ya expuestas en dichos apartados.

Consideremos el siguiente problema de «regla de tres compuesta directa»: 5 *botellas* de 2 *litros* cada una llenas de un líquido pesan 15 *kilogramos*. ¿Cuánto pesan 2 *botellas* de 3 *litros* cada una? Se sobreentiende que el peso es el del líquido sin el envase. Las magnitudes generatrices son la *cantidad de botellas* y el *volumen*. La magnitud generada es el *peso*. Así que las ecuaciones del problema han de tener la forma *botellas*volumen=peso*. El enunciado «5 *botellas* de 2 *litros* cada una llenas de un líquido pesan 15 *kilogramos*» equivale a la siguiente igualdad generatriz:

$$5 \ botella * 2 \ litro = 15 \ kilogramo$$

Análogamente, el enunciado de lenguaje ordinario «¿Cuánto pesan 2 *botellas* de 3 *litros* cada una?», se puede escribir algebraicamente en forma generatriz con la ecuación diádica siguiente:

2 *botella* * 3 *litro* = *x kilogramo* (incógnita generada)

Donde *x* es la medida desconocida del peso de las 2 *botellas* de 3 *litros*. Así, planteadas estas dos ecuaciones, operando con ellas, se transforman en estas dos:

5×2 *botella* *litro* = 15 *kilogramo*

2×3 *botella* *litro* = *x kilogramo*

Podemos dividir diádicamente estas dos ecuaciones miembro a miembro y obtenemos:

$$\frac{5 \times 2\,botella*litro}{2 \times 3\,botella*litro} = \frac{15\,kilogramo}{x\,kilogramo}$$

Así hemos formado dos razones diádicas iguales, por lo que forman proporción no aritmética. Las dos razones son tales que sus numeradores y denominadores constan de díadas con el mismo segundario, por lo que son homogéneas, y su cociente ha de ser igual a la razón aritmética de sus primarios. Así resulta que la proporción diádica anterior justifica asentar la proporción aritmética siguiente:

$$\frac{5 \times 2}{2 \times 3} = \frac{15}{x}$$

Resolviendo, se tiene *x*=9, es decir, que 2 *botellas* de 3 *litros* cada una pesan 9 *kilogramos*, que es la solución al problema.

Observamos que el álgebra diádica permite relacionar unidades tan diferentes como la *botella*, el *litro* y el *kilogramo*. Y este fenómeno es el que se da permanentemente en la Física cuando relaciona unidades fundamentales de longitud, masa o tiempo, como el *metro*, el *kilogramo* y el *segundo*, o cualesquiera otras. En este ejemplo podríamos concebir con relativo valor limitado al supuesto las fórmulas dimensionales del apartado XXVIII, tales como [PESO]=BOTELLA *VOLUMEN. No se nos escapará la incongruencia que puede parecer para la Física que se considere el *peso*, asociado a la *masa*, magnitud fundamental, como unidad generada por el producto generatriz de *botellas* y *volumen*. No obstante lo cual, su validez algebraica es incuestionable.

La libertad en la elección de las magnitudes generatrices queda patente con un primer caso variante inversa: 5 *botellas* de 2 *litros* cada una llenas de un líquido pesan 15 *kilogramos*. ¿De cuántos litros son las *botellas* si 2 pesan 9 *kilogramos*? Las ecuaciones generatrices son en este caso las indicadas a continuación: 5 *botella* ∗ 2 *litro* = 15 *kilogramo* y 2 *botella* ∗ *x litro* = 9 *kilogramo*. La razón aritmética es $(5 \times 2)/(2 \times x) = 15/9$, con lo que $x = 3$. Cada botella tendrá un volumen de *3 litros*, como el primer enunciado.

La segunda variante inversa tendría por incógnita la *cantidad de botellas*. Preguntaríamos ¿cuántas botellas de 3 *litros* pesan 9 *kilogramos*? Las ecuaciones generatrices en este caso son: 5 *botella* ∗ 2 *litro* = 15 *kilogramo* y *x botella* ∗ 3 *litro* = 9 *kilogramo*. Resulta $(5 \times 2)/(x \times 3) = 15/9$, con lo que $x = 2$. La solución son 2 *botellas*, lógicamente, coincidiendo con el primer enunciado.

Veamos otro ejemplo, en este caso de «regla de tres compuesta inversa», como las dos variantes del caso anterior: un sistema de transporte es capaz de trasladar 400 *viajeros* una distancia de 500 *kilómetros* en 8 *horas*. ¿Cuántos *viajeros* se podrán transportar una distancia de 300 *kilómetros* en 12 *horas*? Simplificando la notación mediante el uso de las abreviaturas de unidades *v* para *viajeros*, *km* para *kilómetros* y *h* para *horas*, los dos enunciados anteriores pueden traducirse al lenguaje matemático del álgebra de magnitudes mediante sendas expresiones generatrices:

$$400 \, v \ast 500 \, km = 8 \, h$$

$$x \, v \ast 300 \, km = 12 \, h \text{ (incógnita generatriz)}$$

Operando con el producto diádico, estas dos ecuaciones se transforman en estas otras:

$$400 \times 500 \, v \ast km = 8 \, h$$

$$x \times 300 \, v \ast km = 12 \, h$$

Ambas expresiones tienen en sus primarios la misma unidad compuesta $v*km$ en el primer miembro y el tiempo h en el segundo. Por tanto, dividiéndolas miembro a miembro, resultan dos razones diádicas iguales, es decir, una proporción diádica no aritmética, que se reduce a la proporción aritmética de sus primarios, esto es:

$$\frac{400 \times 500\, v*km}{x \times 300\, v*km} = \frac{8h}{12h} \Rightarrow \frac{400 \times 500}{x \times 300} = \frac{8}{12} \Rightarrow x = 1.000$$

Por tanto, la respuesta al problema es que el número de *viajeros* transportable es de 1.000. En este caso nos encontramos también con la misma observación sobre la relativa validez de la ecuación dimensional que opera en el problema con las magnitudes relacionadas, cuya expresión sería:

[TIEMPO] = VIAJERO * LONGITUD

La incongruencia física de una expresión como esta es manifiesta, toda vez que el tiempo es una magnitud fundamental independiente y en la forma anterior aparece como magnitud compuesta por otras dos, aunque su validez algebraica es plena. Nuevamente nos topamos con la delicadeza necesaria para interpretar las relaciones entre las magnitudes que aparecen en las formulaciones matemáticas de los fenómenos naturales.

Los ejemplos analizados en este artículo pertenecen a los supuestos de «regla de tres compuesta directa e inversa», según la terminología corriente. Las reglas clásicas para resolver estos problemas son todas abstrusas e incompletas, por efecto del vicio de la «aritmetización». Sin embargo, con el álgebra de magnitudes todos los pasos lógicos quedan plenamente justificados por leyes matemáticas y no es preciso recurrir a misteriosas reglas de la cuenta de la vieja ni razonamientos intuitivos, resolviéndose los problemas de este tipo de modo muy directo y con plena facilidad. Y ello sirve muy bien como antesala didáctica en el manejo e interpretación de las leyes y ecuaciones físicas más complejas.

BIBLIOGRAPHY (BIBLIOGRAFÍA)

JOSEPH FOURIER. *Théorie Analitique de la Chaleur*, Gauthier Villars, París, 1888.

DAVID HILBERT. *Grundlagen der Geometrie* (*Fundamentos de la geometría*), 1899.

MAX PLANCK. *Vorlesungen über die Theorie der Wärmestrahlung*, Leipzig, 1906.

R.C. TOLMAN. *Physics Review*, 1914, 1917.

GIOVANNI GIORGI. *Sistemi e unita di mesura*, Enciclopedia delle Matematiche Elementari.

P. W. BRIDGMAN. *Dimensional Analysis*, Yale, University Press (Universidad Nacional de Tucumán, República Argentina).

RICARDO SAN JUAN. *Teoría de las magnitudes físicas y sus fundamentos algebraicos*, Revista de la Real Academia de Ciencias de Madrid, 1947.

P. W. BRIDGMAN. *British Enciclopedia*, edition 1951, article *Dimensional Analysis*.

JULIO PALACIOS. *El lenguaje de la física y su peculiar filosofía*, 1953.

U. STILE. *Messen und Rechnen in der Physic*, Vieweg, Braunschweig, 1961.

JULIO PALACIOS. *Análisis dimensional*, Espasa Calpe, segunda edición, 1964.

P. PUIG ADAM. *Geometría métrica*, Biblioteca Matemática Rey Pastor-Puig Adam, 1970.

SEARS ZEMANSKY. *Física general*, Aguilar, University Physics, 1970.

LUIS A. SANTALÓ. *Vectores y tensores y sus aplicaciones*, Editorial Universitaria de Buenos Aires, 1970.

R. KURTH. *Dimensional Analysis and Group Theory in Astrophysics*, Pergamon, 1972.

F. CATALÁ MORENO, *Álgebra lineal y multilineal*, Academia Iribas, Madrid, 1972.

ANDRÉ LICHNEROWICZ. *Elementos de cálculo tensorial*, Aguilar Sociedad anónima de Ediciones, 1972.

I. CANO DE LA TORRE. *Mecánica Racional*, Academia Luz de Madrid, 1973.

INTERNATIONAL PRACTICAL TEMPERATURE SCALE OF 1968, Amended Edition of 1975, *Metrology*, Comité International des Poids et Mesures, 1976.

R. M. COOKE. *The Algebra of Physical Magnitudes, Foundatios of Physics*, 1980.

JOSÉ CATALÁN CHILLERÓN. *Teoría de las magnitudes físicas*, Instituto Geográfico Nacional, Madrid, 1983.

ISAAC NEWTON. *Principios matemáticos de la filosofía natural*, Alianza Editorial, 2016.

J. M. ARNAIZ. *Matematizar 1 (Fundamentos), Matematizar 2 (Complementos) y Matematizar 3 (Aplicaciones)*, Ediciones Go Beyond, 2016.

BUREAU INTERNATIONAL DES POIDS ET MESURES. *The International System of Units (SI)*.

www.ingramcontent.com/pod-product-compliance
Lightning Source LLC
Chambersburg PA
CBHW071246220526
45468CB00001B/13